WALKING THE TALK?

PROGRESS IN INTERNATIONAL BUSINESS RESEARCH

Series Editors: The European International Business Academy (EIBA)

Recent Volumes:

Volume 1:	Progress in International Business Research – Edited by Gabriel R. G. Benito and Henrich R. Greve
Volume 2:	Foreign Direct Investment, Location and Competitiveness – Edited by John H. Dunning and Philippe Gugler
Volume 3:	New Perspectives in International Business Research – Edited by Maryann P. Feldman and Grazia D. Santangelo
Volume 4:	Research on Knowledge, Innovation and Internationalization – Edited by Jorma Larimo and Tia Vissak
Volume 5:	Reshaping the Boundaries of the Firm in an Era of Global Interdependence – Edited by José Pla-Barber and Joaquín Alegre
Volume 6:	Entrepreneurship in the Global Firm – Edited by Alain Verbeke, Ana Teresa Tavares-Lehmann and Rob van Tulder
Volume 7:	New Policy Challenges for European MNEs – Edited by Rob van Tulder, Alain Verbeke and Liviu Voinea
Volume 8:	International Business and Sustainable Development– Edited by Rob van Tulder, Alain Verbeke and Roger Strange
Volume 9:	Multinational Enterprises, Markets and Institutional Diversity – Edited by Alain Verbeke, Rob van Tulder and Sarianna Lundan
Volume 10:	The Future of Global Organizing – Edited by Rob van Tulder, Alain Verbeke and Rian Drogendijk
Volume 11:	The Challenge of BRIC Multinationals – Edited by Rob van Tulder, Alain Verbeke, Jorge Carneiro and Maria Alejandra Gonzalez-Perez
Volume 12:	Distance in International Business: Concept, Cost and Value – Edited by Alain Verbeke, Jonas Puck and Rob van Tulder
Volume 13:	International Business in the Information and Digital Age – Edited by Rob van Tulder, Alain Verbeke and Lucia Piscitello
Volume 14:	International Business in a VUCA World: The Changing Role of States and Firms – Edited by Rob Van Tulder, Alain Verbeke and Barbara Jankowska
Volume 15:	The Multiple Dimensions of Institutional Complexity in International Business Research – Edited by Alain Verbeke, Rob van Tulder, Elizabeth L. Rose and Yingqi Wei
Volume 16:	International Business in Times of Crisis – Edited by Rob van Tulder, Alain Verbeke, Lucia Piscitello and Jonas Puck
Volume 17:	International Business and Sustainable Development Goals – Edited by Rob van Tulder, Elisa Giuliani and Isabel Álvarez

PROGRESS IN INTERNATIONAL BUSINESS RESEARCH VOLUME 18

WALKING THE TALK? MNEs TRANSITIONING TOWARDS A SUSTAINABLE WORLD

TRIBUTE VOLUME TO ALAIN VERBEKE

EDITED BY

ROB VAN TULDER

RSM Erasmus University Rotterdam, The Netherlands

BIRGITTE GRØGAARD

BI Norwegian Business School, Norway

AND

RANDI LUNNAN

BI Norwegian Business School, Norway

emerald PUBLISHING

United Kingdom – North America – Japan
India – Malaysia – China

Emerald Publishing Limited
Emerald Publishing, Floor 5, Northspring, 21-23 Wellington Street, Leeds LS1 4DL.

First edition 2024

British Library Cataloguing in Publication Data
A catalogue record for this book is available from the British Library

ISBN: 978-1-83549-118-8 (Print)
ISBN: 978-1-83549-117-1 (Online)
ISBN: 978-1-83549-119-5 (Epub)

ISSN: 1745-8862 (Series)

Printed and bound by CPI Group (UK) Ltd, Croydon, CR0 4YY

CONTENTS

PART II
WALKING THE TALK IN SUSTAINABILITY ISSUES

PART III
WALKING THE TALK AT THE INTERFACE WITH GOVERNMENT POLICIES

PART IV
WALKING THE TALK IN GLOBAL VALUE CHAINS

LIST OF FIGURES

LIST OF TABLES

LIST OF CONTRIBUTORS

Matthew M. C. Allen	Manchester Metropolitan University, UK
Martina Barbaglia	Politechnico di Milano, Italy
Roberto Bianchini	Politechnico di Milano, Italy
Vincenzo Butticè	Politechnico di Milano, Italy
Cyntia Vilasboas Calixto Casnici	Leeds University Business School, UK
Luciano Ciravegna	INCAE Business School, Costa Rica; University of Reading, UK
Jeremy Clegg	Leeds University Business School, UK
Cristina Di Stefano	Politecnico di Milano, Italy
Stefano Elia	Politecnico di Milano, Italy
Corina Fehlner	BI Norwegian Business School, Norway
Paola Garrone	Politecnico di Milano, Italy
Birgitte Grøgaard	BI Norwegian Business School, Norway
Lise Lillebrygfjeld Halse	Molde University College, Norway
Guus Hendriks	University of Amsterdam, The Netherlands
Jean-François Hennart	Tilburg University, The Netherlands; Aalborg Business School, Denmark
John Holland	University of Glasgow, UK
Gezim Hoxha	Politechnico di Milano, Italy
Thomas Hutzschenreuter	Technical University of Munich, Germany
Liena Kano	University of Calgary, Canada
Anna Kehl	University of Bremen, Germany
Alessandro Lampo	University of Saint Joseph, China
Gunnar Leymann	University of Bremen, Germany
Randi Lunnan	BI Norwegian Business School, Norway
Jacqueline Mees-Buss	University of New South Wales, Australia
Marina Papanastassiou	Leeds University Business School, UK
Viviana Pilato	HEC Montréal, Canada

Lucia Piscitello	Politecnico di Milano, Italy; Henley Business School, UK
Mohammad B. Rana	Aalborg University, Denmark
Germano Glufke Reis	Federal University of Parana, Brazil
David Schulzmann	Leeds University Business School, UK
Susana C. Silva	Universidade Catolica Portuguesa, Portugal
Ari Van Assche	HEC Montréal, Canada
Rob van Tulder	RSM Erasmus University Rotterdam, The Netherlands
Alain Verbeke	University of Calgary, Canada; University of Reading, UK; Vrije Universiteit Brussel (VUB), Belgium
Stefan Zagelmeyer	The University of Manchester, UK

PREFACE: A TRIBUTE TO PROFESSOR ALAIN VERBEKE

This Progress in International Business Research (PIBR) volume is a tribute to one of the most influential international business (IB) scholars, Professor Alain Verbeke. Professor Verbeke's scholarship transcends four decades where he has successfully advanced IB theory and ensured that scholars remember the importance of managerial relevance. His impact in the IB community is clear. His scholarly work is highly cited by other scholars both within and outside the IB community. His legacy is also clearly visible through his active involvement in the IB community including formal and informal service to the academic community and IB journals. Professor Verbeke is currently a distinguished fellow of European International Business Academy (EIBA) as well as the Academy of International Business and has edited 11 past volumes of PIBR. He recently stepped down as the Editor-in-Chief of the *Journal of International Business Studies (JIBS)*, after serving in this role for two consecutive terms.

In addition, and perhaps most importantly, our decision to make this volume a tribute to Professor Verbeke, is based on his pioneering work related to sustainability and his tireless efforts to support and encourage other scholars to contribute to this field in research, teaching as well as practical engagement. During the 2022 EIBA conference, scholars discussed if firms are really walking the Talk and developing necessary capabilities and collaborative efforts to move toward a sustainable world. Verbeke and co-authors made significant advances to this conversation over 25 years ago. One example of this is their distinction between compliance to environmental policies and developing green capabilities (Rugman & Verbeke, 1998b). Professor Verbeke was also an early and influential contributor linking sustainability issues to IB through research on topics including (but not limited to) collaboration between governments and firms, environmental strategies and stakeholder management, and the effects of environmental policies and taxation (Buysse & Verbeke, 2003; Rugman & Verbeke, 1998a, 2000; Verbeke & Coeck, 1997). His pioneering article on proactive environmental strategies and stakeholder management in the prestigious *Strategic Management Journal* (Buysse & Verbeke, 2003) contributed to bridging IB and sustainability research, and remains his most cited publication to date.

Alain Verbeke's prolific teaching activities have spanned a long period, covering many IB relevant topics, but mostly focused on the strategic management of multinational enterprises (MNEs). He has taught in or has been associated with a large number of prestigious academic institutions (including Dalhousie University, University of Toronto, Université Catholique de Louvain, Tilburg University, Erasmus University, University of Reading, University of Oxford,

University of Cambridge, Skolkovo Institute for Emerging Market Studies, China Europe International Business School, Nanyang Technical University, and University of Antwerp) in the role of associate professor, visiting professor, or MBA program director (Solvay Business School, Vrije Universiteit Brussel). For his teaching activities, he received the PhD Supervisor Excellence award at the University of Calgary in 2018.

Professor Verbeke's societal engagement has involved active participation in projects with stakeholders – for instance, port clusters. He has been a member of the European Science and Technology Assembly, the highest advisory body to the European Commission on the future of European scientific and innovation policy and has served on the board of directors of various educational and scientific research institutions.

Professor Verbeke early recognized the challenges of energy transition, the role of renewable energy, and the importance of integrating sustainability into strategic thinking (Illegems & Verbeke, 2003; Yuan et al., 2011). In the role of editor of JIBS, Verbeke co-organized a special issue on the long-term energy transition with the *British Journal of Management*, emphasizing the importance of actionable research addressing global grand challenges (Verbeke, 2021; Verbeke & Hutzschenreuter, 2021). This special issue is merely one of many examples where Verbeke has championed the importance of making IB research actionable and relevant for business, particularly related to sustainability. Through his exemplary scholarship and service, he continues to support the IB community with a passion and genuine interest to ensure that IB research makes an impact (Santangelo & Verbeke, 2022). This will be increasingly important to support firms, governments, and societies in transitioning toward a sustainable world.

His personal contribution to this volume (Chapter 2) illustrates his continued passion and concern about the relevance of IB scholarship and the critical role played by MNEs as well as by societal stakeholders in creating a more sustainable world. To quote his words:

> Our role as IB scholars is to describe and explain the value creating and value distributing processes put in motion by entrepreneurial parties with complementary resources, willing and able to establish and to nurture business linkages across borders. Acknowledging the role of parties not directly involved in establishing and operating international activities but with some type of decision rights in this realm, is clearly critical. However, an objective and evidence-based analysis is unlikely to conclude that more can and should "obviously" be done by all internationally operating firms beyond their present role in creating and distributing economic value. International entrepreneurial behavior emerges because economic actors chase attractive business opportunities and this opens them up to a wide spectrum of vulnerabilities, which explain the relative absence of global MNEs.

We cannot agree more!
The editors

➔ Photo of Alain Verbeke

2001 – selected McCaig Chair in Management (with wife Juliet)

2020 – with R. Edward Freeman

REFERENCES

Buysse, K., & Verbeke, A. (2003). Proactive environmental strategies: A stakeholder management perspective. *Strategic Management Journal (John Wiley & Sons, Inc.) – 1980 to 2009*, *24*(5), 453.

IIlegems, V., & Verbeke, A. (2003). *Moving towards the virtual workplace: Managerial and societal perspectives on telework*. Cheltenham: Edward Elgar Publishing.

Rugman, A., & Verbeke, A. (2000). Six cases of corporate strategic responses to environmental regulation. *European Management Journal*, *18*(4), 377.

Rugman, A. M., & Verbeke, A. (1998a). Corporate strategies and environmental regulations: An organizing framework. *Strategic Management Journal (John Wiley & Sons, Inc.) – 1980 to 2009*, *19*(4), 363–375.

Rugman, A. M., & Verbeke, A. (1998b). Corporate strategy and international environmental policy. *Journal of International Business Studies*, *29*(4), 819–833.

Santangelo, G. D., & Verbeke, A. (2022). Actionable guidelines to improve 'theory-related' contributions to international business research. *Journal of International Business Studies*, *53*(9), 1843–1855.

Verbeke, A. (2021). The long-term energy transition and multinational enterprise complexity: A BJM–JIBS joint initiative. *Journal of International Business Studies*, *52*(5), 803–806.

Verbeke, A., & Coeck, C. (1997). Environmental taxation: A green stick or a green carrot for corporate social performance? *Managerial & Decision Economics*, *18*(6), 507–516.

Verbeke, A., & Hutzschenreuter, T. (2021). Imposing versus Enacting Commitments for the Long-Term Energy Transition: Perspectives from the Firm. *British Journal of Management*, *32*(3), 569–578.

Yuan, W., Bao, Y., & Verbeke, A. (2011). Integrating CSR initiatives in business: An organizing framework. *Journal of Business Ethics*, *101*(1), 75–92.

CHAPTER 1

INTRODUCTION: SETTING THE SCENE

Rob van Tulder[a], Birgitte Grøgaard[b] and Randi Lunnan[b]

[a] *RSM, Erasmus University Rotterdam, The Netherlands*
[b] *BI Norwegian Business School, Norway*

WHY THE TOPIC OF "WALKING THE TALK"?

Across the globe, concerns escalate about the effects of greenhouse gas emissions on global climate conditions, implications on global trade from pandemics, and the destruction of ecosystems from the exploitation of limited and non-replaceable global resources. Politicians and businesses alike agree that something must be done; the question is what, how, when, and by whom? What should firms do to reduce their carbon footprint while continuing to secure jobs and profits? What should institutions and governments do to regulate and incentivize the road to a more sustainable future? How can firms and individuals reduce waste and secure a better use and reapplication of resources to protect the sustainability of ecosystems? When should firms make decisions to transform their businesses given the uncertainty of markets as well as the lack of clear guidance from global institutions? In this volume, we ask: Are we actually walking the Talk?

Transitioning toward a sustainable world is considered a global grand challenge that transcends national borders (Buckley et al., 2017). Global grand challenges have traditionally not received a lot of attention in international business (IB), which is surprising given the transnational nature of these phenomena. The complex nature of global grand challenges requires research that can capture complexities of cross-border interdependencies, multiple levels of analysis, and the collaboration of different stakeholders and institutions. Addressing these

Walking the Talk? MNEs Transitioning Towards a Sustainable World
Progress in International Business Research, Volume 18, 1–15

ISSN: 1745-8862/doi:10.1108/S1745-886220240000018001

challenges require diversity in methods and theoretical perspectives, which already characterize the IB community (Bass & Grøgaard, 2021; Buckley et al., 2017).

On occasion, business in general, and multinational enterprises (MNEs) in particular, are seen as obstacles to the transition toward a sustainable world. For example, Yu et al. (2023) scrutinized MNEs for taking advantage of cross-border arbitrage opportunities to maximize profits at the expense of the environment. While there may be multiple examples of such destructive behavior, there are also examples of organizations identifying opportunities to contribute positively to a sustainable future, and find ways to address challenges in coalitions with diverse stakeholders.

The 2022 European International Business Academy (EIBA) Annual Conference in Oslo, Norway aimed to highlight some of the issues to consider, and which research questions should be raised within IB to support businesses, industries, and societies when transitioning toward a sustainable world. This book includes select contributions from the 2022 EIBA conference as well as a number of invited contributions. The chapters in this volume illustrate a broad specter of research questions and ways to answer them within the IB community that provide evidence that many types of actors are taking – or can take – steps to actually "Walk the Talk." The contributions also show, however, that MNEs face considerable challenges to make their ambitions real, which in turn presents a challenge for IB scholarship to develop relevant and robust analytical approaches to cover transition problems that MNEs face when confronted with the inevitable gap between the realization of their strategic intentions (Mintzberg et al., 2009) or what in the corporate social responsibility (CSR) discourse has become classified as the "promise-performance gap" (Rasche & Waddock, 2014). In this volume, Verbeke (Chapter 2) and Hennart (Chapter 3) discuss this as the discrepancy between "wishes" and "realities," which has strongly influenced the IB discourse over the past decades.

WHAT DO WE MEAN BY "WALKING THE TALK"?

The main theme of the 2022 EIBA conference was "Transitioning Towards a Sustainable World: Are We Waking the Talk?" This theme raises a number of important research questions. To start, do we know if firms are actually attempting to implement their intended sustainability goals? How do we measure progress and outcomes of actions intended to contribute to a sustainable world? Are we using the right unit of analysis to measure if firms and governments are walking the Talk?

In their recent book, Van Tulder and Van Mil (2023) point to the intention–realization gap present in sustainable business efforts. They present four positions that firms typically take: "don't talk and don't walk," "talk, but don't walk," "talk and walk," and "walk, but don't talk" (pp. 466–467). The "don't talk and don't walk" perspective represents a neo-classical view of the responsibility of businesses to maximize profits for its shareholders without paying attention to actors in the broader society. This perspective is becoming more and more difficult to pursue not only because sustainability increasingly is becoming the most important topic in boardrooms (Eccles & Klimenko, 2019), but also because there is ample evidence that in some areas – in the introduction of environmental solutions – a positive return on investment already exists. So, it already "pays off" to invest in sustainable solutions. Consequently, over the past decades, firms have experienced growing pressures and expectations from governments and societies that have resulted in more attention to environmental and social issues, making "don't talk, don't walk" increasingly challenging for long-term competitiveness. More and more firms have started to talk about sustainability issues but may still struggle with implementing expressed intentions ("talk, but don't' walk"). These firms often get criticized for "greenwashing," although empirical IB research capturing such managerial mindsets as well as the effect on potential implementation obstacles for sustainability strategies remain rare. The third position, "talk and walk," reflects firms that actually develop resources and processes to act on sustainability intentions. Finally, some MNEs choose the position of "walk, but don't talk." While it may seem surprising that firms under-communicate their contributions toward a sustainable world, it may also be a way of managing expectations and avoiding criticisms from particular stakeholders that may still not be satisfied or that have another take on the proper combination of economic–social–environmental sustainability approaches (Van Tulder & Van Mil, 2023).

WHAT FORCES ARE DRIVING THE "WALKING AND TALKING"?

Academic approaches to increased responsibility and sustainability include terms like CSR; health, environment, and safety; and environment, social, and governance (ESG). The essence of these approaches is to broaden the perspective of a firm to take a more ethical approach in its operations. Central definitions capturing this notion include Bowens (1953) definition: "the obligations of businessmen to pursue those policies, to make those decisions or to follow those lines of action which are desirable in terms of the objective and values of our society" (p. 6), and the Brundtland Commission (1987) report's definition of sustainability: "meeting the needs of the present without compromising the ability of future generations to meet their own needs."

These definitions see the firm as more responsible to a broader range of stakeholders than its owners, or shareholders. From a stakeholder perspective, a business can be understood as a set of relationships among groups that have a stake in the activities that make up the business (Freeman, 1984; Parmar et al., 2010). The essence of this perspective is to link value creation to ethics, manage value creation and value capture among its stakeholders, including its employees, suppliers, customers, and the society at large. Thus, forces driving firms to "Walk the Talk" come from broad and diverse stakeholder groups that increasingly are concerned with the overuse of the planet's resources, and that are demanding firms to respond and take their share of the responsibility to mitigate these challenges.

A firm operates under a social contract with society, fulfilling explicit and implicit expectations of behavior (Meyer & Rowan, 1977). These expectations are not similar, nor static, but change over time with changing norms and values of the society. The firm needs to be compliant, that is, fulfill its obligations to society. Demands of compliance may be implicit but are also made explicit in industry standards. The latter demands of compliance have exploded in many industries, creating new jobs, such as compliance officers and firms helping companies fulfill these demands, but also increased the workload for firms. Compliance can be seen as a "myth and ceremony" (Meyer & Rowan, 1977), and lead to "greenwashing," where the firm selects to display only certain characteristics to satisfy core stakeholders while hiding the true nature of its sustainability. Studies show that the higher transparency of a firm's business to other actors, the less viable is a greenwashing strategy (Marquis et al., 2016). We can conclude that diverse forces are increasing the pressure on firms to stay legitimate and compliant, driving the need for firms to "Talk and Walk."

Walking and talking may drive the strategy of a firm if they manage to use sustainability as a way to differentiate their products and services (Bansal & Roth, 2000). Sustainability elements may include raw materials (such as recycled materials, or materials with lower emissions), processes (using less energy or waste in the process), or in the manner the product is delivered or used (transport, packaging, etc.). Schillebeeckx et al. (2022) find that firms are more willing to replace existing inputs with sustainable ones the higher the demonstrated effect on total sustainability (to customers). They also find that immaterial resources are more easily transformed. "Walking and talking" to gain a strategic edge are, therefore, an important force for change.

Individuals are driven by different motivations. Evolutionary theory shows that pro-social behavior may affect performance positively (Wilson et al., 2021). Idealistic owners, employees, and other stakeholders may drive "walking the Talk" discussions in meetings, over lunch, and in boardrooms. Research also shows that the evolution of norms and values are core motivations for higher sustainability actions in both talking and walking (Aguilera et al., 2021).

THE ROLE OF INSTITUTIONS PROMOTING "TALKING AND WALKING"

Both national and international institutions play a vital role when addressing the complexities of a grand challenge such as the transition toward a sustainable world. At the end of 2015, 196 nations agreed to a multilateral treaty on climate change, the Paris agreement. The agreement was signed and ratified in 2016.[1] This agreement recognizes that change cannot take place in isolation and that nations must work together toward a common goal. Another example is the Basel Convention on the control of transboundary movement of hazardous wastes that prevents firms from dumping hazardous materials in other countries. The convention was ratified and has been in operation since 1992. However, agreeing on a common set of rules can be very challenging. For example, the Hong Kong convention for safe and environmentally sound recycling of ships was drafted in 2009 but was not ratified by a sufficient number of countries until 2023 when Liberia and Bangladesh signed. It will not come into effect until 2025, when some will argue that it is already outdated. Others question how it will be implemented and controlled. These examples show the importance of institutions, but also their challenges in gathering sufficient support and making progress.

From an IB research perspective, this highlights the importance of taking a broader and richer institutional perspective into consideration. To date, most institutional research within IB has selected a narrower view, examining how specific formal or informal institutions influence firm-level decisions. The studies typically assume that specific institutional factors homogenously affect (i.e., constrain) firm-level decisions, without taking complexities and interdependencies into account (Aguilera & Grøgaard, 2019).

In a study comparing the climate change strategies to assess if European and North American firms are walking the Talk, Backman et al. (2015) found that European firms were better at communicating intentions and concerns about climate change. This brought attention to processes, government policies, and governance issues. However, the results show that North American firms had similar or better performance related to actual technological investments to address climate change. The authors point to the importance of good public policy that caters to firm heterogeneity.

The institution-induced link between Walk and Talk will be further enforced by more mandatory initiatives around the world on reporting regulation. Whereas in the past, most non-financial reporting was based on voluntary standards such as the Global Reporting Initiative, growing pressure from societal stakeholders, investors, and government agencies have made more transparent reporting on intentions and achievements on sustainability necessary. The practice of sustainability reporting in the past has thereby concentrated around the so-called "single materiality" issues that had direct effect on the operation of the company. Recent initiatives around reporting regulation, make reporting not only mandatory, but also add a "double materiality" perspective to reporting, that is, include outside-in considerations of the broader contribution of companies to societal issues. The most pervasive and far-reaching initiative at the moment concerns the European Corporate Sustainability Reporting Directive issued by the European Union (EU), which from 2024 onwards demands from companies above a specific size – involving more than 50,000 companies that are active in Europe – to publish their double materiality intentions (their Talk) and realizations (their Walk). The increasing comparability (due to harmonization of requirements) of reporting initiatives will provide a rich source of research material for IB scholars. In the past, the study of comparative MNE strategies on the basis of annual CSR reports was criticized for being based on "secondary" (ideologically biased) information. The new wave of corporate reporting will make it possible to compare reported "primary" information (on "Talk" as well as "Walk") on a massive scale for large numbers of MNEs. As a consequence, techniques like narrative analysis will become more feasible and robust for IB scholarship in this area.

"WALKING THE TALK" BY MAKING SUSTAINABILITY PART OF YOUR STRATEGY

"Walking the Talk" can be particularly challenging when sustainability initiatives and actions are detached from core business. Porter and Kramer (2006, 2011) suggested that to truly "Walk the Talk," companies should move away from generic "good citizenship" and instead try to incorporate sustainability thinking into improvements of value chain activities and prioritize social dimensions that improve the firm's competitive context. This creates win–win situations, referred to as shared value, that incentivize firms to commit to a long-term sustainability thinking that benefits both the firm and society-at-large.

According to an article posted by the World Economic Forum, 90% of executives agree that sustainability is important. However, only about 60% of surveyed firms had clear sustainability strategies or had established chief sustainability officer positions that reported directly to the board (Rafi, 2022). Comparable results were found with frontrunner MNEs. A UN Global Compact and Accenture (2019) study found, for instance, that 67% of their corporate signatories were committing to sustainability at the CEO level, yet only 48% had been implementing sustainability into their operations.

There are multiple drivers that contribute to making sustainability part of firms' strategies. As discussed earlier, national and transnational commitments to sustainability goals such as the Paris Agreement and UN SDGs may lead to regulatory changes that in turn put pressure on firms to change. Similarly, investors have increasingly started asking questions about the sustainability orientation of their assets. Some investors such as BlackRock take a proactive role in contributing to developing a strategic focus in firms (BlackRock, 2023). However, having vocal investors that are focused on sustainability is not a guarantee for a more strategic approach to "walking the Talk." A recent study of foreign institutional owners found that foreign institutional investors may indeed push for a stronger communication of environmental performance, but this does not necessarily translate into a willingness to invest into longer-term environmental innovation (Ellimäki et al., 2023).

One particular challenge that MNEs face is how to *internally align* all areas of management across borders. Strategic alignment "ensure that all components of the organization work towards the core goals and the company's vision" (Morrison, 2014), to help ensure that whatever activity takes place, resources are expended only when they contribute to the achievement of the overall business strategy (Birchall, 2004). This is also known as "coupling," the degree to which decisions made in one channel are associated with decisions made in other channels (Ocasio & Joseph, 2005). When the different departments, business units, or functional areas of management all do different things with the same perceived purpose, the firm can become its own major obstacle for realizing change. Overcoming these internal obstacles in a "learning-by-doing" manner triggers processes of "strategic tinkering" or "bricolage," which are extremely difficult to assess from a scientific point of view: are MNEs actually creating the conditions within the organization for higher levels of sustainability to thrive?

Successful transition strategies include both coupling and decoupling decisions. Decoupling for sustainability strategies need to be resolved over time, or otherwise the company will face fragmented strategies that do not surpass an inevitable "muddling through" or "tinkering" phase (Crilly et al., 2012). IB scholarship can add value in research on how to deal with timely coupling and decoupling processes, particularly research related to international management questions focusing on centralization–decentralization and/or examine the "global" versus the "local." Multi-domestic MNEs adapt to each of the local sustainability circumstances (changing issues, stakeholders, and regulation) and will face internal coordination and alignment problems which makes it difficult to couple strategies. MNEs with global strategies have been found to support international/multilateral standards for sustainability (such as Global Compact or ISO standards). The effectiveness of their strategy depends on the degree to which their local units in the rest of the world will adapt to these standards (Van Tulder, 2015).

THE IMPORTANCE OF COLLABORATION TO BE ABLE TO "WALK THE TALK"

Transforming firms and industries to a more sustainable future is challenging, as these issues are complex (characterized by many interactions and associations), uncertain (how they evolve is difficult to forecast), and evaluative (cutting across many jurisdictional boundaries) (Ferraro et al., 2015). Although firms have an important role to play, it is increasingly apparent that stand-alone efforts will not be sufficient to reach global agreements or goals of a more sustainable world. As Equinor's[2] CEO stated, "Tackling climate change will require the largest public-private partnership the world has ever seen…Without a global perspective, we risk national measures with little real impact, reducing the opportunity to reach our common goal" (Equinor, 2016). This suggests that we must look at collaborations beyond dyadic firm relationships and examine collaborations that transcend firms, industries, and nations. These collaborations involve multiple parties with different characteristics, posing increased challenges to run collaborative processes as well as govern them and reaching fair and desired outcomes over time (Gray & Purdy, 2018).

Collaborations may be instigated by firms within the same country and industry, but increasingly there is a need to work together on solutions spanning many industries and contexts. In the maritime sphere, for example, the issue of maritime ocean plastics is a concern to many countries. Plastic from ocean industries, such as fishing, and aquaculture operations and general waste are polluting beaches and seas globally, creating problems for creatures of the sea as well as industries like tourism. Legislation like the extended producer responsibility (directive by EU)[3] makes plastics producers responsible for the entire lifetime of the product, including recycling. These types of initiatives have stimulated collaborations with all actors along the value chain to increase circularity and reduce sea pollution. Marine plastics have been developed as a material that can be used to make products such as furniture, socks, and football nets, innovations that have all taken place in collaborations.

Extending and changing value activities are often driven through collaborative efforts, simply because many actors are needed to create a new ecosystem. For example, the change from diesel to alternative low emission fuels such as hydrogen or ammonia requires infrastructure in all ports if ships are going to change to low emission fuels. Capturing carbon from industrial production or waste sites require the development of new value chains and providers, including designing and constructing new types of ships to transport the carbon from land to deep sea storage. These types of changes require new producers, suppliers, products, services, and a totally new infrastructure. Ecosystems are groups of firms with products and/or services that together comprise a coherent solution (Adner & Kapoor, 2010), consisting of complementary actors that collaborate to create value and compete to capture it (Hannah & Eisenhardt, 2018). Understanding how to enter into collaborative agreements within new ecosystems, develop, and benefit from these collaborations over time, become central for firms that want to Walk the Talk, and constitutes an important research area for IB scholars.

The ethical communication literature (Schoeneborn et al., 2020) adds a formative aspect to the discourse in which various communication strategies can help involve internal and external stakeholders in proactive change strategies. Schoeneborn et al. (2020) discern between three formative approaches at increasing levels of stakeholder involvement: (1) "talking-to-Walk" reflects an "outside-in" approach to communication as "aspirational Talk" in which sustainability ambitions and intent ("Talk") precede practice and performance ("Walk"); (2) "walking-to-Talk" represents an "inside out" approach to CSR communication in which practiced and reported "Walk" – combined with stakeholder dialogues – precedes future-oriented "Talk" and enables extend sustainability ambitions; (3) "T(w)alking" involves the most interactive approach in which stakeholders and companies co-create a sustainability approach which involves dilemma sharing, deliberations on intricate trade-offs, and the like.

CHALLENGES TO MNEs WHEN "WALKING AND TALKING"

Transforming industries will change how MNEs conduct business, and the role they choose to take in this transformation. An MNE can take a leading role, actively engaging with governments and collaboration partners to develop new solutions, or it can take a more passive role, staying compliant. Both strategies, however, take place in a context that is characterized by digitalization and fast technological changes, environmental crises, and geopolitical events (George & Schillebeeckx, 2022), which have fundamental implications for MNEs.

As the demands of stakeholders increase, where do the responsibilities of the MNE stop? And what do these increased responsibilities do for MNE business models? How can MNEs account for responsible actions of its suppliers' sub-suppliers, and their sub-sub-suppliers? Fehlner, in Chapter 15 of this volume, argues that the transformation of value chains to increase circularity will increase transaction costs. Could one implication of this overall transformation to sustainability be more vertical integration within the MNE?

A characteristic and value creating opportunity of the MNE is its ability to utilize location differences. Elia et al. (Chapter 18) find that there is a difference between CSR expectations in mature and emerging economies. Could this difference open for "sustainability reporting" arbitrage, or different sustainability strategies dependent on the maturity of the context? Concerns about "pollution havens" and a "race to the bottom" have been voiced for MNE activities. Bu and Wagner (2016) found that foreign investment may indeed contribute to a race to the bottom, but also to a race to the top since MNEs with strong environmental capabilities invest in context with more stringent environmental policies.

The transformation of industries will lead to new ecosystems and technological innovations within these. Where will these innovations take place, within established MNEs or will they come from new entrepreneurial ventures? How will the MNE implement, organize, and govern innovations together with other actors?

DEVELOPING AND LEVERAGING CAPABILITIES TO "WALK THE TALK"

The question of which capabilities are needed in the transformation toward sustainability, is complex. Which existing capabilities can the firm leverage to become more sustainable, and which capabilities do firms need to develop to succeed in the long term? One of the main challenges for firms is how to balance existing operations that may be less sustainable with new opportunities or new regulations. Managers must determine how to "balance the potential contribution of an increase in resources devoted to environmental performance against the risks involved with such a decision" (Rugman & Verbeke, 1998). This conundrum is particularly challenging for producers and major users of fossil-fuel based energy. For example, multinational oil and gas companies such as BP, Shell, and Equinor act on opportunities that leverage existing capabilities. These companies have made substantial investments into offshore wind, that build on current strengths such as experience with offshore installations, geological, and engineering knowledge, and the experience of running large-scale development projects. While these firms benefit from leveraging existing firm-specific advantages, they must cope with the increased risks related to uncertain market developments and changing regulatory frameworks. The sustainable changes may require new business models and lead to new criteria to assess profitability and return on investments.

In many cases, firms also need to develop green capabilities to support a sustainable transition (Rugman & Verbeke, 1998). This can be challenging in dynamic environments with changing stakeholder needs. Maksimov et al. (2022) suggest that global connectedness help firms develop green capabilities that support a shift toward environmental sustainability in dynamic stakeholder environments.

Increasingly, the focus is not only on leveraging and developing capabilities, but also on attracting the right talent for the future. Younger generations are placing sustainability at the forefront of their career choices. According to a recent survey by Yale University, "51 percent of students said they would accept a lower salary to work for an environmentally responsible company" (Gross, 2021). Attracting the right talent, and keeping them over time, is, therefore, a major challenge.

THE CONTRIBUTIONS OF THIS VOLUME

All chapters in this volume address various aspects of the challenges of "walking the Talk" using different theories and perspectives. The contributions to the volume are clustered along four core themes that all have a bearing on overcoming the gap between "Walk" and "Talk": (1) Walking the Talk in International Business Studies, (2) Walking the Talk in Sustainability Issues, (3) Walking the Talk at the Interface with Government Policies, and (4) Walking the Talk in International Value Chains.

Part I: Walking the Talk in International Business Studies

Part I focuses on fundamental theoretical and analytical challenges that IB scholarship is faced with. The leading article is written by the Alain Verbeke, to whom the whole volume is a tribute. Three other contributions are written on invitation by scholars that have been strongly influenced by the work of Alain Verbeke.

In Chapter 2, Alain Verbeke presents his personal vision on the relevance of the topic of "Walk and Talk." By pointing at the findings of his pioneering research over the years, he argues that commonly held beliefs in IB, in particular the suggestion that many MNEs are "global" players, have never been based on factual evidence. Rather, ideologies and gurus have dominated the field and also influenced IB scholarship. His advice is to make present IB scholarship more relevant by applying "first principles" of evidence-based IB thinking – the most important first principle being that entrepreneurially oriented firms seek to create and capture economic value through cross-border linkages.

Jean-François Hennart, in Chapter 3, argues that the IB literature has by and large overestimated the role played by technology and at the same time underestimated the strategic importance of complementary resources. Complementary resources are needed to exploit intangibles such as government permits, natural resources, or access to end consumers. These are typically owned by governments, firms, and individuals of countries. The way that MNEs deal with the discrepancy between "wishes" and "reality," determine the speed at which MNEs develop their foreign sales and sustainable footprints.

Thomas Hutzschenreuter in Chapter 4 takes another fundamental IB theme into account: distance. More specifically, he considers distance as the main barrier between an MNE's global intentions and its regional realizations. Hutzschenreuter continues the research pioneered by Rugman and Verbeke in several different directions. He argues that further theoretical and empirical development on concepts like "bounded rationality" or "bounded reliability" are needed to create a deeper understanding of internationalization and (multi)regional strategies.

Chapter 5 by Liena Kano and Luciano Ciravegna, take yet another fundamental question into account to characterize Alain Verbeke's key contribution to IB scholarship and practice: governance for sustainability. By arguing that IB theory should be based on a realistic set of assumptions about the behavior of managers, policy-makers and other market and non-market stakeholders, they complement the argument as presented by Verbeke in Chapter 2. Kano and Ciravegna illustrate this point further by looking at the way stated ESG goals of governments should be aligned with bounded rationality and reliability challenges. Misalignment lowers the effectiveness of governance initiatives.

Part II: Walking the Talk in Sustainability Issues

The contributions in Part II focus on a selected number of key environmental and social issues that are relevant for IB scholarship in the coming years. Many of the chapters are exploratory and conceptual in nature and aimed at providing frameworks for further research in those areas where MNEs must develop sustainable strategies.

Jacqueline Mees-Buss, in Chapter 6, takes a strategic and managerial approach to the Walk–Talk challenge. She considers four types of challenges that can be linked to four distinct strategies. She focuses on the discrepancies that senior managers in large MNEs perceive of what is expected of them (as social and environmental responsibilities) and how to implement these. Four theoretical quadrants represent a strategic choice menu for managers. Mees-Buss's main argument is that rather than adjudicate between them, a holistic interpretation of CSR takes all four types into account. Managers that aim at a more holistic approach, have to be able (and willing) to deal with the "nasty trade-offs" between each of these strategies.

Stefan Zagelmeyer in Chapter 7 considers what is needed to enhance the legitimacy of MNEs by strengthening their approach to Human Rights. The chapter aims at developing an analytical framework on corporate human rights reporting. Special emphasis to the IB discourse is provided by paying attention to external environmental factors influencing the strategies of business organizations that operate internationally.

In Chapter 8, Viviana Pilato and Ari van Assche consider "cross-border adjustment mechanisms" as one of the strategic instruments that MNEs operating in the EU have adopted to prevent carbon leakage. Cross-border adjustment mechanisms are relevant for MNEs that transfer carbon-intensive production activities to countries with laxer environmental constraints (pollution havens). The chapter analyzes the logic and the influence on sectoral MNE activities of a new policy initiative called the carbon border adjustment mechanism that the EU introduced in May 2023 to fight carbon leakage.

The final contribution to this part is by John Holland (Chapter 9) who reports about his effort to help large financial firms to go green in an authentic way – and consequently withstand what he calls "organized hypocrisy." He designs a conceptual framework that can support a green "Behavioural Theory of the Financial Firm' (green BTTF). The BTTF develops an international version of the resource-based view of the firm. His largely explorative contribution aims at overcoming the gap between "talking" and "walking" in the financial sector. It defines testable firm-level hypotheses for "Green Finance," "Net Zero action" as well as – by implication – tests for "greenwashing."

Part III: Walking the Talk at the Interface with Government Policies

Part III focuses on the way sustainability strategies materialize at the intersection between governments and firms.

In Chapter 10, Guus Hendriks examines Chinese foreign-aid public–private partnerships in Africa, comparing projects undertaken by Chinese firms versus government agencies. He identifies differences between projects that raise both ethical and theoretical questions, pointing to the complexities of public–private partnerships and the need for more research on the role of for-profit firms in these projects.

In Chapter 11, Martina Barbaglia, Roberto Bianchini, Vincenzo Butticè, and Stefano Elia point to the dynamic interface of governments and firms, and how strict home country regulations can re-direct perceptions of opportunities in the home country. They find that sustainability-oriented firms have a higher propensity to de-internationalize. Furthermore, the stringency of environmental regulations in the home country increases the likelihood of de-internationalization. The study provides important nuances to our knowledge of factors influencing de-internationalization.

In Chapter 12, Alessandro Lampo and Susana C. Silva examine the role of early-adopters of electric vehicles to understand their influence on the innovation and broader diffusion of technology to global consumers. The study is situated in Macau, where the government has proactively pushed to a transition to sustainable transportation. Their study offers an interesting illustration of interrelationships between governments, firms, and consumers.

In Chapter 13, Gunnar Leymann and Anna Kehl provide a thorough overview of concepts, drivers, and strategies identified in IB research on carbon performance. They point to the complexities of institutional environments and develop a conceptual framework to guide future research. Their framework takes spatial diversity and temporal dynamism into consideration. This chapter is an excellent starting point for scholars interested in research on MNEs strategic reactions to climate policies.

Part IV: Walking the Talk in Global Value Chains

Part IV relates challenges of sustainability to global value chains (GVC). How are these chains affected by sustainability demands from major stakeholders?

Mohammad B. Rana and Matthew M. C. Allen take a broad approach to climate change strategies and sustainability and address these topics from IB and GVC perspectives in Chapter 14. Using illustrative cases from the global north and south, they discuss the changing nature of institutions, markets, and technologies and the implications of these changes for how firms develop their climate change strategies.

In Chapter 15, Corina Fehlner explores the lens of transaction cost theory on circular economy. Her analysis shows that rising transaction costs may pose a barrier for MNEs to transform their business models from linear to circular. This transformation leads to different changes in the MNEs value activities which may have implications for efficient costs allocations, location, and organization. With this, the chapter demonstrates the value of applying an efficiency lens to circular business transformations.

Cristina Di Stefano, Stefano Elia, Paola Garrone, and Lucia Piscitello examine principles of circular economy on activities of GVCs in Chapter 16. Circularity can be obtained by reducing, reusing, or recycling materials, and each of these principles may have different implications for the GVC activities, their localization, as well as governance. This insight opens for various interesting strategies for firms exploring circularity in a global context.

Lise Lillebrygfjeld Halse examines the relation between sustainability and backshoring strategies in Chapter 17. The chapter shows how the motives for backshoring may be more complex than for offshoring and may include other motives than pure costs but be a decision with broader implications for the GVC. Using illustrative cases from the maritime industry, she finds sustainability to be an emergent motive, and uses institutional theory to develop a comprehensive model linking sustainability motives, backshoring decisions, and benefits of sustainability.

In Chapter 18, Stefano Elia, Gezim Hoxha, and Lucia Piscitello examine the role that country-specific factors play on shareholder expectations from CSR and irresponsibility. Building on stakeholder and legitimacy perspectives they argue that shareholders in emerging economies have lower expectations to firms' social behavior, and therefore they reward positive CSR actions and punish negative CSR incidents to a lesser extent than shareholders in developed markets. They find support for these hypotheses, and thus manage to demonstrate a very interesting contextual and global business-related implication of CSR strategies.

In Chapter 19, Cyntia Vilasboas Calixto Casnici, Germano Glufke Reis, David Schulzmann, Marina Papanastassiou, and Jeremy Clegg point to the important roles that non-governmental organizations (NGOs) can take in shaping GVCs. Using a case study of an international NGO, they explore the influence of NGOs on innovation and the transition toward a sustainable GVCs with the agri-food industry. It is an excellent contribution to the GVC literature, showing that NGOs can also play a leading role, which has so far often been attributed MNEs.

We hope that you will enjoy these scholarly contributions as much as we have, and that they will inspire many more studies on sustainability in IB and "walking the Talk" in the future.

NOTES

1. The United States under the Trump administration withdrew from the agreement in 2020 to return to it in 2022 under the Biden administration. As of June 2023, 194 nations plus the EU have ratified the agreement, representing almost all global greenhouse emissions.
2. Equinor is a global energy company. It is also Norway's largest multinational company (measured by market capitalization).
3. https://www.compliancegate.com/eu-extended-producer-responsibility/

REFERENCES

Adner, R., & Kapoor, R. (2010). Value creation in innovation ecosystems: How the structure of technological interdependence affects firm performance in new technology generations. *Strategic Management Journal, 31*(3), 306–333.

Aguilera, R., Aragon Correa, J. A., Marano, V., & Tashman, P. A. (2021). The corporate governance of environmental sustainability: A review and proposal for more integrated research. *Journal of Management, 47*(6), 1468–1497.

Aguilera, R. V., & Grøgaard, B. (2019). The dubious role of institutions in international business. A road forward. *Journal of International Business Studies, 50*(1), 20–35.

Backman, C. A., Verbeke, A., & Schulz, R. A. (2015). The drivers of corporate climate change strategies and public policy: A new resource-based view perspective. *Business & Society, 56*, 545–575.

Bansal, P., & Roth, K. (2000). Why companies go green: A model of ecological responsiveness. *Academy of Management Journal, 43*, 717–736.

Bass, E. A., & Grøgaard, B. (2021). The long-term energy transition: Drivers, outcomes, and the role of the multinational enterprise. *Journal of International Business Studies, 52*(5), 807–823.

Birchall, J. (2004). *Cooperatives and the millennium development goals*. ILO.

BlackRock. (2023). *About us*. https://www.blackrock.com/ca/investors/en/about-us/investment-stewardship

Bowen, H. R. (1953). Social Responsibilities of the Businessman. University of Iowa Press, Iowa City. Federal Council of Churches of Christ in America.

Brundtland Commission. (1987). *Our common future: World commission on environment and development*. UN.

Bu, M., & Wagner, M. (2016). Racing to the bottom and racing to the top: The crucial role of firm characteristics in foreign direct investment choices. *Journal of International Business Studies, 47*(9), 1032–1057.

Buckley, P., Doh, J., & Benischke, M. (2017). Towards a renaissance in international business research? Big questions, grand challenges, and the future of IB scholarship. *Journal of International Business Studies, 48*(9), 1045–1064.

Crilly, D., Zollo, M., & Hansen, M. (2012). Faking if or muddling through? Understanding decoupling in response to stakeholder pressures. *Academy of Management Journal, 55*(6), 1429–1448.

Eccles, R. G., & Klimenko, S. (2019). The investor revolution. *Harvard Business Review, 97*, 106–116.

Ellimäki, P., Aguilera, R. V., Hurtado-Torres, N. E., & Aragón-Correa, J. A. (2023). The link between foreign institutional owners and multinational enterprises' environmental outcomes. *Journal of International Business Studies, 54*(5), 910–927.

Equinor. (2016). https://www.equinor.com/news/archive/well-positioned-low-carbon-future

Ferraro, F., Etzion, D., & Gehman, J. (2015). Tackling Grand Challenges pragmatically: Robust action revisited. *Organization Studies, 36*(3), 363–390.

Freeman, R. E. (1984). *Strategic management: A stakeholder approach*. Pitman.

George, G., & Schillebeeckx, S. J. (2022). Digital transformation, sustainability, and purpose in the multinational enterprise. *Journal of World Business, 57*(3), 101326.

Gray, B., & Purdy, J. (2018). *Collaborating for our future: Multistakeholder partnerships for solving complex problems*. Oxford University Press.

Gross, J. (2021, November, 14). Business schools respond to a flood of interest in E.S.G. *New York Times*.

Hannah, D. P., & Eisenhardt, K. M. (2018). How firms navigate cooperation and competition in nascent ecosystems. *Strategic Management Journal, 39*(12), 3163–3192.

Maksimov, V., Wang, S. L., & Yan, S. (2022). Global connectedness and dynamic green capabilities in MNEs. *Journal of International Business Studies, 53*(4), 723–740.

Marquis, C., Toffel, M. W., & Zhou, Y. (2016). Scrutiny, norms, and selective disclosure: A global study on greenwashing. *Organization Science, 27*(2), 483–504.

Meyer, J. W., & Rowan, B. (1977). Institutionalized organizations: Formal structure as myth and ceremony. *American Journal of Sociology, 83*(2), 340–363.

Mintzberg, H., Ahlstrand, B., & Lampel, J. (2009). *Strategy safari: A guided tour through the wilds of strategic management* (2nd ed.). FT Press.

Morrison, J. (2014). *The social license to operate. How to keep your organization legitimate*. Palgrave MacMillan.

Ocasio, W., & Joseph, J. (2005). An attention-based theory of strategy formulation: lining micro- and macro-perspectives in strategy processes. In *Strategy process* (pp. 39–61). Emerald Group Publishing.

Parmar, B. L., Freeman, R. E., Harrison, J. S., Wicks, A. C., Purnell, L., & De Colle, S. (2010). Stakeholder theory: The state of the art. *Academy of Management Annals*, *4*(1), 403–445.

Porter, M. E., & Kramer, M. R. (2006). Strategy & society: The link between competitive advantage and corporate social responsibility. *Harvard Business Review*, *84*(12), 78–92.

Porter, M. E., & Kramer, M. R. (2011). Creating shared value. *Harvard Business Review*, *89*(1/2), 62–77.

Rafi, T. (2022). *Why sustainability is crucial for corporate strategy*. World Economic Forum. https://www.weforum.org/agenda/2022/06/why-sustainability-is-crucial-for-corporate-strategy/

Rasche, A., & Waddock, S. (2014). Global sustainability governance and the UN Global Compact: A rejoinder to critics. *Journal of Business Ethics*, *122*(2), 209–216.

Rugman, A. M., & Verbeke, A. (1998). Corporate strategies and environmental regulations: An organizing framework. *Strategic Management Journal*, *19*(4), 363–375.

Schillebeeckx, S. J., Kautonen, T., & Hakala, H. (2022). To buy green or not to buy green: Do structural dependencies block ecological responsiveness? *Journal of Management*, *48*(2), 472–501.

Schoeneborn, D., Morsing, M., & Crane, A. (2020). Formative perspectives on the relation between CSR communication and CSR practices: Pathways for walking, talking, and t(w)alking. *Business & Society*, *59*(1), 5–33.

UN Global Compact & Accenture. (2019). *The decade of delivery. A call to business action* (The United Nations Global Compact – Accenture Strategy CEO Study on Sustainability). UNGC.

Van Tulder, R. (2015). Getting all the motives right. A holistic approach to internationalization motives of companies. *The Multinational Business Review*, *23*(1), 36–56.

Van Tulder, R., & Van Mil, E. (2023). *Principles of sustainable business*. Routledge.

Wilson, D. S., O'Brien, D. T., & Sesma, A. (2021). Human prosociality from an evolutionary perspective: variation and correlations at a city-wide scale. *Evolution and Human Behavior*, *30*, 190–200.

Yu, H., Bansal, P., & Arjaliès, D.-L. (2023). International business is contributing to environmental crises. *Journal of International Business Studies*, *54*, 1151–1169.

PART I

WALKING THE TALK IN INTERNATIONAL BUSINESS STUDIES

CHAPTER 2

EVIDENCE-BASED INTERNATIONAL BUSINESS SCHOLARSHIP

Alain Verbeke

Haskayne School of Business, University of Calgary, Canada

ABSTRACT

"First principles" of international business (IB) thinking should be applied systematically when assessing the functioning of internationally operating firms. The most important first principle is that entrepreneurially oriented firms seek to create, deliver and capture economic value through cross-border linkages. Such linkages invariably require complementary resources from a variety of parties with idiosyncratic vulnerabilities to be meshed. Starting from first principles allows bringing to light evidence-based insight. For instance, most companies are not global and even the world's largest firms rarely change the location of key strategic functions. International new ventures (INVs), emerging economy multinational enterprises (MNEs) and family firms face unique vulnerabilities but also command resources that can be used to create value across borders. The quest for "optimal" international diversification appears to be a futile academic exercise, and in emerging economies with institutional voids, relational networks – and more broadly, informal institutions – are unlikely to function as scalable substitutes for formal institutions. In global value chains (GVCs), many lead firms and their partners have been able to craft governance mechanisms that reduce bounded rationality and bounded reliability challenges, and it is also critical for them to use governance as a tool

Walking the Talk? MNEs Transitioning Towards a Sustainable World
Progress in International Business Research, Volume 18, 19–41

ISSN: 1745-8862/doi:10.1108/S1745-886220240000018002

to create entrepreneurial space. Finally, many of the world's largest companies have been on successful trajectories toward reducing their climate change footprint for a few decades. But these firm-specific trajectories are fraught with challenges and cannot just be imposed via unilateral, macro-level targets decided upon by individuals and institutions lacking a clear understanding of innovation and capital expenditure processes in business.

Keywords: Multinational enterprise; firm-specific advantages; regional strategy; formal institutions; climate change strategies; global value chains; entrepreneurial space; bounded rationality; bounded reliability

1. INTRODUCTION

Evidence-based international business (IB) scholarship reflects the purposive and judicious use of the best available intelligence to describe and explain the unfolding of IB activities, with a prime focus on the parties directly involved in these activities. For instance, when discussing the effects of multinational enterprise (MNE) entry, it is important to acknowledge that MNE activity is often actively sought by host country economic actors, or by host country public policy makers and economic development agencies. It is even more important to compare the entire spectrum of effects linked to MNE entry with real-world alternatives and with the past "status quo" situation.

IB scholarship's role is to disseminate factual knowledge on what firms engaged in IB actually do, as opposed to what some believe they should do. In this brief contribution, I describe what I view as the first principles of evidence-based IB scholarship, and then provide a variety of illustrations thereof, building upon my own research.

2. FIRST PRINCIPLES OF IB SCHOLARSHIP

Language and parsimony matter in any field of scientific inquiry, and IB scholarship is no exception to this rule (Santangelo & Verbeke, 2022). Any field requires a limited number of concepts that – as building blocks and in combination – permit describing with some accuracy a wide range of sometimes complex phenomena. Much empirical evidence suggests that entrepreneurially oriented business firms engage in cross-border transactions and operations to create and capture (or to put it more neutrally, "distribute") economic value, meaning also that customers are able and willing to pay for the goods and services produced. If more net economic value could be created and captured by operating domestically rather than through international linkages, there would be no clear incentive for these firms to establish such linkages. I use the term "entrepreneurially oriented" in the sense that few international transactions and operations are exact copies of what a firm may already have done domestically: pure repeatability plays are rare in IB contexts because of various "distance" dimensions, and therefore entrepreneurial judgment will be required to identify IB opportunities and to enact international transactions and operations (Narula & Verbeke, 2015; Verbeke & Ciravegna, 2018; Verbeke & Lee, 2021; Verbeke & Yuan, 2022). Any deviation from a pure repeatability play demands new resource recombination to create economic value, and international activities therefore mostly need to enlist complementary resources held by foreign parties (Hennart & Verbeke, 2022).

The full set of "relevant parties" involved in IB includes both the parties actively engaged in international transactions and operations, and other parties in a position to affect – or that can be affected by – these business activities, for instance government agencies (even if only for granting permits to operate) and activist stakeholders demanding more attention to societal goals. Who should be considered a relevant party changes over time and across space, and what constitutes value and how it is distributed can also be affected by these relevant parties. As one example, much of the environmental, social and governance (ESG) agenda, meaning that firms should meet some level of ESG performance standards, largely originated with parties other than the main players engaged directly in cross-border business. Expectations as to what constitutes acceptable or superior ESG performance, and how the additional "societal value" created affects economic value, have also tended to vary over time and across space (Bebchuk et al., 2022; Bebchuk & Tallarita, 2020; Edmans, 2023). ESG metrics in some shape or form have now become part of assessments by relevant parties as to expected and realized value creation from international activities, and they can also be important in the realm of value distribution (Kaplan & McMillan, 2021; Linnenluecke, 2022).

More generally, both, what could reasonably be included in measuring value and who should be considered the relevant parties to business transactions and operations, have continuously been broadened in many jurisdictions during the past decades, especially since the publication of the 1987 Brundtland Report, *Our Common Future*, from the United Nations World Commission on Environment and Development. Internationally operating firms have faced rising pressures to internalize perceived externalities, thereby adding costs for the parties directly involved in transactions and operations, and often reducing the net economic value thereof accruing to them. Especially the largest and most highly visible MNEs have also been pressured to pursue ambitious and wide-ranging societal value targets, determined by a variety of parties outside the narrow realm of those directly involved in the unfolding of IB activities. Assumed contributions to these societal goals, for instance through developing and exploiting renewable energy resources or by bringing into the market goods such as electric vehicles (EVs) with immediately visible environmental impacts that are comparatively low, have been viewed as worthy of receiving government subsidies and tax incentives (Ramkumar & Eckert, 2023). As a result, the relative attractiveness and viability of many IB value propositions can be affected, for instance in energy-intensive and energy-producing industries (cf. Temple-West, 2021; Verbeke & Hutzschenreuter, 2021a).

Value distribution is also increasingly viewed by "new parties" as an instrument to pursue presumed societal goals. This can occur at the expense of acceptable value distribution among directly involved parties and even make some projects non-viable, though some recent academic literature on GVCs appears to prescribe a stronger focus on value distribution. A positive response to such pressures can become a precondition to receiving and maintaining a social license to operate, especially for foreign owned and controlled companies but can have unintended effects (cf. Van Assche et al., 2023; Van Assche & Narula, 2023).

The distinct feature of international transactions and operations – as opposed to domestic ones – is that foreign firms typically start their activities as relative outsiders facing additional hurdles and burdens, as compared to domestic companies in their home environment. Internationally operating firms almost by definition need to consider more parties relevant to value creation and distribution than would be the case in a purely domestic setting. This holds for both the world's largest MNEs and for the smallest exporters (Verbeke et al., 2019; Verbeke & Ciravegna, 2018). Foreign suppliers of intermediate inputs, employees, distributors, government regulators, societal activists, etc., typically all differ from home country counterparts. All these relevant parties somehow bring complementary resources to the table – including decision rights on a foreign firm's social license to operate – that must be meshed with extant resources for mutually agreed upon value creation and distribution to materialize.

Much guru- and activist talk has focused on the resources-superiority, exploitative power and global scope of internationally operating companies, the books by Naomi Klein – including *No Logo* – being a case in point (Klein, 2000). The empirical evidence mostly does indeed suggest that internationally operating companies command some form of "firm-specific advantages" (*FSAs*) when entering foreign countries (Kano & Verbeke, 2019; Verbeke & Kano, 2016). But in parallel, the reality is often also one of great vulnerability as exemplified by Elon Musk – the world's richest man and founder of Texas-based EV and clean energy company Tesla – agreeing in July 2023 to promote "core socialist values" in China, clearly as an imposed precondition for Tesla's retaining a social license to operate and for further successful business activities in that country (Huang & Kubota, 2023). Pioneering technology as a non-location bound FSA, on its own does not guarantee the commercial viability of international operations.

In my own 35 years of engagement with IB scholarship, most of what I have observed are firms and their senior leadership trying to cope with multifaceted vulnerabilities and attempting to craft efficient governance structures and practices when operating abroad. Even digital firms with FSAs that in theory are easily and cheaply transferable globally suffer from immense vulnerabilities when leaving their dominant positions in the home country (Verbeke & Hutzschenreuter, 2021b). How exactly value is defined, how much of it can be created and how it will be distributed, turns on the interdependencies at hand among the relevant parties' resources, with timing and spatial elements looming large. But the prerequisites for efficient governance – as an intermediate output permitting feasible value creation and distribution – are universal.

Specifically, all parties to business transactions and operations suffer from *bounded rationality* and *bounded reliability* challenges (Kano & Verbeke, 2015). In the IB context, these two challenges are compounded: more parties are involved than in domestic settings, and these parties are dispersed across geographic space, with each party having its own idiosyncratic views on how best to create and distribute value.

Bounded rationality refers to information problems such as the absence of full and accurate information, limited information processing capacity, uneven focus on various information facets and diverging assessments among parties on the meaning of the same information facets. Bounded rationality thus reflects "scarcity of mind" and it is omnipresent at the individual level, the level of units within an organization, the level of organizations in their entirety and in interorganizational settings (Verbeke & Yuan, 2005).

Bounded reliability refers to human frailty and includes opportunism, benevolent preference reversal (especially reprioritization and scaling back on overcommitment) and identity-based discordance (most notably divided engagement and regression). Bounded reliability thus reflects "scarcity of effort towards making good on open-ended promises." It can also be observed at a variety of levels of analysis (Kano & Verbeke, 2015; Verbeke & Greidanus, 2009; Verbeke & Kenworthy, 2008).

The time- and place-based situational context, as well as the specificity of spatial interactions among relevant parties, will influence the scope and levels of bounded rationality and bounded reliability problems that must be addressed, as well as the mechanisms to be deployed for alleviating these challenges. Given the unavoidable presence of bounded rationality and bounded reliability, the nature of the resources brought to the table by the various parties to jointly create value will also affect how these challenges can best be handled.

What can be considered efficient *structural* governance such as the "optimal" mix of entry modes for an MNE and its partners (Benito et al., 2009; Buckley & Casson, 1976) and efficient *strategic* governance, such as the best available managerial practices to be adopted by the parties to coordinate their activities and to mitigate internal conflicts, indeed depends on the characteristics of the resources involved and the interdependencies among them. Characteristics of one set of resources, such as proprietary knowledge or valuable brand names, cannot be viewed in isolation from other resources instrumental to creating and capturing value, such as production capacity or a distribution apparatus (cf. Hennart, 2009). This perspective is closely aligned with David Teece's (1986, 2018) influential work on complementary assets in the realm of profiting from innovation.

Important characteristics to consider in the context of meshing *complementary resources* include: (1) specific interdependencies among these resources (such as pooled, sequential and reciprocal interdependencies to carry out a value creating activity); (2) the extent to which the various resources can (or cannot) be accessed on open markets, and in case of difficult access whether these could be replaced by other resources; (3) the resources' respective levels of tacitness, and the ease with which they can be appropriated or imitated and exploited by other parties; and (4) the value attributed to the resources by customers of final products, etc.

What can reasonably be labeled efficient governance in a real-world sense, thus, depends on how well parties of international transactions and operations organize their resource interdependencies as compared to feasible alternatives, in terms of reducing bounded rationality and bounded reliability challenges, and ultimately creating an organizational context within which their respective resources can be usefully meshed to make desirable value creation and distribution a reality. Here, initial resource ownership and control (for instance private vs public sector ownership or family vs managerial control in firms), and decision rights over some of the resources in play (including both *de jure* rights such as those held by a government agency and *de facto* rights such as those exercised by a non-governmental organization or activist group), can greatly affect the governance setup of international transactions and operations.

The IB context adds two more critical features, as compared to domestic settings, to the processes of establishing a feasible value proposition and negotiating acceptable shares in the ensuing value distribution. *First*, a firm's resources reservoir at home that constitutes its FSAs must be unbundled into *location bound* and *non-location bound FSAs*, with only the latter resource bundles being transferable, deployable and (potentially) profitably exploitable in foreign markets (Rugman & Verbeke, 1992). As noted above, even when commanding the latter, a firm will typically still need to make substantial resource recombination efforts, involving parties commanding complementary resources abroad. And even if the requisite complementary resources are accessible, there will in most cases also be the need for some degree of national or regional responsiveness abroad, that is, a *resource recombination process* leading internationally operating firms (and sometimes their business partners) to develop new, location bound FSAs in each host environment. International transactions and foreign operations can also be the source of new, non-location bound FSAs, and sometimes even subsidiary-specific advantages (SSAs) commanded by affiliates abroad and closely aligned with the local context, for instance when a foreign affiliate assumes – or is assigned – the role of center of excellence within the MNE (Rugman & Verbeke, 2001; Verbeke & Yuan, 2013).

Second, internationally expanding firms mostly capitalize on *location advantages* accessible in the home base and in the host environment selected. Extending Michael Porter's classic – albeit deeply flawed – analysis of the country-level environment (the so-called national diamond of factor conditions, demand conditions, related and supporting industries and firm strategy, structure and rivalry), these internationally operating firms are "diamond connectors" (Rugman et al., 2011; Rugman & Verbeke, 1993; Verbeke & Lee, 2021). The main flaw of the original diamond model is that it assumes a sequence of successful, entirely home-based growth of firms, followed by their easy international expansion. In reality, international linkages are often a key ingredient of early success, even in the home base. A difficulty in this context is that international partners can learn from each other at a different pace, for instance in the realm of international joint ventures and strategic alliances, thereby potentially leading to unexpected capability losses by one of the partners that can then affect initial agreements (cf. the so-called "smiling curve" phenomenon first described by Stan Shih, the founder of the Taiwanese technology company Acer, see Buckley & Verbeke, 2016). It is therefore important to pay great attention to *ex post* governance in cross-border agreements to create and distribute value.

In conclusion, the first principles of IB scholarship include the following:

(1) *Entrepreneurially oriented firms*, ranging from individual entrepreneurs to large MNEs, take the initiative to establish IB linkages (whether at the input or output side, or both), with a view to create and capture economic value. The successful completion of such initiatives requires these firms to command resource bundles that amount to *FSAs*.
(2) In most cases, only a fraction of a company's FSAs from its home location can be transferred, deployed and profitably exploited abroad (whether through trade, licensing/production contracts or investment), that is, represents *non-location bound FSAs*, thereby creating the need for *novel resource combinations* when establishing international linkages.
(3) Each party directly involved in IB transactions and operations commands *complementary resources* that are meshed together to create and capture value, and the nature and interdependencies among the resources involved will affect the structural and strategic governance of the international transactions and operations.
(4) When creating a value proposition, each of the parties involved will build upon the *location advantages* of its *home environment* and will attempt to benefit from the *host environment's location advantages* where other parties reside.
(5) A value proposition acceptable to all parties involved in IB transactions and operations, and implementable in practical terms, requires efficient governance. Mitigating challenges of *bounded rationality* and *bounded reliability* constitute the essence of efficient governance. Governance can be called *efficient* if it represents an organizational regularity, meaning that interdependent parties freely decide, over and over again, to use the same governance structure and fine-grained systems when facing the same challenges of bounded rationality and bounded reliability, and similar resource interdependencies.
(6) The international activities and operations initiated can be associated with developing *new FSAs* by the parties directly involved, both location bound and non-location bound ones. But there can be an asymmetry in learning benefits (and also other types of asymmetries such as differential access to market intelligence) among the various partners that can in turn affect the viability of the initial value proposition, thereby making early attention to *ex post* governance a must.
(7) A broad spectrum of parties, not directly involved in the core value proposition which led to international activities and operations, can affect value creation and value distribution. Whether the interventions of these parties and the impacts thereof should be considered positively or negatively is an empirical question and subject to value judgments. But both the commonly observed inadequacy of the internationally operating firm's initial reservoir of FSAs as described above and the need to accommodate a larger set of parties – including especially those who hold the keys to permitting the firm to operate in host environments – create additional layers of vulnerability not experienced by domestic companies operating exclusively at home and with

home-based value chains. Coping effectively with the added vulnerabilities resulting from IB transactions and operations can be instrumental to new FSAs, but these will require effort and time, for instance in addressing exogenous crises (Mithani et al., 2022; Van Tulder et al., 2022), and more generally when operating large internal and external MNE networks (Rugman et al., 2011; Rugman & Verbeke, 2003).

In the following sections, I use these first principles to address a variety of IB phenomena, based on my own research.

3. HOW GLOBAL ARE THE WORLD'S LARGEST MNEs?

The evidence shows that the globalization of even the largest firms, and therefore also their global market power, is often exaggerated. In my work with Thomas Hutzschenreuter and Johannes Voll, based on the expansion of German listed companies, we demonstrated that "adding distance" to an existing internal network of multinational operations is no small feat (Hutzschenreuter et al., 2011). When an MNE adds substantially more distance in one period, that is, enters markets very differently from the ones in which the MNE operates already, this slows down the rate of further international expansion during the next 3–5 years. Distance needs to be absorbed, meaning that extant firm-level resources need to be recombined with new resources abroad, and these resource combination processes require time to come to fruition. We also found that the extant diversity of the MNE's portfolio of international operations had a negative impact on adding further distance, thus suggesting that bounded rationality and bounded reliability challenges in managing internal networks limit the geographic scope of MNEs.

Wenlong Yuan and I investigated the differential roles that subsidiaries can play in these internal MNE networks, which tend to become very complex for large firms. Economic activities are often fine sliced, with foreign subsidiaries that can become very interdependent, and with substantial bounded rationality and bounded reliability challenges haunting the relationships between the head office and foreign subsidiaries (Rugman et al., 2011; Verbeke & Yuan, 2005, 2007, 2013, 2020).

My work with the late Alan Rugman and several other colleagues has unambiguously shown that both the sales and the asset base of most large MNEs have remained concentrated in their home region or just a limited number of segments of the world economy, with these MNEs typically having a much smaller market share abroad than at home, and a clearly non-dominant position in one or more of the world's main economic regions (Rosa et al., 2020; Rugman & Verbeke, 2004, 2005, 2007, 2008a). This evidence is the clearest proof that many MNE FSAs are not globally transferable, deployable and profitably exploitable. Most MNEs are dominant only in their home region. Their location bound FSAs lead to an asymmetrical spatial footprint, with a home-regional concentration of activities and not a more balanced "global" dispersion. Given this spatial asymmetry, a review of prevailing organizational structures shows that many MNEs have injected regional components in their organizational structure and systems (regional head offices, regional divisions; regional and interregional coordination mechanisms) to reduce challenges of bounded rationality and bounded reliability (Rugman & Verbeke, 2008b; Verbeke et al., 2016; Verbeke & Kano, 2012a).

A prediction I voiced with Christian Asmussen is that most MNEs will likely continue to experience differential success across geographic space, and especially across regions (Verbeke & Asmussen, 2016). For most firms, an asymmetrical footprint is there to stay, further amplified by macro-level events driving deglobalization (Verbeke, 2020). From a research perspective, adding regional elements to the analysis can improve the methodological soundness and sophistication of much IB research, especially in an era of supposed "global decoupling" and "regional blocks," see for instance the entry mode analysis performed in Hillemann et al. (2019) in the context of the European Monetary Union. Resource recombination and access to complementary resources in host regions become more challenging as the result of animosity among macro-level parties that can in turn influence international transactions and multinational operations at the micro-level.

Suggestions that a small number of large firms dominate their industry globally and thereby wield enormous power across the world are in most cases simply not evidence-based.

4. ARE THE WORLD'S LARGEST FIRMS REALLY FOOTLOOSE?

It is often argued that MNEs lack "loyalty" to their home country and can easily shift activities to other nations. With Regis Coeurderoy, I analyzed the mobility of head offices of the largest companies in the world. The evidence showed only 33 head office relocations or 1.46% of the Forbes 2,000 largest firms (over 2008 and 2009) since their inception (Coeurderoy & Verbeke, 2016). Head office activities are partly related to the legal domicile of the firm (e.g., accounting- and information disclosure and external governance-related functions). They also include central monitoring and control functions (e.g., aggregated performance assessment of worldwide operations). And in terms of the entrepreneurial function, the head office establishes the boundaries of the firm and takes the most critical resource allocation decisions.

The absence of head office mobility is relatively easy to explain. A high gross domestic product per capita, institutional stability and the presence of a well-functioning capital market contribute to explaining the presence of at least one very large firm with a domestic head office. Furthermore, institutional quality proxies contribute much to explaining a high head office density, especially regulations that facilitate contract enforcement, trading across borders and closing a business.

The evidence suggests that to build home grown MNEs with their head office in the home nation, institutional quality is critical, much in line with Ghoshal et al.'s (1999) analysis. Direct incentives from governments to attract large firms can lead to the increased location of footloose activities in these countries, but such incentives are no substitute for institutional quality critical to the presence of head offices. Paradoxically, many large MNE FSAs as far as head office strengths are concerned, are at least partly location bound, and embedded in the home country institutional system. Here, macro-level institutional quality reflects a reliable and predictable environment for business that can be leveraged in the context of both domestic and international transactions and operations, while at the same time reducing the MNE's vulnerability to less reliable and less predictable foreign institutions.

5. WHAT FSAs DO INVs HAVE?

INVs' distinctive characteristic is that they supposedly internationalize at inception or shortly thereafter, thereby often having (too) limited time to build up non-location bound FSAs resulting from R&D and patents at the upstream side or advertising and brand name creation at the downstream side. They thus have additional vulnerabilities as compared to firms that have reached a larger scale domestically before internationalizing.

In their early years, new venture companies have little firm-level experience by definition, beyond the experience of their owners-managers, that is, the founding entrepreneurs. Therefore, these founding entrepreneurs' influence is critical to new venture success, whether in the home country or abroad.

My analysis with Oleksi Oseyievski and Amin Zargarzadeh based on Kauffman data showed that requisite resources accumulation and recombination may actually have been undertaken at the individual level, by the founding entrepreneurs themselves, in the pre-stage of the new venture and would therefore remain "invisible" when focusing solely on the organizational experience built up in the new venture (Verbeke et al., 2014). Particular founding entrepreneurs' characteristics can thus function as FSAs. The evidence showed that especially immigrant founders with a high education level appear to engage in early new venture internationalization. The status of immigrant makes it easier to establish linkages with foreign-based actors to develop and act upon international value propositions and to engage in new resource combinations. Irrespective of technical knowledge skills, a higher education level of these founders could at least in part be interpreted as a non-location bound FSA, helpful in reducing bounded rationality challenges.

6. WHAT FSAs DO EMERGING ECONOMY FIRMS HAVE?

It has been argued that emerging economy MNEs represent a new class of firms, requiring dedicated theory, supposedly because their home country location *dis*advantages hinder them in developing non-location bound FSAs at the outset. Emerging economy MNEs must therefore supposedly go international to access FSAs available elsewhere on the strength of their home country location advantages. This line of thinking is debatable as a general case. If emerging economy companies did not command some type of resources that would be attractive to foreign parties as the basis for a joint value proposition, no international transactions and operations would ensue, regardless of their home country location advantages.

As with INVs, many emerging economy companies may simply have a different configuration of FSAs than established MNEs from developed countries. Liena Kano and I studied 60 case studies covering 11 emerging economy MNEs (Verbeke & Kano, 2015). We found these firms to have the following location bound and non-location bound FSAs. The location bound FSAs included privileged access to skilled labor, privileged access to low-cost production, embeddedness in local networks (including distribution networks), government connections and knowledge of local markets. More importantly, these emerging economy MNEs also commanded the following non-location bound FSAs: entrepreneurial agility, commitment to innovation, FSAs in local responsiveness/flexibility, adversity advantages, access to global distribution networks and higher-order management capabilities/sophisticated routines.

The last two elements are especially noteworthy. *First*, commanding resources that can complement those held by foreign parties are useful only insofar as the various parties are informed about each other's resources, which constitutes the foundation of a value proposition for international transactions and operations. *Second*, international expansion requires the disciplined execution of projects that result from entrepreneurial judgment by senior executives. Higher-order managerial capabilities precisely reflect this dual requirement, that is, the ability to exercise sound judgment and to be organized for disciplined execution, especially when creative resource combination across borders will be required.

As to state-owned MNEs, their FSAs were of a particular nature, and included various resource types conferred or enabled by parties not always directly involved on an ongoing basis themselves in international transactions and operations, such as government agencies: access to cheap capital, increased latitude for risk and access to decision-making bodies/ability to obtain exemptions from stringent regulations and policies (e.g., human rights codes and environmental regulations). These elements highlight the sometimes-critical role that parties with decision rights over the social license to operate can perform. In this case, non-business parties did not threaten to withhold such license from foreign entrants, but extended or elevated such license to support domestic firms' international growth.

However, state-involvement can also be detrimental in the processes of developing non-location bound FSAs. Shelter-based rather than efficiency-based behavior can sometimes be observed in emerging economies, when the emphasis is on national champions, infant industry protection and quasi-public monopolies. Successful international activity breaks down if associated with excessive waste of resources, and this holds for any economy promoting shelter-based strategies (Rugman & Verbeke, 1990, 2017).

We concluded that the empirical challenge when studying any "new type" of MNE is to find appropriate proxies to measure these enterprises' FSAs. In accordance with Aharoni (1993), we concluded that it may be difficult to capture the FSAs of new MNE types by using industry averages and performing statistical tests. We therefore suggested complementing conventional approaches to appraise FSAs with qualitative and longitudinal studies of individual firms.

7. DO FAMILY FIRMS FACE SPECIFIC INTERNATIONALIZATION CHALLENGES?

Do family firms internationalize differently (vis-á-vis non-family firms) in terms of organizing cross-border transactions and operations? Much evidence suggests that family firms intrinsically suffer from a bifurcation bias (Verbeke & Kano, 2012b). This means they make a "one versus zero" assessment of resources, that is, view the resources they command either as very valuable and with the firm for the long term or as commodity-like and easy to dispose of. This bifurcation bias is a special case of the well-known "affect bias," and it can be interpreted as a form of bounded rationality, in the sense of imperfect information processing with the value of resources and potential resources being wrongly assessed. The outcome can be an increase of bounded reliability, for instance if a family member is elevated beyond her or his competence to run international family operations, thereby also demotivating more competent, non-family managers better suited for this job.

More generally, there can be an overestimation of family-based resources' international transferability, deployability and exploitation potential, combined with a simultaneous underestimation of the value of extant non-family-based resources. The bifurcation bias will also increase the likelihood of underestimating the value of – and need for – complementary resources abroad, thereby leading to sub-optimal mixes of resources when selecting foreign operating modes (Kano & Verbeke, 2018). Perhaps most importantly, the erroneous assessment of both family-based resources and complementary resources abroad will make it less likely that value propositions for foreign projects will come to fruition.

The empirical evidence suggests that family firms can do many things to reduce the impact of bifurcation bias on internationalization decisions, especially in the sphere of human resources management and governance (Almodóvar et al., 2016).

For example, recent research shows that non-family members serving on the Board of Directors, including directors with international experience, can reduce bounded rationality challenges and be valuable to internationalization efforts (Pongelli et al., 2023). Here, the combination of a family CEO (supposedly with superior knowledge of the firm's family heritage resources) with a more active Board composed in part of non-family members with international experience can be instrumental to pursuing a presence in international activities. Such Board composition reduces bounded rationality and bounded reliability challenges and makes it easier to engage with cross-border value propositions and to craft requisite new resource combinations.

8. IS HIGHER INTERNATIONAL DIVERSIFICATION ALWAYS BENEFICIAL?

The multinationality–performance relationship has been the subject of a particularly rich IB literature. This literature's ambition is high. If we knew in a general sense how multinationality affects firm-level performance, clear guidance for managers would ensue to move toward an "optimal" level of multinationality. And if we discovered this optimal level of multinationality, we would really also understand the firm's optimal level of diversification, and we would therefore have an empirically validated, core ingredient of a general theory of the firm.

However, the question immediately arises whether it makes sense to search for a firm's "optimal" level of multinationality. Similarly, does it make sense to search for a firm's optimal size, optimal R&D level, optimal level of women in top management or optimal CEO salary? My view is that the situational context will largely determine any answers as to "optimality" and that empirical findings based on past data are unlikely to provide much guidance for the future. Multinationality is an intermediate outcome that follows from multiple decisions by parties to create value based on their complementary resources (including entry mode decisions).

In addition, past empirical research has often suffered from a series of weaknesses (Verbeke & Brugman, 2009; Verbeke & Forootan, 2012). When measuring multinationality, many studies have adopted a rather reductionist perspective on the value chain activities considered relevant (for instance the share of foreign sales in total sales, or foreign assets in total assets, or foreign employment in total employment). We can also often observe the absence of distinguishing between internationalization (which could be in a single host nation) and international diversification (whereby the number of countries is accounted for and the share of each in the internationalization level), and between related and unrelated diversification. As to the latter point, the "distance" vis-á-vis the home country or existing operations matters and there could be regional effect, such as an European Union (EU)-based firm investing in another EU country versus outside of the EU. The impact of home country operations on multinationality levels also tends to be neglected. The share of home country operations will typically be larger for firms with a larger home market.

The measurement of performance for the purpose of establishing a multinationality–performance relationship has also been somewhat problematic. For instance, many large-scale empirical studies neglect the expected impact of strategic investment motives on performance (e.g., strategic asset seeking vs market seeking). Each motive will clearly lead to a very different expected stream of future costs and benefits as these materialize over time.

The issue of alternative performance measures also looms large. For instance, the return-on-assets parameter typically neglects R&D, and this parameter also measures something very different than Tobin's q. Dynamic elements, especially external shocks, also appear to have been excluded from most performance assessments. There is mostly the implicit assumption made that no large-scale shocks occurred that affected at least some firms in the sample. These dynamic elements would include shocks as diverse as changes in the EU's membership and its level of integration, China's entry into the World Trade Organization, the Great Recession, significant exchange rate changes, etc.

As to the linkages between multinationality and performance, the performance of foreign operations was mostly not isolated from overall, firm-level performance, so that one does not know whether any linkage identified resulted from the international operations, versus merely from domestic operations, or from a combination of both. Sometimes, cross-sectional data or panel data covering a relatively short time span were used instead of longitudinal data, so that there was no comprehensive assessment of the "real-life" trajectory of "real" firms. Many past studies did not address fully problems of reverse causality (whereby prior performance might explain changes in multinationality) and endogeneity. As to the latter point, extant FSAs held by an internationally operating company and its ability to access complementary resources abroad and to recombine resources, are likely the most critical elements to explain multinationality. The timing of international expansion can also matter a lot, with early versus late entrants to foreign markets likely to experience different performance effects. And finally, internationally expanding firms may face host country incumbents who will strategically respond to new entrants, thereby affecting the performance effects of the expansion moves (Hutzschenreuter et al., 2014).

From an evidence-based perspective, the "good" is the usage of increasingly sophisticated statistical analyses (with many control variables) and the testing for a variety of possible relationships between multinationality and performance. The "bad" is that many, even sophisticated, empirical pieces on the multinationality–performance relationship fail some of the baseline tests for methodological soundness that I outlined above. Especially important is neglecting the characteristics of the home country market (including especially its size and the investment opportunities in that market), assuming erroneously the impact of large external shocks, and finally discounting the importance of each individual firm's history.

This type of research should build upon actual longitudinal data, covering firms' entire historical trajectories. Analysis should focus on highly specific samples with companies facing similar strategic drivers and motivations for international expansion, a common and well-understood environmental context such as a mature home market and considering both critical FSA-bundles and opportunities for accessing complementary resources and for realizing novel resource combinations abroad. Testing should occur of well-grounded propositions, richly informed by business history cases and by credible theory, rather than by providing long shopping lists of alleged costs and benefits of multinational activity, with the impact of these variables not tested directly.

9. WHAT MACRO-LEVEL COMPLEMENTARY RESOURCES DO MNEs NEED WHEN DOING BUSINESS IN EMERGING ECONOMIES?

It is now commonplace for activist scholars to prescribe that MNEs should do more to serve the world's poorest people, especially in the Global South or at the base of the pyramid – BOP. This prescription is closely aligned with the normative message that MNEs can "do well by doing good" (Hillemann & Verbeke, 2014).

The focus of much of the related literature has been on drastically rethinking business models and on relational contracting and networks, that is, informal institutions, to substitute for non-existing or poorly functioning formal institutions.

MNEs should compensate for institutional voids. Is this prescription wishful thinking in terms of what is expected from MNEs? What are feasible governance approaches to support BOP market creation? There is scant empirical evidence beyond isolated and highly situation-dependent examples that corporate actions favoring BOP market creation through informal means and grand business model redesign alleviate poverty. It should also be mentioned that a great challenge in many emerging economies is often not institutional voids, but various expressions of institutional overkill (Eddleston et al., 2020; Verbeke & Kano, 2013).

As a general point, the evidence in a country like India, the world's largest democracy, has been one of low creditworthiness and default risk of BOP customers (Verbeke et al., 2023). There has been an absence of credible safeguards that can be provided by BOP entrepreneurs and consumers to establish enforceable contracts. This situation has affected transactions and value creating propositions with lending institutions and providers of products and services, including foreign MNEs. High lending costs, including the poverty premium, reflect default risk, even in the absence of intentional deceit.

The governance solution to this perverse interaction of bounded rationality and expected bounded reliability hindering BOP market creation consists of introducing safeguards to protect the BOP customers, lenders and MNEs involved against failure to execute contracts. One critical point is access to correct creditworthiness information permitting MNEs and retailers to segment the BOP consumer base and to extend credit at reasonable rates, thereby reducing the poverty premium for BOP consumers with high expected reliability.

The evidence, at least for India, suggests that a scalable, practical solution does not consist of promoting relational networks but of introducing an arsenal of formal information technology-related systems, operating at the macro-level but with an unlimited number of "micro-level plug-ins," where MNEs can be proactive (Verbeke et al., 2023).

The massive introduction of these systems in India, starting in 2010 has been built on three components: identity validation (in the context of India through "Aadhaar"-based authentication since 2010); a generalized, real-time mobile payments system (in India through the Unified Payments Interface – UPI since 2016); and an efficient ownership structure and usage of data on income and payments (in progress). As a result of this three-legged system, bounded rationality and bounded reliability challenges can be addressed: various parties to transactions can better understand and monitor BOP customer behavior (thereby moving away from a "BOP average" to "a BOP unique customer"). Moreover, individual creditworthiness can be assessed immediately through documentation of all prior relevant transactions and by tracking these systematically (e.g., payments due and the fulfilling of contractual obligations). Financial institutions, MNEs and retailers can credibly judge whether a potential BOP customer is "willing and able" to purchase (and pay for) goods on credit, based on experiential data.

More generally, when formal institutional voids at the macro-level can be observed that hinder international transactions, the scalable solution for internationally operating actors is not to attempt filling voids with informal institutions, but to participate through micro-level plug-ins in developing or leveraging novel, properly functioning formal institutions. Here, host country parties, including government agencies and commercial players, rather than foreign MNEs must perform a lead role in establishing macro-level governance systems that permit inclusion of BOP customers.

10. WHAT GOVERNANCE MECHANISMS DO MNEs AND THEIR INTERNATIONAL GVC PARTNERS ADOPT TO MITIGATE VULNERABILITIES?

Substantial recent literature laments the functioning of GVCs and calls for the MNEs acting as lead firms to treat their GVC partners better than they have in the past, and to contribute to sustainable development. But prescription is cheap. It is factually correct that many lead firms in GVCs benefit from a size and information asymmetry advantage vis-á-vis smaller, international partners. However, as with all other IB transactions and operations, GVCs will only survive in the longer run subject to credible and enforceable agreements to create and distribute value.

In Verbeke et al. (2021), we addressed the bounded rationality and bounded reliability challenges facing GVCs. Our analysis of several case studies suggests that various parties in GVCs have often been able to anticipate and solve potential governance problems through a mix of creative safeguards. Perhaps the most interesting insight from this analysis is that some GVC parties have started to focus on creating entrepreneurial space in GVCs. There has been extensive prior work about various dimensions of entrepreneurial space in the IB literature, centered around MNE actions such as national responsiveness, insiderization, subsidiary attention, subsidiary initiatives, centers of excellence, SSAs and more generally distributed resource recombination. However, entrepreneurial space is a broader concept, critical to governance mechanisms, both inside MNE internal networks and their GVCs.

In Verbeke et al. (2021), we discussed the challenge of boosting entrepreneurial space in GVCs through entrepreneurship-infusing governance tools viewed as a complement to the conventional formal and informal (or relational) tools used in many types of contracting. The main recommendation from this analysis for large MNEs managing external and often international partners is to avoid a myopic focus on cost-efficiencies-only, and to support developing the entrepreneurial capacity of partners. We suggested that MNEs should search for contracting partners with proven entrepreneurial and innovative capabilities. All parties directly engaged in GVC activities should re-direct their own and their partners' self-interest seeking toward mutually beneficial innovation outcomes through creating a cooperative environment (which can be linked to formal property rights and incentives). Here, lead MNEs should support especially their "strategic" partners by paying particular attention to these parties' unique areas of expertise, and they should co-develop innovations with these partners. All parties should be involved in jointly crafting creativity-oriented goals, practices, performance measures, etc. Finally, MNEs can in many cases orchestrate knowledge transfers to partners, for example, through providing access to the MNE's global supply networks and to network-level innovation in partners' areas of greatest competence.

Strategy scholars have often erroneously argued that entrepreneurial action and attention to governance are antithetical (see Barney et al., 2023), but in the real world of GVCs these should be closely aligned. Governance-related research in IB would benefit greatly from explicitly studying the entrepreneurship-infusing dimension of contracting tools such as supplier contracts, distribution contracts, labor contracts, etc. All these contracts are expressions of complementary resources being meshed, with entrepreneurial space in these contracts in many instances being the driver for creating and distributing economic value. Every contract should thus be viewed as an opportunity to be more entrepreneurial and innovative. MNE head offices and subsidiaries should design their contracts with GVC partners to create entrepreneurial space, meaning innovation-oriented behavior, moving far beyond conventional formal safeguards and relational facets of contracts, thereby increasing the economic value that can be created in the future.

11. DO THE WORLD'S LARGEST FIRMS ONLY TALK AND NOT WALK WHEN ADDRESSING CLIMATE CHANGE?

The issue of climate change impact mitigation has been high on the agenda of many strategy and IB scholars during the past decade. Prescriptions for large, internationally operating firms abound. Climate change is a challenge of global reach with large-scale consequences and material financial implications. Climate change has had uneven impacts across geographies, industries, firms within a single industry, and their stakeholders. There have also been chaotic response processes and unclear system linkages. Climate change impact mitigation can be viewed as a shared responsibility for causes and remedies among governments, business firms, consumers, and a variety of societal stakeholders. Perhaps the main problem is that climate change impact mitigation is largely situated in the realm of "governance without government."

With my colleagues, Charles Backman and Robert Schulz, I developed a conceptual framework to analyze how large (and mostly internationally operating) firms from Europe and North America have addressed the climate change issue in terms of reducing their greenhouse gas (GHG) emissions. For this purpose, we used carbon disclosure project data and followed the impact mitigation strategies of 117 large MNEs during the 2004–2020 period (Backman et al., 2017; Schulz et al., 2022). We found that firms can engage in impact mitigating actions along four resource domains, which we call the GISTe model. Firms can do this by improving *G*overnance (referring to the level in the firm responsible for the climate change file and to the presence of incentives to reduce impacts), *I*nformation gathering and processing on the firm's climate change impacts (including the level of disaggregation thereof), *S*ystems of target setting and performance monitoring and finally investments in *Te*chnological innovation, both on the process and product side.

Our evidence brought to light the fallacy that European firms would be structurally performing better than North American ones. We found that many firms, whether European or North American ones, have not just "talked" about climate change impact mitigation, but actually "walked the talk." Our longitudinal data showed that European firms were initially better than their North American counterparts in what we call "walk I," denoting attention to governance, information management and systems, that is, they tried to establish a reputation for seriously addressing climate change through managerial tools, thereby also securing their license to operate, given the prevailing institutional environment in Europe. But when it came down to "walk II," meaning actual technology-related investments, North American firms' performance has often been equal or superior to that of the European companies, at least until 2012, in this case more driven by market forces.

Importantly, the trajectories over time of the firms included in the analysis did suggest a convergence of European and North American ones in terms of outputs achieved, even though the processes to get there may have been very different. We concluded that "governance without government" can still be associated with significant efforts at climate change impact mitigation, but also that adopting advanced managerial tools is not necessarily a direct driver of superior mitigation outcomes in terms of having products and processes with a lower impact.

What perhaps matters most and became visible from the empirical evidence were the various, equifinal trajectories enacted by the firms we studied. We found that energy companies became proactive earlier than firms in other sectors, undoubtedly because of the unrelenting scrutiny from regulators and other societal stakeholders with decision rights to withhold these firms' license to operate. Energy firms thus made irreversible technological investments in GHG reductions earlier than other companies.

Over time, there appeared few meaningful differences in outcomes as to the investments in the various resource domains relevant to climate change impact mitigation among European and North American firms, but in 2016, the North American firms appeared to lag somewhat the European ones in terms of process innovation, and in 2020, there was a lag in product innovation. Finally, firms in B2B sectors were the slowest to adjust, likely because they were not directly influenced by their customers, and they are often not considered explicitly as sources of carbon by societal stakeholders.

The main point of the above is to recognize that firms need time to develop sustainability-oriented strategies, because this amounts to material changes in their investment projects aimed at value creation and value distribution. Artificially imposed, strict deadlines, in this instance to achieve particular reductions in GHG emissions, cannot act as substitutes for firm-level trajectories that need to consider the requirements for sustained economic value creation and acceptable value distribution among the parties involved. Unfortunately, there regularly appears to be a missing link between imposed GHG emission targets at the macro-level and the enactment thereof at the firm level. Public policy, when formulated by people with little understanding of business practice, tends to ignore firm-level constraints and feasible trajectories to innovate and to make capital investment decisions (Verbeke & Hutzschenreuter, 2021a). The unreliability of policy makers in the form of changing the rules of the game when firms have already made large, irreversible investments with a particular carbon footprint, in turn incentivizes these firms to become unreliable themselves and to engage in virtue signaling, political rent-seeking and greenwashing or ceremonial adoption of imposed rules. A vicious cycle of unreliable behavior ensues.

12. CONCLUSION

IB scholarship has much to offer to cumulative knowledge development in the social sciences (Beugelsdijk, 2022). As the above account based on some of my own research has shown, IB scholars should always start from first principles when studying empirically a range of business phenomena that unfold over time and across geographic space. It is especially important to forego the urge to prescribe simple solutions for solving complex problems. Scholars should refrain from spinning elegant narratives that are not evidence-based. Or in the words of the late Oliver Williamson (1993, p. 485): 'The world of commerce is reorganized in favor of the cynics, as against the innocents, when social scientists employ user-friendly language that is not descriptively accurate-since only the innocents are taken in."

Our role as IB scholars is to describe and explain the value creating and value distributing processes put in motion by entrepreneurial parties with complementary resources, willing and able to establish and to nurture business linkages across borders. Acknowledging the role of parties not directly involved in establishing and operating international activities but with some type of decision rights in this realm, is clearly critical, as powerfully advocated by van Tulder and van Mil (2022) in the context of the United Nations' Sustainable Development Goals. However, an objective and evidence-based analysis is unlikely to conclude that more can and should "obviously" be done by all internationally operating firms beyond their present role in creating and distributing economic value. International entrepreneurial behavior emerges because economic actors chase attractive business opportunities and this opens them up to a wide spectrum of vulnerabilities, which explain the relative absence of global MNEs (Verbeke et al., 2018).

If other relevant parties are prepared to alter (and potentially heighten) these vulnerabilities of the main actors engaged in international transactions and operations, the geographic configuration of these activities may change (Wong, 2023), or firms may try to endogenize public policy (Rugman & Verbeke, 1998). In the worst-case scenario, the opportunities for creating economic value across borders may simply fade away (Hennart, 1989; Verbeke, 2021). IB scholars' main pedagogical goal in this realm should not be to defend any and all behavior manifested by internationally operating firms, but to educate all relevant parties on the first principles of IB functioning in the real world.

REFERENCES

Aharoni, Y. (1993). In search for the unique: Can firm-specific advantages be evaluated? *Journal of Management Studies*, *30*(1), 31–44.

Almodóvar, P., Verbeke, A., & Rodríguez-Ruiz, Ó. (2016). The internationalization of small and medium-sized family enterprises: The role of human asset quality. *Journal of Leadership & Organizational Studies*, *23*(2), 162–174.

Backman, C. A., Verbeke, A., & Schulz, R. A. (2017). The drivers of corporate climate change strategies and public policy: A new resource-based view perspective. *Business & Society*, *56*(4), 545–575.

Barney, J. B., Arikan, A. M., & Arikan, I. (2023). Does IB need a strategic theory of the firm? *Journal of International Business Studies*, *54*(6), 1105–1114.

Bebchuk, L. A., Kastiel, K., & Tallarita, R. (2022). *Does enlightened shareholder value add value?* [Working Paper #643]. Harvard Law School (28 pp.).

Bebchuk, L., & Tallarita, R. (2020). The illusory promise of stakeholder capitalism. *Cornell Law Review*, *106*, 91–177.

Benito, G. R., Petersen, B., & Welch, L. S. (2009). Towards more realistic conceptualisations of foreign operation modes. *Journal of International Business Studies*, *40*(9), 1455–1470.

Buckley, P. J., & Casson, M. (1976). *The future of the multinational enterprise*. Palgrave Macmillan.

Beugelsdijk, S. (2022). Capitalizing on the uniqueness of international business: Towards a theory of place, space, and organization. *Journal of International Business Studies*, *53*(9), 2050–2067.

Buckley, P. J., & Verbeke, A. (2016). Smiling and crying curves in international business. *International Business Review*, *25*(3), 749–752.

Coeurderoy, R., & Verbeke, A. (2016). The unbalanced geography of the world's largest MNEs: Institutional quality and head office distribution across countries. *Global Strategy Journal*, *6*(2), 127–148.

Eddleston, K. A., Banalieva, E. R., & Verbeke, A. (2020). The bribery paradox in transition economies and the enactment of 'new normal' business environments. *Journal of Management Studies*, *57*(3), 597–625.

Edmans, A. (2023). The end of ESG. *Financial Management*, *52*(1), 3–17.

Ghoshal, S., Hahn, M., & Moran, P. (1999). Management competence, firm growth and economic progress. *Contributions to Political Economy*, *18*(1), 121–150.

Hennart, J. F. (1989). Can the "new forms of investment" substitute for the "old forms"? A transaction costs perspective. *Journal of International Business Studies*, *20*(2), 211–234.

Hennart, J. F. (2009). Down with MNE-centric theories! Market entry and expansion as the bundling of MNE and local assets. *Journal of International Business Studies*, *40*(9), 1432–1454.

Hillemann, J., & Verbeke, A. (2014). An internalization theory perspective on the bottom of the pyramid. In R. Van Tulder, A. Verbeke & R. Strange (Eds.), *International business and sustainable development* (Vol. 8, pp. 69–90). Emerald Group Publishing Limited.

Hennart, J. F., & Verbeke, A. (2022). Actionable and enduring implications of Oliver Williamson's transaction cost theory. *Journal of International Business Studies*, *53*(8), 1557–1575.

Hillemann, J., Verbeke, A., & Oh, W. Y. (2019). Regional integration, multinational enterprise strategy and the impact of country-level risk: The case of the EMU. *British Journal of Management*, *30*(4), 908–925.

Huang R., & Kubota Y. (2023, July 6). Tesla joins Chinese carmakers in pledge to end EV price war. *Wall Street Journal*. https://www.wsj.com/articles/tesla-joins-chinese-carmakers-in-pledge-to-end-ev-price-war-47223bd9

Hutzschenreuter, T., Kleindienst, I., Groene, F., & Verbeke, A. (2014). Corporate strategic responses to foreign entry: Insights from prospect theory. *The Multinational Business Review*, *22*(3), 294–323.

Hutzschenreuter, T., Voll, J. C., & Verbeke, A. (2011). The impact of added cultural distance and cultural diversity on international expansion patterns: A Penrosean perspective. *Journal of Management Studies*, *48*(2), 305–329.

Kano, L., & Verbeke, A. (2015). The three faces of bounded reliability: Alfred Chandler and the micro-foundations of management theory. *California Management Review*, *58*(1), 97–122.

Kano, L., & Verbeke, A. (2018). Family firm internationalization: Heritage assets and the impact of bifurcation bias. *Global Strategy Journal*, *8*(1), 158–183.

Kano, L., & Verbeke, A. (2019). Theories of the multinational firm: A microfoundational perspective. *Global Strategy Journal*, *9*(1), 117–147.

Kaplan, R. S., & McMillan, D. (2021, February 3). Reimagining the balanced scorecard for the ESG era. *Harvard Business Review Digital Articles*. https://hbr.org/2021/02/reimagining-the-balanced-scorecard-for-the-esg-era?ab=hero-main-text

Klein, N. (2000). *No logo*. Vintage Canada.

Linnenluecke, M. K. (2022). Environmental, social and governance (ESG) performance in the context of multinational business research. *The Multinational Business Review*, *30*(1), 1–16.

Mithani, M. A., Narula, R., Surdu, I., & Verbeke, A. (Eds.). (2022). *Crises and disruptions in international business: How multinational enterprises respond to crises*. Springer Nature.

Narula, R., & Verbeke, A. (2015). Making internalization theory good for practice: The essence of Alan Rugman's contributions to international business. *Journal of World Business*, *50*(4), 612–622.

Pongelli, C., Majocchi, A., Bauweraerts, J., Sciascia, S., Caroli, M., & Verbeke, A. (2023). The impact of board of directors' characteristics on the internationalization of family SMEs. *Journal of World Business*, *58*(2), 101412.

Ramkumar, A., & Eckert, N. (2023, June 22). Ford venture gets record $9.2 billion government loan for EV batteries. *Wall Street Journal*. https://www.wsj.com/articles/ford-venture-gets-record-9-2-billion-government-loan-for-ev-batteries-b0ed8445

Rosa, B., Gugler, P., & Verbeke, A. (2020). Regional and global strategies of MNEs: Revisiting Rugman A.M. & Verbeke A. (2004). *Journal of International Business Studies*, *51*(7), 1045–1053.

Rugman, A. M., & Verbeke, A. (1990). Strategic trade policy is not good strategy. *Hitotsubashi Journal of Commerce and Management*, *25*(1), 75–97.

Rugman, A. M., & Verbeke, A. (1992). A note on the transnational solution and the transaction cost theory of multinational strategic management. *Journal of International Business Studies*, *23*(4), 761–771.

Rugman, A. M., & Verbeke, A. (1993). Foreign subsidiaries and multinational strategic management: An extension and correction of Porter's single diamond framework. *Management International Review*, *33*(2–1), 71–85.

Rugman, A. M., & Verbeke, A. (1998). Multinational enterprises and public policy. *Journal of International Business Studies*, *29*(1), 115–136.

Rugman, A. M., & Verbeke, A. (2001). Subsidiary-specific advantages in multinational enterprises. *Strategic Management Journal*, *22*(3), 237–250.

Rugman, A. M., & Verbeke, A. (2003). Extending the theory of the multinational enterprise: Internalization and strategic management perspectives. *Journal of International Business Studies*, *34*(2), 125–137.

Rugman, A. M., & Verbeke, A. (2004). A perspective on regional and global strategies of multinational enterprises. *Journal of International Business Studies*, *35*(1), 3–18.

Rugman, A. M., & Verbeke, A. (2005). Towards a theory of regional multinationals: A transaction cost economics approach. *Management International Review*, *45*(1), 5–17.

Rugman, A. M., & Verbeke, A. (2007). Liabilities of regional foreignness and the use of firm-level versus country-level data: A response to Dunning et al. *Journal of International Business Studies*, *38*(1), 200–205.

Rugman, A. M., & Verbeke, A. (2008a). A new perspective on the regional and global strategies of multinational services firms. *Management International Review*, *4*(48), 397–411.

Rugman, A. M., & Verbeke, A. (2008b). A regional solution to the strategy and structure of multinationals. *European Management Journal*, *26*(5), 305–313.

Rugman, A. M., & Verbeke, A. (2017). *Global corporate strategy and trade policy*. Routledge (25th year anniversary edition).

Rugman, A. M., Verbeke, A., & Yuan, W. (2011). Re-conceptualizing Bartlett and Ghoshal's classification of national subsidiary roles in the multinational enterprise. *Journal of Management Studies*, *48*(2), 253–277.

Santangelo, G. D., & Verbeke, A. (2022). Actionable guidelines to improve 'theory-related' contributions to international business research. *Journal of International Business Studies*, *53*(9), 1843–1855.

Schulz, R., Backman, C., Verbeke, A., & Heger, H. (2022). Firm adaptation to climate change: Applying the GISTe model to a cohort of FT500 firms 2004–2020. *Academy of Management Proceedings* (Vol. 2022/1, p. 17504). Briarcliff Manor, NY: Academy of Management.

Teece, D. J. (1986). Profiting from technological innovation: Implications for integration, collaboration, licensing and public policy. *Research Policy*, *15*(6), 285–305.
Teece, D. J. (2018). Profiting from innovation in the digital economy: Enabling technologies, standards, and licensing models in the wireless world. *Research Policy*, *47*(8), 1367–1387.
Temple-West, P. (2021, January 26). The ESG investor's dilemma: To engage or divest. *Financial Times*. https://www.ft.com/content/814cbd2c-00db-41b7-91af-28435301a8a2
Van Assche, A., & Narula, R. (2023). Internalization strikes back? Global value chains, and the rising costs of effective cascading compliance. *Journal of Industrial and Business Economics*, *50*(1), 161–173.
Van Assche, A., Soundararajan, V., Bapuji H., & Ertug G. (2023). *Global supply chain practices and inequality* [Mimeo] (38 pp.).
van Tulder, R., & van Mil, E. (2022). *Principles of sustainable business: Frameworks for corporate action on the SDGs*. Taylor & Francis.
van Tulder, R., Verbeke, A., Piscitello, L., & Puck, J. (2022). International business in times of crisis: What perspective to take? In R. van Tulder, A. Verbeke, L. Piscitello & J. Puck (Eds.), *International business in times of crisis: Tribute volume to Geoffrey Jones* (Progress in International Business Research (PIBR), Vol. 16, pp. 3–24). Emerald Publishing Limited.
Verbeke, A. (2020). Will the COVID-19 pandemic really change the governance of global value chains? *British Journal of Management*, *31*(3), 444–446.
Verbeke, A. (2021). The long-term energy transition and multinational enterprise complexity: A BJM–JIBS joint initiative. *Journal of International Business Studies*, *52*(5), 803–806.
Verbeke, A., & Asmussen, C. G. (2016). Global, local, or regional? The locus of MNE strategies. *Journal of Management Studies*, *53*(6), 1051–1075.
Verbeke, A., & Brugman, P. (2009). Triple-testing the quality of multinationality–performance research: An internalization theory perspective. *International Business Review*, *18*(3), 265–275.
Verbeke, A., & Ciravegna, L. (2018). International entrepreneurship research versus international business research: A false dichotomy? *Journal of International Business Studies*, *49*(4), 387–394.
Verbeke, A., Ciravegna, L., Lopez, L. E., & Kundu, S. K. (2019). Five configurations of opportunism in international market entry. *Journal of Management Studies*, *56*(7), 1287–1313.
Verbeke, A., Coeurderoy, R., & Matt, T. (2018). The future of international business research on corporate globalization that never was.... *Journal of International Business Studies*, *49*(9), 1101–1112.
Verbeke, A., & Forootan, M. Z. (2012). How good are multinationality–performance (M-P) empirical studies? *Global Strategy Journal*, *2*(4), 332–344.
Verbeke, A., & Greidanus, N. S. (2009). The end of the opportunism vs trust debate: Bounded reliability as a new envelope concept in research on MNE governance. *Journal of International Business Studies*, *40*(9), 1471–1495.
Verbeke, A., & Hutzschenreuter, T. (2021a). Imposing versus enacting commitments for the long-term energy transition: Perspectives from the firm. *British Journal of Management*, *32*(3), 569–578.
Verbeke, A., & Hutzschenreuter, T. (2021b). The dark side of digital globalization. *Academy of Management Perspectives*, *35*(4), 606–621.
Verbeke, A., Hutzschenreuter, T., & Pyasi, N. (2021). The dark side of B2B relationships in GVCs–Micro-foundational influences and strategic governance tools. *Journal of Business Research*, *135*(C), 816–828.
Verbeke, A., & Kano, L. (2012a). An internalization theory rationale for MNE regional strategy. *Multinational Business Review*, *20*(2), 135–152.
Verbeke, A., & Kano, L. (2012b). The transaction cost economics theory of the family firm: Family–based human asset specificity and the bifurcation bias. *Entrepreneurship Theory and Practice*, *36*(6), 1183–1205.
Verbeke, A., & Kano, L. (2013). The transaction cost economics (TCE) theory of trading favors. *Asia Pacific Journal of Management*, *30*(2), 409–431.
Verbeke, A., & Kano, L. (2015). The new internalization theory and multinational enterprises from emerging economies: A business history perspective. *Business History Review*, *89*(3), 415–445.
Verbeke, A., & Kano, L. (2016). An internalization theory perspective on the global and regional strategies of multinational enterprises. *Journal of World Business*, *51*(1), 83–92.

Verbeke, A., Kano, L., & Yuan, W. (2016). Inside the regional multinationals: A new value chain perspective on subsidiary capabilities. *International Business Review, 25*(3), 785–793.
Verbeke, A., & Kenworthy, T. P. (2008). Multidivisional vs metanational governance of the multinational enterprise. *Journal of International Business Studies, 39*(6), 940–956.
Verbeke, A., & Lee, I. I. (2021). *International business strategy: Rethinking the foundations of global corporate success*. Cambridge University Press.
Verbeke, A., Simoes, S. & Grøgaard B. (2023). *The role of multinational enterprises and formal institutions in base of the pyramid markets* [Mimeo].
Verbeke, A., & Yuan, W. (2005). Subsidiary autonomous activities in multinational enterprises: A transaction cost perspective. *Management International Review, 45*(2), 31–52.
Verbeke, A., & Yuan, W. (2007). Entrepreneurship in multinational enterprises: A Penrosean perspective. *Management International Review, 2*(47), 241–258.
Verbeke, A., & Yuan, W. (2013). The drivers of multinational enterprise subsidiary entrepreneurship in China: A new resource-based view perspective. *Journal of Management Studies, 50*(2), 236–258.
Verbeke, A., & Yuan, W. (2020). The tyranny of the head office? Revisiting corporate headquarters'(CHQs) role in MNE subsidiary initiatives. *Journal of Organization Design, 9*(1), 1–11.
Verbeke, A., & Yuan, W. (2022). Rethinking intrapreneurship in the established MNE. *Global Strategy Journal, 12*(4), 738–758.
Verbeke, A., Zargarzadeh, M. A., & Osiyevskyy, O. (2014). Internalization theory, entrepreneurship and international new ventures. *Multinational Business Review, 22*(3), 246–269.
Williamson, Oliver E. (1993) Calculativeness, trust, and economic organization. *The Journal of Law and Economics, 36* (1 Part 2), 453–486.
Wong, J. (2023, June 26). The U.S. and Modi rewire Indian defense and tech. *Wall Street Journal.* https://www.wsj.com/articles/the-u-s-and-modi-rewire-indian-defense-and-tech-dfabec47

CHAPTER 3

TAKING WISHES FOR REALITY: THE NEGLECTED ROLE OF COMPLEMENTARY RESOURCES IN THE CHOICE OF FOREIGN MARKET ENTRY, THE INTERNATIONALIZATION OF DIGITAL VENTURES, AND CATCHING-UP BY EMERGING MARKET FIRMS

Jean-François Hennart[a,b]

[a]*Tilburg University, The Netherlands*
[b]*Aalborg Business School, Denmark*

ABSTRACT

Why is it that, despite repeated claims that digital-content firms and internet-based businesses can internationalize everywhere almost instantly, many seem unable to profitably expand outside their home markets? Why have emerging market firms (EMNEs) caught up with established developed-country multinationals (DMNEs) so much faster than expected? In this chapter, the

Walking the Talk? MNEs Transitioning Towards a Sustainable World
Progress in International Business Research, Volume 18, 43–60

ISSN: 1745-8862/doi:10.1108/S1745-886220240000018003

author argues that the clue to these two puzzles lies in the realization that, contrary to the dominant view in the international business (IB) literature that focuses only on the intangibles exploited by DMNEs and assumes that these firms are free to unilaterally decide on their mode of entry and operation, doing business in a foreign country is only possible if intangibles are bundled with complementary local resources, usually held by local firms. Taking into account these complementary local resources and their owners makes it clear that DMNEs are not always free to choose their entry mode but must enlist the cooperation of local resource owners. The need of digital-content and internet-based firms for local complementary resources also explains why they sometimes experience problems when expanding abroad. Lastly, control of complementary local resources provides EMNEs with a home advantage against DMNEs competing with them in their home market. The author shows how EMNEs can capitalize on this advantage to obtain the intangibles they lack and need. The fact that these advantages are available on efficient global markets, while complementary local resources are not, explains the surprising speed of EMNE catch-up.

Keywords: Bundling model; complementary local resources; modes of foreign market entry; born digitals; internet platforms; emerging market multinationals; theories of the multinational enterprise; OLI paradigm; transaction cost theory; springboard model

INTRODUCTION

Imagine a world where multinational enterprises (MNEs) based in developed markets (DMNEs) can unilaterally decide how much control they will have over their foreign operations, where high-tech Silicon Valley entrepreneurs can capture markets all over the globe without having to deal with foreign firms and governments, and where DMNEs can rest assured that their superior technology will allow them to continue to dominate their emerging country rivals. Shades of such an utopic vision have permeated the IB field. In this chapter, I argue that the IB literature has by and large overestimated the role played by technology and at the same time underestimated the strategic importance of complementary resources. By complementary resources, I mean those necessary to exploit intangibles such as government permits, natural resources, access to final consumers, and so on. These are typically owned by the governments, firms, and individuals of the countries that DMNEs are trying to enter.[1] I will show that the strategic role of suppliers of complementary resources has been overlooked or at best minimized by the IB literatures on (1) the choice of modes of foreign market entry; (2) the impact of digitalization and internet platforms on internationalization; and (3) the technological catch-up of EMNEs.

I begin with a quick review of internalization theory and its focus on the exploitation of intangibles, then outline the crucial role of complementary resources in that exploitation. Following that, I show that the IB literature has largely ignored the role of such resources in its approach to the choice of modes of foreign market entry, in its discussion of the internationalization of digital firms, and in its treatment of the competition between DMNEs and EMNEs. I conclude with some recommendations for further research.

THE INTERNALIZATION THEORY OF THE MNE

A fundamental insight behind the internalization theory of MNEs (Buckley & Casson, 1976; Hennart, 1977, 1982; Rugman, 1981; Rugman & Verbeke, 2001) is that one of the reasons they arise is to facilitate the exploitation of intangibles across borders.[2] Intangibles are various types of knowledge as well as reputation. They have two specific properties: (1) they are public goods in the sense that their use by one person does not diminish their utility to another, and (2) their production is characterized by high fixed costs but low marginal costs.[3] Consequently, scaling up production, once the initial fixed costs have been made, incurs very low marginal costs, and so it makes sense to sell to as many customers as possible in order to cover fixed costs and generate a profit. Especially if the domestic market is small, this may well mean selling to foreign customers and in some cases investing in foreign production facilities, that is, making FDIs (Buckley & Casson, 1976; Hennart, 1982). The same logic applies to reputation. Once a favorable one has been established with one group of customers or with one type of product, it can be extended to new customers and to new products at very low marginal cost (Hennart, 1982).

Another connection between intangibles and FDI is that it is generally more costly for foreign firms to operate in a country than it is for native ones. Foreign firms are often subject to a liability of foreignness because they do not know local operating conditions, are not integrated into local networks, are sometimes discriminated against by host-country consumers and governments, and operate at a distance from their headquarters, which increases management costs (Hymer, 1976; Hennart, 1982; Zaheer, 1995). Firms operating abroad must therefore have some kind of advantage, often based on intangibles, that allows them to compensate for the disadvantages of foreignness. That advantage must be specific to the firm, for otherwise it would be imitated by rivals, and so would no longer serve as an advantage. Hence, ownership advantages (Dunning, 1977) have been called firm-specific advantages (FSAs) (Rugman, 1981).

As was established early on, the fact that a firm has intangibles it seeks to exploit does not mean that it will be an MNE (Dunning, 1977). An MNE is a firm that has employees outside its own country (Hennart, 1982). This is likely to occur under very specific circumstances. We need to remember that firms or individuals who want to exploit their intangible assets can do it in five different markets, that for labor, that for intellectual property rights (IPs), that for the services of IPs (technology and franchising licenses), that for firms in which intangibles are embedded, and that for goods and services in which the intangibles are embedded. A firm or an individual will use whichever market is most efficient. MNEs only arise when intangibles are exploited on the market for firms and on that for goods and services, and when their exploitation involves having employees abroad. This occurs (1) when individuals or teams develop intangibles within a firm and get a return on their R&D efforts by selling the firm, partly or fully, to foreign firms – in that case the acquirer ends up with employees abroad and (2) when owners of intangibles exploit them in the market for goods and services by producing them in an owned facility in a foreign location (Hennart, 2009b). In that case, intangible owners skip the market for intellectual property and integrate vertically into the market for goods and services in which intangibles are embedded. In doing so, they are transferring intangibles internally from conception to exploitation, that is, they are internalizing its transfer. Hence the term internalization to define this theory (Rugman, 1981).

COMPLEMENTARY RESOURCES

Consider the ubiquitous laser printer. Manufacturers can only successfully sell them in a foreign country if they manage to efficiently line up all the factors of production needed to procure and assemble the parts, to take orders, to manage payments, to handle physical delivery, and to provide after-sales service and repairs. Consumers themselves also contribute complementary services. They must have electric power, be able to install printer software on their computer and make it communicate with the printer and provide suitable toner and paper. A printer is of little use without all these complementary resources. Monetizing intangibles by embedding them in products and services is therefore a strategy that requires lining up complementary resources (Hennart, 2009a; Teece, 1986). These complementary resources, if they are available, are generally supplied by local governments, firms, and consumers.

Their range is quite wide. It goes from government permits, to finance, land, employees, raw materials, utilities, and consumer access, which in turn requires communication, transportation, and payment infrastructures. Access to complementary resources determines the feasibility, cost, and profitability of foreign operations. The ideal situation is one where they are available and can be purchased at competitive prices, thus allowing the intangible owner to capture most of the gains generated by the intangible-complementary resources bundle (Teece, 1986). If they are not available, then intangible-exploiting firms must persuade others to provide them, or they must supply them themselves. Complementary resources may be available, but on narrow markets. In that case, their owners may be able to hold up owners of intangibles and to capture a very large share of the profits generated by their exploitation. A firm is likely to find lining up complementary resources especially challenging when it requires such resources in a foreign location. There, and, as we will see, contrary to the assumptions of Dunning's OLI model (Dunning, 1977), its lack of knowledge of local resource providers and of their reliability, as well as potential discrimination by local government authorities, may cause problems.

Two factors are important in this regard: first, there is considerable variation in the relative importance of foreign complementary resources in the intangible-complementary services bundle; in some cases, profitable exploitation of intangibles abroad in the product and service markets requires a significant amount and variety of complementary services, in others few. For example, and as described more fully below, some firms, like Atlassian in its early days, sell computer programs downloadable from the Internet and destined for a niche clientele of experts; such firms can operate with very few, if any, complementary resources in target foreign markets. Platforms like Uber or Jumia, on the other hand, need considerably more. The greater the types and diversity of complementary resources needed, the more likely some will be missing or, if present, will be such as to demand substantial adaptation of the firm business model (see the case of Jumia below). Firms that need substantial complementary resources will find it more difficult to exploit their FSAs abroad, may take longer to do so, and may extract less profit from the process. A related concept is the distinction made by Rugman and Verbeke (2001) between location-bound and non-location-bound FSAs. Non-location-bound FSAs can be exploited abroad with varying degrees of recombination with local complementary resources (Verbeke, 2009; Verbeke et al., 2018; Verbeke & Kano, 2015). Location-bound FSAs, on the other hand, are, as the term makes clear, specific to a particular location and hence can only be exploited in other locations if substantial adaptations to the business model and recombinations with local complementary resources are made. These will slow down expansion abroad, and may even make it uneconomic (Bohnsack et al., 2021; Verbeke et al., 2014). These considerations are applicable to all types of firms and all technologies. Sellers of digitalized products and services and internet-based platforms are no exception, as I show below.

The extent to which complementary services can be obtained in competitive markets also varies across countries. Most developed Western countries have efficient markets for labor, IP, goods and services, and firms. In many emerging markets, on the other hand, exchange is governed by relationships, and this puts outsiders at a considerable disadvantage (Hennart & Sutherland, 2022; Li, 2003; Li et al., 2004). As I will show below, these factors influence (a) the modes of entry and operation firms choose to enter a foreign country; (b) the feasibility of selling abroad and the speed at which foreign sales can be developed; and (c) the relative bargaining power of owners of intangibles and of those of complementary resources, and more specifically of that of EMNEs relative to DMNEs.

COMPLEMENTARY RESOURCES AND FOREIGN ENTRY MODES

In Hennart (2009a), I show that the IB literature has ignored complementary resources or taken them for granted. Theories of modes of entry and operation, based on transaction cost theory or on organizational learning, have ignored the fact that profitable exploitation of FSAs requires access to complementary resources and that this affects the optimal mode of entry. Dunning's OLI paradigm (Dunning & Lundan, 2008) has not ignored them, but has taken them for granted by assuming that they are accessible on equal terms by all firms present in a foreign country.

For Dunning, whether to serve foreign customers through exports or through foreign production, and where to locate foreign production facilities, depends on a comparison of the quantity and quality of complementary resources available in domestic and foreign locations and between locations. Dunning calls complementary local resources "location advantages." In contrast to ownership advantages, which are firm-specific, "location advantages are specific to a particular country but available to all firms" (Dunning & Lundan, 2008, p. 6). Accordingly, because domestic and foreign firms can access them on an equal basis, they do not play any role in determining whether a foreign firm can profitably operate in a foreign market and the optimal form its entry will take.

While OLI takes access to complementary resources for granted, most theories of modes of entry ignore them altogether. In their recent review of the empirical literature on an MNE's choice between full and partial control of its foreign subsidiaries, Chhabra et al. (2021, p. 244) summarize it in this way:

> The literature argues that equity ownership stake for MNEs is a key strategic decision (Anderson & Gatignon, 1986; Chari & Chang, 2009) involving a trade-off between relative costs (e.g. commitment of resources, risks in the foreign market) and benefits (e.g. degree of control in the foreign subsidiary).

In other words, the decision on whether to enter and operate in a foreign market through a wholly owned subsidiary or through a joint venture depends only on the MNE's own desire for control and on its own evaluation of the risks involved, which, if large, will lead to sharing its foreign plant with other firms, that is, to an equity joint venture. No consideration is given to the complementary resources the MNE requires, and to the potential need to incentivize their owners to supply them by giving them a stake in the venture.[4] Similarly, the literature on entry mode based on organizational learning sees the choice between entering foreign markets through greenfields and acquisitions as depending on a firm's international experience or on its experience with a particular entry mode, with Barkema and Vermeulen (1998) arguing that firms with greater foreign market experience will unilaterally opt for greenfields, while Padmanabhan and Cho (1999) hypothesize that they will opt for greenfields if they have used that mode of entry in the past. There is no recognition that the choice between greenfields and acquisition may be influenced by the need to access complementary resources and that in some cases an internationally experienced MNE, or one with past experience with greenfields, will prefer to enter through acquisitions because it is more efficient to access complementary resources through the market for firms (i.e., through acquisitions) than through that for inputs (i.e., through greenfields).

If foreign market entry and operation require the bundling of intangibles and of local complementary resources, then it is clear that the modes of foreign market entry that should be used by MNEs are those that align the incentives of all parties involved in that entry, that is, the MNE and the suppliers of complementary resources. For example, in the case of market-seeking FDIs, whether the MNE will enter by licensing its intangibles to local firms or by locating in the foreign market the production of goods and services integrating its intangibles – hence establishing a foreign production subsidiary – and how much control it will take in that subsidiary will hinge on the relative efficiency of the market for intangibles compared to that for the products incorporating the intangibles. If the market for the latter is more efficient, the MNE will choose to enter through an FDI. However, if in that case it is faced with a monopolized market for complementary resources, it may need to incentivize their local owners by giving them part of the equity of the foreign plant. Hence the choice of mode of entry depends on the efficiency of the various possible markets available to exploit intangibles, as well as those local owners of complementary resources can use to sell their complementary resources. When both markets are inefficient, the MNEs and local suppliers of complementary resources may find that sharing the profits of the venture is the most efficient solution. In that case, MNEs will enter the foreign market with equity joint ventures not because they choose to, as suggested by most authors, but because they must (Hennart, 2009a).

We follow this approach in Hennart et al. (2015) when considering the choice US MNEs entering Brazil made between full or partially owned subsidiaries. Our expectations were that US firms would opt for equity joint ventures when the complementary resources they needed were not available on efficient markets. We therefore looked at the number of potential Brazilian distributors for the products manufactured by US firms in Brazil if they sold to Brazilian customers, and the number of potential marine shipping firms if they sold to foreign ones. As predicted, we found that, keeping all other factors constant, US investors tended to opt for equity joint ventures when there were few potential distributors and maritime shippers for their products.

DIGITALIZATION AND COMPLEMENTARY RESOURCES

Many authors writing on the impact on IB of recent advances in digitalization, electronic commerce, and artificial intelligence also underestimate the role played by complementary resources. That literature (e.g., Autio et al., 2021; Coviello et al., 2017; Nambisan et al., 2019) holds that advances in information technology have drastically increased the ability to sell abroad. Mihailova (2023, p. 1), for example, writes that digital technologies "have triggered unprecedented possibilities for international scalable expansion by decreasing transportation and production costs and developing platforms enabling instant global expansion and connectivity." Digital technologies are deemed to have specific attributes that make international expansion easy (Stallkamp et al., 2023). They are said to be scale-free (Adner et al., 2019), which is another way of saying that, like intangibles, they are public goods. They are also described as fungible, that is, their value remains the same in all contexts. This contrasts with the conventional view that intangibles developed at home may lose value when transferred to different country contexts (Rugman & Verbeke, 2001).

While it is true that digitalization has drastically reduced the cost of sharing information and of distributing content, and that it has made it possible in some cases to sell to foreign customers without a foreign footprint, there are reasons to believe that in many – indeed perhaps most – cases the exploitation of digital intangibles in foreign markets requires some local complementary resources, and that these resources tend to be country-specific, and hence not cheaply re-deployable across markets. This may make the exploitation of digital-content in foreign countries more difficult and slower than that predicted by the digitalization and electronic platform literature, and in some cases unprofitable.

Digitalization is the act of

> converting "things" (information, sound, shapes, etc.) into digital data that can be stored, processed, reprocessed and redeployed an infinite number of times, at low or zero marginal cost. It involves digital data, technologies, infrastructure, and business models, all of which represent some form of digital assets. (Verbeke & Hutzschenreuter, 2021, p. 607)

Digitalization has spawned at least three types of business models (Hennart, 2019): (1) firms that sell digital products through digital means, such as software, information (search engines), music, and entertainment distributed through the internet – these firms have been named "born digitals" (Monaghan et al., 2020); (2) firms that operate B2B, B2C, and C2C electronic brokerage internet platforms which connect individuals to other individuals (C2C), firms to other firms (B2B), and individuals and firms (B2C and C2B) for social contact, employment, the sale of goods and services, and other purposes; and (3) firms that use platforms to distribute their own physical goods and services, for example Amazon.

Born digitals are firms "with a market offer that is both digital in nature and distributed by digital technology" (Monaghan et al., 2020, p. 13). Banalieva and Dhanaraj (2019), Birkinshaw (2022), Brouthers et al. (2022), and Nambisan et al. (2019), among others, claim that born digitals can internationalize without owning physical assets in foreign markets. For instance, Brouthers et al. (2022) write that such firms can sell to foreign customers through "virtual presence," that is without setting up "operations of any kind (including cooperative modes like export agents, distributors, or export sales subsidiaries) in the foreign market, instead relying on digital technologies to reach customers/users" (p. 2095). According to them, such novel ways of entering foreign countries show the need for "employing new theoretical perspectives to understand the complex phenomena of internationalization" (p. 2088). Likewise, digital platforms and ecosystems are said to create worldwide marketplaces that facilitate cross-border transactions and allow platform firms to "rapidly enter, scale, and grow their operations in foreign markets" (Nambisan et al., 2019, p. 1474). According to Singh et al. (2023, p. 1), they can even "provide a virtual field, unconstrained by space and time limitations, to achieve rapid internationalization and value creation."

I have discussed elsewhere (Hennart, 2019, 2022; Hennart & Verbeke, 2022) claims that governance structures brought about by digitalization are new and as such require new theories. I do not believe that is the case, and indeed argue that existing theory is in fact particularly useful in making sense of these supposedly new governance structures. Here, I want to focus instead on the claims that digitalization and internet platforms have made it possible to expand abroad without complementary local resources, thus abolishing most if not all the familiar problems caused by geographic and cultural distance.

Along with Verbeke and Hutzschenreuter (2021) and Stallkamp et al. (2022, 2023), I believe these claims are somewhat exaggerated. Verbeke and

Hutzschenreuter (2021) argue that the positive impact of digitalization on international expansion is somewhat countered by constraints at the macro- and micro-levels. At the macro-level, both host customers and governments can be expected to react to some of the negative social consequences of digital globalization. At the micro-level, they point out that the literature has often downplayed the location-boundedness of digital FSAs, that is, the extent to which they lose value when exploited in a foreign country. While demand differences and consumer and governmental reaction to the penetration of foreign digital companies are important constraints, I focus here on how the need for local complementary resources affects foreign expansion. I first discuss born digitals, and then look at digital platforms.

I start with a simple question: What determines the speed at which firms develop their foreign sales and their foreign footprint? I try to provide an answer in a 2014 article published in *Entrepreneurship Theory and Practice* (Hennart, 2014). My approach is to identify the tasks a firm might have to perform to successfully sell abroad, starting with identifying potential customers; understanding their preferences; modifying the marketing mix in response; persuading them to buy; then providing after-sales education, advice and repair, billing; extending credit and collecting debt; and finally physical delivery of products and services. I argue that there are large differences between firms in the cost and time needed to perform these tasks, and that the differences depend on the business model used. Mass-market B2C firms expanding abroad must spend considerable time and resources to identify likely customers and devise ways to persuade them to buy. Customers of mass-market goods, such as automobiles and food, also tend to have country-specific tastes, which foreign firms need to understand and to which they must adapt. Lastly, because their products are weakly differentiated, and hence have local substitutes, exporting from a home base makes their products too expensive relative to locally based ones. Hence sellers of mass-market products must locate production close to their foreign customers, a task that takes a considerable amount of time and money. This expensive and time-consuming process is the one described by the Uppsala model: a slow, timid, step-by-step progression from culturally and institutionally close to more distant countries (Johanson & Vahlne, 1977).

In contrast, firms that focus on specialized products with global appeal – global niche products – can immediately sell to foreign customers. Most, if not all the tasks listed above, can, in their case, be done quickly and cheaply. To make my point, I chose the example of a born digital, Atlassian, an Australian-based developer of specialist software aimed at professional programmers. Atlassian's program, Jira, which can be downloaded from the firm's website, is priced low enough that programmers can put it on their credit card without involving higher-ups. Jira created a buzz in the programmer community, and two years after its founding Atlassian had some 3,500 customers in close to 40 countries (Zalan et al., 2009). There were no marketing costs as programmers learned about it by word-of-mouth and sought Atlassian out, there was no product adaptation, little after-sales service, zero transportation costs, and no need to physically locate production close to the consumer. No wonder international expansion was instantaneous and cheap. Atlassian could exploit its FSAs abroad without foreign-based complementary resources. In the words of Rugman and Verbeke (2001), Atlassian's FSAs were not location-bound. Empirical studies on a sample of Italian SMEs shows that this business-model explanation is the best predictor of internationalization speed, and accounts for more of the variance in that speed than the international experience of founders which some have postulated is the key to internationalization (Hennart et al., 2021).

Concentrating on a concrete case like Atlassian is useful because it makes it possible to showcase the factors that allowed it to expand abroad so quickly. One of them is that it was selling a low-priced but high-tech product to highly knowledgeable and networked professionals. Atlassian did not have to find them, persuade them to buy, and hold their hands afterwards. Hence it did not need a local salesforce. This is not, however, the case for all sellers of digital products. Consider the internationalization history of HD-Furniture, as reported by Ojala et al. (2020). HD-Furniture sells software that allows furniture manufacturers and retailers to show their products in 3D. Its offering, software sold over the internet to business customers, is the same as Atlassian's. But as is evident from Ojala et al.'s (2020) description, the devil is in the details.

> In theory, HD-Furniture was able to distribute its service to any geographical location; however, while its digital service itself was borderless … internationalization was slowed down by the marketing and sales processes that required lengthy face-to-face negotiations with prospective customers. In addition, HD-Furniture's customers were not able to add their physical furniture into the digital service as doing so required extensive technical knowledge. HD-Furniture's employees conducted this task until they found partners to take over the task. This slowed down the internationalization process as customers were not able to adopt the service in a self-service form. (Ojala et al., 2020, p. 4656)

In other words, to operate efficiently HD-Furniture required a key local complementary resource, that is, a local sales force, something Atlassian did not need. Because the local sales force had to be deployed in each target market, HD's international expansion was slower and more costly than Atlassian's. More generally, the speed and extent to which a born digital will expand abroad will depend on its need for local complementary resources, a need that will vary in subtle ways depending on its business model.[5] Even born digitals, the most likely candidates for foreign asset-free virtual internationalization (Brouthers et al., 2022), may require local complementary resources, and the need for lining up such resources is likely to reduce their ability to scale up their business internationally as well as the speed at which they can do it. Recent empirical findings support this view. Stallkamp et al. (2022) hypothesize that the extent to which a born digital relies on non-digital complementary resources affects its foreign sales ratio and find support for the hypothesis when looking at 169 US-based born digitals. In later work (Stallkamp et al., 2023), they find that 132 out of the 169 US born digitals in their sample had foreign subsidiaries, and that the probability a born digital would establish a subsidiary in a foreign country was greater the more culturally and geographically distant that country was from the USA.

The need for complementary local resources is likely to be even greater for platforms like Alibaba or Amazon. Consider Jumia, an e-commerce platform incorporated in Nigeria in 2012 with financial and technical support from Rocket Internet, a Berlin-based incubator. The plan was to create an African Amazon targeting the poorly served African middle class in countries other than South Africa (Peprah et al., 2021). In 2019, the company did an IPO on the New York Stock Exchange and its stock price peaked at US$49.77. By then it was operating in 14 African countries. Since then, the company has not done well. It has yet to turn a profit, has retreated to 11 countries, and has seen its two founders resign (Kene-Okafor, 2022). Its stock, as of 5 May 2023, was down to US$2.85. As I will show, the problems faced by Jumia seem to come from an underestimation of the importance of complementary resources in Amazon's business model which Jumia sought to imitate.

Amazon's business model relies on (1) consumers familiar with the kinds of products it sells; (2) comfortable ordering and paying online; (3) third party payment and logistic providers; (4) a reliable communication infrastructure; and (5) a transportation infrastructure and a system of unique physical addresses. These complementary resources often do not exist in the countries where Jumia operates (Peprah et al., 2021). Because many customers do not have computers and are not used to ordering online, Jumia has had to hire employees who go to public places with computer tablets and help in the placing of orders. Many of Jumia's customers do not trust online platforms and some lack bank accounts, so Jumia has had to allow them to pay cash-on-delivery. This means that delivery personnel carry cash, increasing the risk of theft and robbery. Many customers also insist on having a look at the merchandise before buying, and then change their mind, resulting in a high return rate. Another reason for a high return rate is consumer inexperience with some of the products Jumia sells, for example, PCs, TVs, and mobile phones, so it has had to hire personnel to help customers install and use them. Jumia had initially to do its own delivery because no reliable third-party provider was available. To remedy the lack of street addresses, it hires neighborhood representatives at strategic locations who hold packages for customers and collect payment. These additional expenses undertaken to remedy the lack of local complementary resources are undermining Jumia's initial business model and making it unprofitable, indeed in 2019 its fulfillment expenses were higher than its gross profits: in short, Jumia paid more to serve its customers than it earned from them (Madowo, 2023). While Jumia is perhaps an extreme case, almost all platforms rely on host-country complementary resources, some of them provided by governments, some by firms, and some by consumers themselves. The extent to which these resources are available on competitive markets determines their internationalization prospects.

CATCHING UP BY EMNEs

The benign view of complementary resources taken by many Western IB researchers, business leaders, and politicians explains the surprise at the speed with which EMNEs are catching up with MNEs. Keep in mind that OLI theory stresses the need for foreign investors to own FSAs, typically based on intangibles, but posits that local complementary resources are freely available to one and all. According to such a worldview, one would expect firms based in emerging markets to face a hard challenge in developing internally their own FSAs, and some have argued that doing that would take a substantial amount of time. Rugman (2009, p. 53), for example, predicted that it would take "ten to twenty years" for China to "generate its own world-class MNEs." Nolan (2004) was equally pessimistic. The consensus in the beginning of the new century was that EMNEs might at best be able to survive in their home markets or to sell to countries with similar levels of income and institutional development (Cuervo-Cazurra & Genc, 2008), but not to be able to compete with DMNEs in Western markets.[6]

This rather unrealistic view of the superior competitive abilities of DMNEs comes, I argue, from an overestimation of the strategic importance of intangibles coupled with a systematic underestimation of that of complementary resources. Once one abandons the fiction that FSAs are the only drivers of international competition, with complementary resources playing a passive role, it become clear that the dominant position of DMNEs was bound to be temporary. The challenge for DMNEs eager to exploit their intangible assets in a foreign country is to access local complementary resources on competitive terms. Some complementary resources in some emerging markets may not be available, so DMNEs will have to provide them. This is not new. United Fruit built railroads and harbors in Central America before the turn of the last century to facilitate banana exports. Complementary resources may be available but controlled by local firms. In fact, as I have argued in Hennart (2012, 2018), a successful strategy for EMNEs competing with DMNEs in their home market is to seek to monopolize the supply of complementary resources that DMNEs need. They can then use their monopoly power to accumulate funds to purchase the intangible needed to compete with DMNEs, or they can enter into local equity joint ventures in which the complementary resources they control are traded for DMNE intangibles. In Hennart (2012, 2018), I document how this strategy has been used by several EMNEs, such as Lenovo, Haier, and CRRC in China, Grupo Bimbo in Mexico, and CVRD-Vale in Brazil. Lenovo, for example, started as a Chinese distributor for Western computer firms at a time when they were prevented by the Chinese government from setting up their own distribution network in the country. This allowed Lenovo to have a close look at Western products and at how Chinese customers responded to them. While continuing to build a formidable distribution network in China, Lenovo integrated into computer manufacture, and pursued a strategy of incorporating cutting-edge components from global suppliers into products sold at competitive prices. Based on this strategy of control over distribution, Lenovo became the dominant player in the Chinese PC market and started to compete with Western firms such as Dell and HP in developed countries. Today Lenovo is the world's largest PC seller.

Lenovo, Haier, and Grupo Bimbo are private firms, but the strategy they have used to catch up with DMNEs has also been employed by state-owned ones. CRRC provides an interesting example of how the Chinese government leveraged access to its internal market to obtain cutting-edge Western technology for national firms. In 2004, the Chinese government decided to build a very large high speed train network and when domestic firms proved unable to provide the necessary equipment, it solicited bids from Western high speed railway companies.[7] The RFPs required most of the train sets to be built in Sino-foreign joint ventures in which the foreign partner could hold no more than a 49% stake. By pitting against each other Siemens, Bombardier, Alstom, and Kawasaki, the four major Western manufacturers of high-speed train sets, the Chinese government managed to get them to transfer their latest technology to Chinese train makers CSR and CNR and to train their engineers (Hout & Ghemawat, 2010; Massie, 2011).[8] It also insisted that the train sets be trademarked under the Chinese partner's brand (Cory, 2021). By 2011, CSR and CNR had fully digested Western technology and were manufacturing train sets more technologically advanced than those of their foreign teachers; these days they are competing with DMNEs and winning orders for high-speed train projects outside China.[9]

The Chinese government has not been equally successful in its attempt to trade market access for technology in other sectors such as automobiles and semiconductors (The Economist, 2017a), nevertheless, the high-speed railway case highlights the strategic importance of local complementary resources, in that case market access.[10] In fact, it is not hard to show that complementary resources are often more strategic than FSAs. As I show in Hennart (2012, 2018), the competition between DMNEs and EMNEs in emerging markets is one where DMNEs need to line up complementary resources in foreign markets while EMNEs need to access or acquire intangibles similar or superior to those held by their DMNE competitors. I advance in Hennart (2012) that doing the former can be much more challenging than doing the latter, and events since appear to have confirmed that insight.

Why has acquiring intangibles proven to be easier for EMNEs than lining up complementary local resources has been for DMNEs? Bear in mind that knowledge can be accessed in the markets for the services of individuals, for IP (patents and trademarks), for goods and services (machines, products, and consulting), and for firms. One can efficiently access intangibles by uncompensated appropriation (patent infringement and counterfeiting) or by purchase on efficient markets. EMNEs, especially Chinese ones, have widely infringed upon the IP of their Western competitors (Brander et al., 2017). Until a recent shift in policy in the USA and in Europe toward curbing the acquisition of native high-tech firms by Chinese firms and the export of dual-use products and equipment to China, the markets for high-tech goods and services and that for intangible-rich firms have been relatively efficient in Western countries. For example, EMNEs have widely tapped the Western labor market and made use of Western consultants. In Hennart (2018), I use the example of Nio, a Chinese automotive startup, which has been able to recruit experts from 40 countries, including veterans from Ford and Volkswagen (The Economist, 2017b). Huawei took advantage of the bankruptcy of Nortel and of layoffs by Ericsson to hire engineers to staff newly established greenfield R&D centers in Ottawa (where Nortel was located) and in Stockholm and San Diego (where it hired laid off Ericsson engineers) (Schaefer, 2020). Chinese and Indian firms have been able to tap into the large number of their nationals who have graduated from US universities. EMNEs have acquired advanced Western inputs and machinery, for example computer chips and semiconductor equipment (Allen & Benson, 2023), and been able to take advantage of the efficient market for corporate control in

Western countries to purchase small and medium Western firms with cutting-edge technology, and in the process obtain a full set of technological and marketing capabilities. CSR, the Chinese train equipment company mentioned earlier, through its subsidiary Times Electric, was able to purchase the British firm Dumex, a maker of key components for the electric drive systems of high-speed train sets, components that CSR had not been able to develop on its own (He et al., 2018). Likewise, Geely, a Chinese automaker, accessed expert knowledge of automobile transmissions by acquiring Drivetrain Systems International, an Australian specialist maker of gearboxes. The knowledge thus gained allowed Geely to open three new transmission plants in China and to launch its own advanced six speed gearboxes (Ivarsson & Petersen, 2019).

Contrast these examples to DMNEs attempting to acquire complementary resources in emerging markets. Many complementary resources in emerging markets, for instance natural resources, are owned by governments, which also control access to the market for goods and services. Many emerging countries put restrictions on foreign FDI.[11] State-owned and family firms dominate the economy, while the number of private firms with dispersed ownership is limited. While in developed countries firms are usually highly specialized, the inefficiency of markets for intermediate products in emerging countries forces EMNEs to be vertically integrated into their production. A Western acquirer needs therefore to sell-off a good part of the acquired assets. Add to this that regulatory supervision of EMNEs is often lax, which makes due diligence difficult.[12] For all these reasons, it has proven more difficult for DMNEs to access complementary local resources than for EMNEs to acquire intangibles. The outcome was predictable: many EMNEs, operating in an opaque environment and often protected by their governments have quickly achieved technological parity with DMNEs, shrunk DMNE market share in emerging markets, and started to compete with DMNEs on their home turf. Only those who have bought into the mistaken notion of irrelevance of complementary local resources will have been surprised.[13]

CONCLUSION

I have argued that the neglect of complementary resources in much of IB research has led to the mistaken view that Western MNEs are free to unilaterally decide on their mode of entry, that digitalized and internet platform-based firms can sell to foreign customers without a foreign footprint, and that control of local complementary resources does not provide advantages to their owners because such resources are freely available to all on competitive markets. In the real world, MNEs cannot always use their preferred mode of foreign entry as they must enlist the cooperation of local resource owners. The need for local complementary resources is also a constraint for digital firms. While digital technologies may appear on the surface to offer the promise of instant and costless internationalization, in actuality the firms exploiting them frequently require local – often human – complementary resources to sell, implement, and adapt them. This poses the usual problems of lining up local complementary resources and possibly reconfiguring their business models.

These considerations suggest that future research on modes of foreign market entry should focus on the dependence of the entrant's business model on local complementary resources, on whether such resources are available locally, and on how to access them. This is likely to depend on the match between the foreign entrant's business model and the characteristics of the target foreign market. Since business models vary even within an industry, a more micro-approach may therefore be advisable. The same goes for studies of the impact of digital technologies on internationalization. Digitalization has affected many different business configurations (Hennart, 2019), each with its own dependence on foreign-based complementary resources, so its impact on internationalization is also likely to vary. Lastly, the strategic importance of complementary resources, for example, access to customers, has been ignored for far too long in the IB and strategic management literatures. Yet, it is clear that firms such as Amazon and Apple have considerable bargaining power that is due to their hold on consumers.

To conclude, things would be a lot easier for DMNEs if they could exploit their intangibles without having to deal with local suppliers of complementary resources. Intangible-rich DMNEs would continue to dominate markets worldwide if indeed we were in an OLI world where complementary resources are freely available on competitive markets and their owners have no bargaining power. Some IB scholars seem to have taken such wishes for reality.

NOTES

1. By DMNEs, I mean MNEs based in high-income countries such as Australia, Canada, most of Europe, Japan, South Korea, New Zealand, and the USA.

2. A more general statement is that MNEs arise when coordination of international interdependencies through hierarchical governance (employment contracts) is more efficient than coordination through the market. This makes it possible to also explain foreign direct investment (FDI) motivated by access to resources, both real and intangible, and access to customers (forward vertical integration). See Hennart (1982, 2010).

3. For instance, once a speaker has spent the time to prepare a talk, the costs of scaling up and reaching a thousand attendees are only marginally higher than that of delivering it to just a few.

4. For exceptions, see Delios and Beamish (1999), Hennart (1988), Hennart et al. (2015), and Luo (2001).

5. A business model approach is useful in this context because it goes beyond product to also consider types of customers targeted and ways to capture value. For an IB application, see Hennart et al. (2021).

6. Some (e.g., Zeng & Williamson, 2007) had a different view.

7. By 2020, the Chinese had a network of 35,740 km of high-speed track, then 68% of the world's total (Cory, 2021, p. 10).

8. CSR and CNR were merged into CRRC in 2015.

9. CRRC is providing the train sets for the new Jakarta–Bandung high-speed line in Indonesia (Artymiuk, 2022).

10. Naturally, a strategy of leveraging access to complementary resources is more effective for firms and governments based in countries with a large domestic market or which possess unique natural resources like oil or minerals.

11. With the exception of Brazil, BRIC countries ranked at the top of the 2020 OECD FDI Regulatory Restrictiveness Index which measures restrictions on incoming FDI. Russia scored 0.26, and China and India 0.21. In contrast, the OECD average was 0.06. See OECD (2023).

12. The low level of regulatory supervision in China, for example, explains the accounting scandals surrounding some Chinese firms registered on US stock exchanges.

13. Surprisingly, the importance of control of complementary local resources and the strategies of EMNEs and governments to leverage it seem to have been overlooked by most, if not all, researchers trying to explain technological catch-up by EMNEs. Kotabe and Kothari (2016) is a good example of this omission.

REFERENCES

Adner, R., Puranam, P., & Zhu, F. (2019). What is different about digital strategy? From quantitative to qualitative change. *Strategy Science*, *4*(4), 253–261.

Allen, G., & Benson, E. (2023, March). *Clues to the U.S. – Dutch–Japanese semiconductor export controls deal are hiding in plain sight*. Center for Strategic and International Studies.

Anderson, E., & Gatignon, H. (1986). Modes of foreign entry: A transaction cost analysis and propositions. *Journal of International Business Studies*, *17*(3), 1–26.

Artymiuk, S. (2022, September 7). First high-speed trains exported from China arrive in Indonesia. *International Railway Journal*. Retrieved on May 6, 2023 from https:/railjournal.com/fleet/first-high-speed-trains-expected-from-China-arrive-in-indonesia/

Autio, E., Mudambi, R., & Yoo, Y. (2021). Digitalization and globalization in a turbulent world: Centrifugal and centripetal forces. *Global Strategy Journal*, *11*(1), 3–16.

Banalieva, E., & Dhanaraj, C. (2019). Internationalization theory for the digital economy. *Journal of International Business Studies*, *50*(8), 1372–1387.

Barkema, H., & Vermeulen, F. (1998). International expansion through start-up or acquisition: A learning perspective. *Academy of Management Journal*, *41*(1), 7–26.

Birkinshaw, J. (2022). Move fast and break things: Reassessing IB research in the light of the digital revolution. *Global Strategy Journal*, *12*(4), 619–631.

Bohnsack, R., Ciulli, F., & Kolk, A. (2021). The role of business models in firm internationalization: An exploration of European electricity firms in the context of the energy transition. *Journal of International Business Studies*, *52*(5), 824–852

Brander, J., Cui, V., & Vertinsky, I. (2017). China and intellectual property rights: A challenge to the rule of law. *Journal of International Business Studies*, *48*(7), 908–921.

Brouthers, K., Chen, L., Li, S., & Shaheer, N. (2022). Charting new courses to enter foreign markets: Conceptualization, theoretical framework, and research directions on non-traditional entry modes. *Journal of International Business Studies*, *53*(9), 2088–2115.

Buckley, P., & Casson, M. (1976). *The future of the multinational enterprise*. McMillan.

Chari, M., & Chang, K, (2009). Determinants of the share of equity sought in cross-border acquisitions. *Journal of International Business Studies*, *40*(8), 1277–1297.

Chhabra, A., Popli, M., & Li, Y. (2021). Determinants of equity ownership stake in foreign entry decisions: A systematic review and research agenda. *International Journal of Management Reviews*, *23*(2), 244–276.

Cory, N. (2021). *Heading off track: The impact of China's mercantilist policies on global high-speed rail innovation*. Information Technology and Innovation Foundation.

Coviello, N., Kano, L., & Liesch, P. (2017). Adapting the Uppsala model to a modern world: Macro context and microfoundations. *Journal of International Business Studies*, *48*(9), 1151–1164.

Cuervo-Cazurra, A., & Genc, M. (2008). Transforming advantages into disadvantages: Developing country MNEs in the least developed countries. *Journal of International Business Studies*, *39*(6), 957–979.

Delios, A., & Beamish, P. (1999). Ownership strategy of Japanese firms: Transactional, institutional and experience influences. *Strategic Management Journal*, *20*(10), 915–933.

Dunning, J. (1977). Trade, location of economic activity and the multinational enterprise: a search for an eclectic approach. In B. Ohlin, P. Hesselborn, & P. Wijkman (Eds.), *The international allocation of economic activity* (pp. 395–418). Palgrave.

Dunning, J., & Lundan, S. (2008). *Multinational enterprises and the global economy*. Edward Elgar.

He, S., Khan, Z., & Shenkar, O. (2018). Subsidiary upgrading under emerging market acquirers. *Journal of World Business*, 53(2), 248–262.

Hennart, J.-F. (1977). *A theory of foreign direct investment* [Ph.d. dissertation, University of Maryland, College Park, MD].

Hennart, J.-F. (1982). *A theory of multinational enterprise*. University of Michigan Press.

Hennart, J.-F. (1988). A transaction cost theory of equity joint ventures. *Strategic Management Journal, 9*(4), 361–374.

Hennart, J.-F. (2009a). Down with MNE-centric theories! Market entry and expansion as the bundling of MNE and local assets. *Journal of International Business Studies, 40*(9), 1432–1454.

Hennart, J.-F., (2009b). Theories of the multinational enterprise. In A. Rugman (Ed.), *Oxford handbook of international business* (2nd ed., pp. 125–145). Oxford University Press.

Hennart, J.-F. (2010). Transaction cost theory and international business. *Journal of Retailing, 86*(3), 257–269.

Hennart, J.-F. (2012). Emerging market multinationals and the theory of the multinational enterprise. *Global Strategy Journal, 2*(3), 168–187.

Hennart, J. F. (2014). The accidental internationalists: A theory of born globals. *Entrepreneurship Theory and Practice, 38*(1), 117–135.

Hennart, J.-F. (2018). Springing from where? How emerging market firms become multinational enterprises. *International Journal of Emerging Markets, 13*(3), 568–585.

Hennart, J.-F. (2019). Digitalized service multinationals and international business theory. *Journal of International Business Studies, 50*(8), 1388–1400.

Hennart, J.-F. (2022). How much is new in Brouthers et al.'s new foreign entry modes, and do they challenge the transaction cost theory of entry mode choice? *Journal of International Business Studies, 53*(9), 2116–2132.

Hennart, J.-F., Majocchi, A., & Hagen, B. (2021). What's so special about born globals: Their entrepreneurs or their business model? *Journal of International Business Studies, 52*(9), 1665–1694.

Hennart, J.-F., Sheng, H., & Pimenta, G. (2015). The drivers of entry and expansion modes of US-based MNEs in Brazil. *International Business Review, 23*(3), 466–475.

Hennart, J.-F., & Sutherland, D. (2022). International business research: The real challenges are data and theory. *Journal of International Business Studies, 53*(9), 2068–2087.

Hennart, J.-F., & Verbeke, A. (2022). Actionable and enduring implications of Oliver Williamson's transaction cost theory. *Journal of International Business Studies, 53*(8), 1557–1575.

Hout, T., & Ghemawat, P. (2010, December). China vs. the world: Whose technology is it? *Harvard Business Review*, pp. 94–103.

Hymer, S. (1976). *The international operations of national firms: A study of foreign direct investment*. MIT Press.

Ivarsson, I., & Petersen, B. (2019). *Market-political ambidexterity and resource-asset bundling: A recipe for competitive advantage for emerging market firms?* [Paper presentation] European International Business Academy.

Johanson J., & Vahlne, J-E. (1977). The internationalization process of the firm –A model of knowledge development and increasing foreign market commitment. *Journal of International Business Studies, 8*(1), 23–32.

Kene-Okafor, T. (2022, November 18). Jumia to cut products and overhead as new management chases profits. *Techcrunch*. Retrieved May 6, 2023, from https://techcrunch.com/2022/11/18/jumia-to-cut-products-and-overhead-as-new-management-chase-profits/

Kotabe, M., & Kothari, T. (2016). Emerging market multinational companies' evolutionary paths to building a competitive advantage from emerging markets to developed countries. *Journal of World Business, 51*(5), 729–743.

Li, J.S. (2003). Relation-based *vs*. rule-based governance: An explanation of the East Asian miracle and crisis. *Review of International Economics, 11*(4), 651–673.

Li, S., Park, S-H., & Li, S. (2004). The great leap forward: The transition from relation-based governance to rule-based governance. *Organizational Dynamics, 33*(1), 63–78.

Luo, Y. (2001). Equity sharing in international joint ventures: And empirical analysis of strategic and environmental determinants. *Journal of International Management, 7*(1), 31–58.

Madowo, L. (2023, April 25). Jumia: The e-commerce start-up that fell from grace. *BBC News*. Retrieved May 6, 2023, from www://bbc.com/news/world-africa-52439546

Massie, J. (2011). Digestion and re-innovation: A lesson learned from China's high-speed rail technology-transfer agreements. *Intellectual Property Brief*, *25*, 25–26.
Mihailova, I. (2023). Business model adaptation for realized international scaling of born digitals. *Journal of World Business*, *58* (2), 101418.
Monaghan, S., Tippmann, E., & Coviello, N. (2020). Born digitals: Thoughts on their internationalization and a research agenda. *Journal of International Business Studies*, *51*(1), 11–22.
Nambisan, S., Zahra, S., & Luo, Y. (2019). Global platforms and ecosystems: Implications for international business theories. *Journal of International Business Studies*, *50*(9), 1464–1486.
Nolan, P. (2004). *China at the crossroads*. Polity Press.
Ojala, A., Rollins, M., Fraccastoro, S., & Gabrielsson, M. (2020). The internationalization of B2B digital platform providers: The role of cross-national distance and digital characteristics. In *Proceedings of the 53rd Hawaii international conference on systems science* (pp. 4652–4661). https://doi.org/10.24251/HICSS.2020.571
OECD. (2023). *OECD FDI regulatory restrictiveness index*. Retrieved May 31, 2023, from Stats.oecd.org/Index.aspx?Datasetcode=FDI_Flow_Industry#
Padmanabhan, P., & Cho, K. (1999). Decision-specific experience in foreign ownership and establishment strategies: Evidence for Japanese firms. *Journal of International Business Studies*, *30*(1), 25–44.
Peprah, A., Giachetti, C., Larsen, M., & Rajwani, T. (2021). How business models evolve in weak institutional environments: The case of Jumia, the Amazon.com of Africa. *Organization Science*, *33*(1), 431–463.
Rugman, A. (1981). *Inside the multinationals*. Columbia University Press.
Rugman, A. (2009). Theoretical aspect of MNEs from emerging countries. In R. Ramamurti & J. Singh (Eds.), *Emerging multinationals in emerging markets* (pp. 42–63). Cambridge University Press.
Rugman, A., & Verbeke, A. (2001). Subsidiary-specific advantages in multinational enterprises. *Strategic Management Journal*, *22*(3), 237–250.
Schaefer, K. (2020). Catching up by hiring. *Journal of International Business Studies*, *51*(9), 1500–1515.
Singh, N., Munjal, S., & Kundu, S. (2023). Marketplace platforms as game changers: Internationalization of smaller enterprises. *Journal of International Management*, *29*(4), 101035. https://doi.org/10.1016/j.intman.2023.101035
Stallkamp, M., Chen, L., & Li, S. (2023). Boots on the ground: Foreign direct investment by born digital firms. *Global Strategy Journal*, 1–25. https://doi.org/10.1002/gsj.1474
Stallkamp, M., Hunt, R., & Schotter, A. (2022). Scaling, fast and slow: The internationalization of digital ventures. *Journal of Business Research*, *146*, 95–106.
Teece, D. (1986). Profiting from technological innovation: Implications for integration, collaboration, licensing and public policy. *Research Policy*, *15*(6), 285–305.
The Economist. (2017a, September 23). China sets its sights on dominating sunrise industries. Retrieved on May 11, 2023 from https://economist.com/finance-and-economics/2017/09/23/china-sets-its-sights-on-dominating-sunrise-industries
The Economist. (2017b, September 23). China's audacious and innovative generation of entrepreneurs. Retrieved on May 11, 2023 from https://economist.com/briefing/2017/09/23/chinas-audacious-and-inventive-new-generation-of-entrepreneurs
Verbeke, A. (2009). *International business strategy*. Cambridge University Press.
Verbeke, A., Coeurderoy, R., & Matt, T. (2018). The future of international business research on corporate globalization that never was…*Journal of International Business Studies*, *49*(9), 1101–1112.
Verbeke, A., & Hutzschenreuter, T. (2021). The dark side of digital globalization. *Academy of Management Perspectives*, 35(4), 606–621.
Verbeke, A., & Kano L. (2015). The new internalization theory and multinational enterprises from emerging economies: A business history perspective. *Business History Review*, *89*(3), 415–445.
Verbeke, A., Zargarzadeh, M., & Osiyevskyy, O. (2014). Internalization theory, entrepreneurship, and international new ventures. *Multinational Business Review*, *22*(3), 246–269.
Zaheer, S. (1995). Overcoming the liability of foreignness. *Academy of Management Journal*, *38*(2), 341–363.
Zalan, M., Muzychenko, O., & Burshtein, S. (2009). Atlassian: Supporting the world with legendary service. *Journal of International Business Education*, *4*, 27–46.
Zeng, M., & Williamson, P. (2007). *Dragons at your door: How Chinese cost innovation is disrupting global competition*. Harvard Business School Press.

CHAPTER 4

DISTANCE, (MULTI)REGIONAL STRATEGIES, AND LIMITATIONS TO BRIDGE DISTANCE

Thomas Hutzschenreuter

Technical University of Munich, Germany

ABSTRACT

Distances between contexts that a multinational enterprise (MNE) is familiar with and those that it is not familiar with do not hinder the MNE from becoming global. In this sense, distance serves as a barrier between an MNE's global intentions and its regional realizations. (Multi)regional strategies are a possible expression of an MNE's inability to approach distances. Considering the managerially relevant distances is essential to analyze the internationalization and the resulting (multi)regional strategies. A deeper understanding of how distance works requires linking distance to micro-foundations. Besides bounded rationality, bounded reliability opens the conceptual pathway toward a deeper understanding of internationalization and (multi)regional strategies.

Keywords: Distance; added distance; MNEs; (multi)regional strategies; micro-foundations; bounded reliability; outsidership; experiential learning

Walking the Talk? MNEs Transitioning Towards a Sustainable World
Progress in International Business Research, Volume 18, 61–69

ISSN: 1745-8862/doi:10.1108/S1745-886220240000018004

INTRODUCTION

The internationalization process of firms has been the subject of considerable research in the field of international business (IB). In parallel, the concept of distance has become one of the cornerstones of IB. The consensus is that firms crossing borders are exposed to liabilities of outsidership (Johanson & Vahlne, 2009). To overcome the liabilities of outsidership, firms need to become an insider in local networks, learn, and accumulate knowledge. A higher degree of outsidership means the firm needs to close a more significant knowledge gap. The knowledge gap, as the difference between experience/knowledge possessed and that required in a new/target context, is a consequence and expression of the distance between the contexts in which the firm is an insider and the new/target context the firm is an outsider. Thus, the challenge of successful internationalization is becoming an insider, or in other words, overcoming distances.

The expression of internationalization efforts is the resulting portfolio of domestic and non-domestic (international) business activities. Firms significantly differ in the degree of how global or regional their portfolio of business activities is. According to their portfolio of business activities, firms can be global or multiregional but not global, multinational but single-regional, or domestic. Among the reasons why some firms are global and others are (only) multinational but single-regional is that firms differ in their abilities to overcome distances. In other words, the extent to which a firm's (multi)regional strategy differs from being global could be seen as an expression of the difference between intentions to be global and "regional" realizations. The theory of the multinational firm discusses those limitations under the term micro-foundations, hinting at human limitations in handling challenges that arise from doing business in unfamiliar environments. In this chapter, I will establish a connection between the distance being a relevant factor when firms internationalize, the resulting portfolio of business activities placing the firm into a category of (multi)regional strategies, and the micro-foundations of the theory of the MNE.

FIRM INTERNATIONALIZATION AS A CORE SUBJECT

Internationalization is a topic at the very core of IB. Hymer's (1960, 1968) starting point while explaining internationalization is that firms entail disadvantages in competing with indigenous firms in foreign markets ("disadvantage of foreignness"). Therefore, internationalization is about overcoming this inherent disadvantage. Jan Johanson and Jan-Erik Vahlne highlighted in their seminal paper (Johanson & Vahlne, 1977) that the firm's internationalization should be interpreted as an incremental learning process. As expressed by Forsgren and Johanson (1992, p. 10): "International expansion is inhibited by the lack of knowledge about markets and such knowledge can mainly be acquired through experience from practical operations abroad." The critical problem in internationalization is "the lack of knowledge" and embeddedness. This knowledge has to be acquired through learning-by-doing processes when conducting business abroad, and the embeddedness has to be reached through interacting with stakeholders in the host environment.

DISTANCE TO BE BRIDGED

The lack of knowledge results from dissimilarities between the contexts a focal firm is familiar with and those into which firm expands. In the field of IB, such context dissimilarities are called distances. Thus, the critical problem in internationalization is to bridge the distance between contexts by acquiring relevant knowledge and embedding it in host contexts.

What is distance when a firm internationalizes? Distance refers to the inverse degree of closeness between two entities concerning one or several dimensions (Deza & Deza, 2006). For the phenomenon of firm internationalization: (1) what are the two entities to be considered? and (2) what are the dimensions to be considered? In IB, distance is usually defined as the difference between two countries (Håkanson & Ambos, 2010). It typically relies on the distance between an MNE's home country and the new foreign (host) country. The home–host country distance has recently been substantially questioned. Tung and Verbeke (2010, pp. 1268–1269) have reasoned that what

> matters in the case of a new entry is not so much the distance between the home country A and the host country B, but between the newly entered host country B and the host country C, where the firm already has substantial experience and that shows the lowest distance to the newly entered country B.

To decide what concept of distance regarding the selection of the two entities that are compared matters is crucial from which sources and how the company learns to overcome the liabilities of outsidership in the new/target context.

WHICH DISTANCE IS RELEVANT?

Outsidership can be overcome by direct experiential learning, that is, learning through own experience, and experiential indirect learning, that is, learning from the experiences of others (Bingham & Davis, 2012). Learning from the experience of others includes indirect learning from an external source (i.e., imitative learning from competitors, or through the acquisition of another firm, or by making use of the knowledge of network partners) and the possibility of indirect learning from MNE-internal sources, in the sense that subsidiaries can learn from other units of the MNE. Experiential indirect learning allows the firm to benefit from knowledge stocks previously accumulated in the MNE's portfolio of subsidiaries. An MNE is a firm that has already gained international experience and thus has accumulated knowledge on how to compete in foreign contexts. Those stocks of knowledge differentiate MNEs from purely domestic firms and one another. A global MNE differs from a local MNE in the volume and quality of knowledge about international activities and the environments within which those activities are successfully conducted (Rugman & Verbeke, 2001, 2004). When MNEs internationalize further, they can build upon already accumulated knowledge stocks inside the firm.

Given the possibility of learning from different sources inside the portfolio of subsidiaries of an MNE, the resulting *added distance* that matters for the relevant outsidership in internationalizing is the smallest distance between the new/

target context and a source of learning from inside the MNE. The added distance could be equal to the home–host country distance or smaller. Added distance is smaller if something relevant for the internationalization step can be learned from a source inside the MNE that is closer to the new/target context than the home country. A host country's understanding of distance treats distance as static and exogenous. Instead, if MNEs can learn from experiences inside and outside, distance in the form of added distance becomes dynamic and endogenous. In other words, distance endogenously manifests as a dynamic consequence of the "MNEs' Walk" of further internationalization.

The impact of added distance has been introduced to the literature and empirically been tested from a country-level perspective in the form of the "closest neighbor" distance (Hutzschenreuter et al., 2011; Hutzschenreuter & Voll, 2008), which measures the distance between the new/target country and the most similar country the MNE has already been active in. However, alternatives to the closest neighbor approach would be to use the MNE subsidiary portfolio or the relevant regional cluster as a distance reference point (Hutzschenreuter & Matt, 2017). In addition, the added distance implementations are not restricted to a country-level perspective but could also use a sub- or supranational perspective (Hutzschenreuter et al., 2020).

As mentioned, a second fundamental question is which dimensions distance is being determined. The most used dimension in the literature is cultural distance. However, as Håkanson and Ambos (2010) have found, although cultural distance influences perceptions about distance, it is not the strongest determinant of such perceptions. Similarly, Ghemawat (2001) reasoned that the distance between two countries might manifest along four basic dimensions: cultural, administrative, geographic, and economic. Hutzschenreuter et al. (2014) tested for the differentiated effects of these dimensions and found that added administrative distance seems to explain the most significant part of the variation in firm performance, followed by added cultural distance.

(MULTI)REGIONAL STRATEGIES AS OUTCOMES OF INTENTIONAL DISTANCE-BRIDGING INTERNATIONALIZATION

> Most large multinational enterprises (MNEs) are not global. This statement simply means that these firms have been unable to emulate fully their domestic and home region success – in terms of achieved sales volumes or size of activity base – in equivalently sized markets outside of their home country or region (Rugman & Verbeke, 2004, 2005, Rugman et al., 2007, Rugman & Verbeke, 2008a, Rugman & Verbeke, 2008b). (Verbeke & Kano, 2016)

The extent of an MNEs (multi)regional strategy is an outcome of internationalization activities (expansion of activity base) and achievements in terms of market success in host markets (i.e., achieved sales) relative to host markets' importance. A world economy without distances would let us expect a successful focal firm in its home market to become global by transferring its firm-specific advantages

(Rugman, 1980; Rugman & Verbeke, 1992) to other countries. In this sense, distances serve as barriers to global strategies and opportunities for firms familiar with contexts that are distant from those our focal firm is familiar with. Or put differently, distances are barriers between an MNEs' intentions to become and be global and its actual regional realizations.

How far does a company get overcoming these barriers, bridging the distances? First, whether or not a company grows by expanding internationally is a matter of intentionality. Executives, and consequently firms, could significantly differ in their intentions to enter new host markets (Hutzschenreuter et al., 2007, 2010). Different levels of the intention to internationalize (how much), different intentions to which host countries to internationalize (where), and varying priorities to internationalize along a corporate lifetime (when) can result in entirely different approaches to becoming global or staying regional. Consequently, (multi)regional strategies must be interpreted, considering intentions as the starting point for any potentially deliberate managerial and firm action. This argument is similar to that of Hennart (2009), who argues that when interpreting the choice of governance modes if MNEs internationalize, it is too narrow-sided to assume that MNEs can select from the full range of governance modes independent of the conditions of host markets for ownership rights on firms or unbundled resources. Thus, firms can only attempt to internationalize to those host countries they intend to.

Second, assuming that the focal firm intends to become global or multiregional, or multinational but single-regional, it is a matter of bridging distances between the contexts our focal firm is familiar with and the new contexts the firm still has to expand to. In this sense, that is, the firm intends to make it to the next level, a (multi)regional but un-global strategy indicates that the company has been able to bridge distances successfully but cannot bridge the remaining additional distances so far. Given this cascade of reasoning, interpreting (multi) regional strategies becomes a discussion on limited capabilities and a discussion on the difference between "talking" global strategy and "walking" a (multi) regional strategy.

EXPLAINING (MULTI)REGIONAL STRATEGIES BY LIMITED CAPABILITIES TO BRIDGE DISTANCE

What capabilities does the firm need to bridge distances to "Walk the global strategy Talk"? As we have seen, bridging distances requires closing a knowledge gap and embedding in local contexts. Closing a knowledge gap involves identifying the need for additional knowledge, knowledge acquisition, and processing and applying the relevant knowledge. Fulfilling these tasks is typically associated with rationality. Unbounded rationality would imply an unlimited ability to new knowledge creation and knowledge application to new, so far unfamiliar contexts. In reality, executives and firms are boundedly rational and thus limited in their abilities for knowledge creation and application (Simon, 1961, 1997).

Embedding the firm in local contexts requires coping with bounded reliabilities outside the firm. The concept of bounded reliability has been established as a second envelope concept along with bounded rationality to explain MNE governance choices, that is, coping with bounded reliabilities within the firm, as part of the internalization theory of the MNE (Kano & Verbeke, 2015, 2019; Verbeke, 2009; Verbeke & Greidanus, 2009). The concept of bounded reliability extends or substitutes the much narrower concept of opportunism. In addition to opportunism (intentional deceit), it contains benevolent preference reversal and identity-based discordance as micro-foundational human behavioral elements to explain human failing commitments.

If it is true that "bounded reliability is dispositional" (Verbeke & Greidanus, 2009, p. 1489) and that their implications in forms of inefficiencies, ineffectiveness, and required governance choices are "largely situational" (Verbeke & Greidanus, 2009, p. 1490) then bounded reliability is also a core behavioral characteristic of (potential) stakeholders in (new) host countries and requires situational approaches/choices to deal with. Thus, embedding in (new) local contexts to bridge distances requires coping with bounded reliabilities of potentially new local stakeholders. Analyzing the added distance between the respective contexts offers insights into dispositional bounded reliabilities and situational approaches/choices to cope with them. This chapter does not aim to comprehensively discuss how distances potentially translate into dispositional bounded reliabilities and situational choices. Instead, the following gives exemplary indications of this relationship.

Benevolent preference reversal addressing "delaying a commitment to a course of action in favor of pursuing other opportunities" (Kano & Verbeke, 2015) in the form of postponed or canceled commitments might differ according to differences in cultural background. Simply, the meaning of commitments might be different across cultures. In some cultures, commitments are strict behavioral instructions; in others, commitments might serve as general guidelines. In addition, the legal perspective on commitments might significantly differ due to administrative distance. Depending on legal enforcement, commitments are to be understood as stronger or weaker. Economic differences can result in differing consequences for pressures/incentives to reprioritization. In contexts with higher economic turbulences, commitments could be understood as less strict and more relative than in those with more economic stability. Similar reasoning can be made concerning overcommitment. Economic differences could lead to dissimilarities in how much overcommitment pays off. In addition, if institutions differ, for example, regarding the insolvency law, it could well be that this affects overcommitment. Finally, in some cultures, overcommitment is seen as entrepreneurial, whereas in others, it is seen as unserious.

Depending on the context an MNE becomes active, stakeholders might differ toward identity-based discordance. Identity-based discordance "refers to commitment non-fulfillment due to conflict between 'what one promises' (in good faith) and 'what one represents' or 'values' in terms of one's identity" (Kano & Verbeke, 2015). In some cultures, "clan thinking" might play a more critical role than in others, leading to a more or less dominant role of regression as an expression of identity-based discordance. Or, the cultural background might influence whether stakeholders act as "mercenaries" versus "ambassadors" when interacting with a foreign MNE. In other words, cultural dissimilarities could significantly impact the level of divided engagement. Last but not least, from an administrative distance perspective, the nature of the political system might significantly impact how strongly stakeholder engagement is divided. The priorities of allegiance could be diametrically different.

LIMITATIONS, IMPLICATIONS, AND CONCLUSION

The chapter draws a line connecting the emergence of (multi)regional strategies via internationalization, added distances that must be bridged along the internationalization path, and the capabilities to cope with bounded rationality and bounded reliabilities. An appropriate understanding of the MNE involves considering the relevant distance for a focal MNE depending on its knowledge stock and accumulated practices embedding itself in host contexts. Most empirical IB research lacks such appropriateness by (often conceptually implicit) and empirically explicitly applying a host–home country understanding of distance. Home–host might reflect managerially relevant challenges, but in most cases, not. If the further internationalizing MNE benefits from experience in host countries where it is already active, home–host is inappropriate.

An MNE that intends to become global has to become an insider in different local contexts. Since each MNE is limited in its capabilities to cope with distances, it is of uppermost importance to use all available "help" inside and outside the MNE. This is impressively underlined when considering the bounded reliabilities of stakeholders in host contexts. As the exemplary analysis of determinants of bounded reliabilities in relationship to cultural, administrative, and economic distances shows, the "devil" is really in the distance "detail." Future research will further and in detail explore this link between (added) distance and bounded reliabilities. In this regard, Alain Verbeke's work impressively shows how important it is to consider the difference between home and host orientation and thus understand distance as an endogenous and dynamic phenomenon; and lights the way for further conceptually and empirically investigating the nature of distance. In addition, by emphasizing bounded reliabilities, Verbeke's work highlights one of the most important explanations for the difference between "talking global strategy" and "walking (multi)regional strategies." By employing the envelope concept of bounded reliabilities, future research will further investigate the link between intentions to be or become global and actual realizations in (multi) regional strategies.

However, it must also be noted that analyzing the connection between bounded reliabilities and distances has its dangers. One critical trap is the reinforcement of cultural stereotypes and biases. Self-evidently, a scientific discussion has strictly to avoid such hazards and must do everything to explore the respective links sensitively. While going deeper into the micro-foundations of how distances work, it is also necessary to reaggregate and look at interrelationships between different distance sources. Verbeke and Asmussen (2016) suggest considering the compounded effect of multiple distances that an outsider MNE encounters when expanding to a new host location. A holistic view of the distance puzzle in IB will go both directions: down the micro-foundational road and up the compounded distance path.

REFERENCES

Bingham, C. B., & Davis, J. P. (2012). Learning sequences: Their existence, effect, and evolution. *Academy of Management Journal, 55*(3), 611–641.

Deza, E., & Deza, M. M. (2006). *Dictionary of distances*. Elsevier.

Forsgren, M., & Johanson, J. (1992). Managing in international multi-centre firms. In M. Forsgren & J. Johanson (Eds.), *Managing networks in international business* (pp. 19–31). Gordon and Breach.

Ghemawat, P. (2001). Distance still matters. The hard reality of global expansion. *Harvard Business Review, 79*(8), 137–147.

Håkanson, L., & Ambos, B. (2010). The antecedents of psychic distance. *Journal of International Management, 16*(3), 195–210.

Hennart, J. F. (2009). Down with MNE-centric theories! Market entry and expansion as the bundling of MNE and local assets. *Journal of International Business Studies, 40*(9), 1432–1454.

Hutzschenreuter, T., Han, U.-S., & Kleindienst, I. (2010). Exploring the role of managerial intentionality in international business. *Advances in International Management, 23*, 113–136.

Hutzschenreuter, T., Kleindienst, I., & Lange, S. (2014). Added psychic distance stimuli and MNE performance: Performance effects of added cultural, governance, geographic, and economic distance in MNEs' international expansion. *Journal of International Management, 20*(1), 38–54.

Hutzschenreuter, T., & Matt, T. (2017). MNE internationalization patterns, the roles of knowledge stocks, and the portfolio of MNE subsidiaries. *Journal of International Business Studies, 48*, 1131–1150.

Hutzschenreuter, T., Matt, T., & Kleindienst, I. (2020). Going subnational: A literature review and research agenda. *Journal of World Business, 55*(4), 1–14.

Hutzschenreuter, T., Pedersen, T., & Volberda, H. W. (2007). The role of path dependency and managerial intentionality: A perspective on international business research. *Journal of International Business Studies, 38*(7), 1055–1068.

Hutzschenreuter, T., & Voll, J. C. (2008). Performance effects of "added cultural distance" in the path of international expansion: The case of German multinational enterprises. *Journal of International Business Studies, 39*(1), 53–70.

Hutzschenreuter, T., Voll, J. C., & Verbeke, A. (2011). The impact of added cultural distance and cultural diversity on international expansion patterns: A Penrosean perspective. *Journal of Management Studies, 48*(2), 305–329.

Hymer, S. (1960). *The international operations of national firms: A study of direct investment* [Ph.d. dissertation, Sloan School of Management, Massachusetts Institute of Technology] (Published by MIT Press under the same title in 1976).

Hymer, S. (1968). La Grande Firme Internationale. *Revue Economique, 14*(b), 949–973.

Johanson, J., & Vahlne, J.-E. (1977). The internationalization process of the firm: A model of knowledge development and increasing foreign market commitments. *Journal of International Business Studies, 8*(1), 23–32.

Johanson, J., & Vahlne, J.-E. (2009). The Uppsala internationalization process model revisited: From liability of foreignness to liability of outsidership. *Journal of International Business Studies, 40*(9), 1411–1431.

Kano, L., & Verbeke, A. (2015). The three faces of bounded reliability: Alfred Chandler and the microfoundations of management strategy. *California Management Review, 58*(1), 97–122.

Kano, L., & Verbeke, A. (2019). Theories of the multinational firm: A microfoundational perspective. *Global Strategy Journal, 9*, 117–147.

Rugman, A. (1980). A new theory of the multinational enterprise: Internationalization versus internalization. *Columbia Journal of World Business, 15*(1), 23–29.

Rugman, A., & Verbeke, A. (1992). A note on the transnational solution and the transaction cost theory of multinational strategic management. *Journal of International Business Studies*, *23*(4), 761–771.

Rugman, A. M., & Verbeke, A. (2001). Subsidiary-specific advantages in multinational enterprises. *Strategic Management Journal*, *22*(3), 237–250.

Rugman, A. M., & Verbeke, A. (2004). A perspective on regional and global strategies of multinational enterprises. *Journal of International Business Studies*, *35*(1), 3–18.

Rugman, A., & Verbeke, A. (2005). Towards a theory of regional multinationals: A transaction cost economics approach. *Management International Review*, *45*(1), 5–17.

Rugman, A., & Verbeke, A. (2007). Liabilities of regional foreignness and the use of firm-level versus country-level data: A response to Dunning et al. (2007). *Journal of International Business Studies*, *38*(1), 200–205.

Rugman, A., & Verbeke, A. (2008a). The theory and practice of regional strategy: A response to Osegowitsch and Sammartino. *Journal of International Business Studies*, *39*(2), 326–332.

Rugman, A., & Verbeke, A. (2008b). A new perspective on the regional and global strategies of multinational services firms. *Management International Review*, *48*(4), 397–411.

Simon, H. A. (1961). *Administrative behaviour* (2nd ed.). Macmillan.

Simon, H. A. (1997). *Models of bounded rationality, Volume 3: Empirically grounded economic reason*. MIT Press.

Tung, R. L., & Verbeke, A. (2010). Beyond Hofstede and GLOBE: Improving the quality of cross-cultural research. *Journal of International Business Studies*, *41*(8), 1259–1274.

Verbeke, A. (2009). *International business strategy: Rethinking the foundations of global corporate success*. Cambridge University Press.

Verbeke, A., & Asmussen, C. G. (2016). Global, local, or regional? The locus of MNE strategies. *Journal of Management Studies*, *53*(6), 1051–1075.

Verbeke, A., & Greidanus, N. S. (2009). The end of the opportunism vs trust debate: Bounded reliability as a new envelope concept in research on MNE governance. *Journal of International Business Studies*, *40*(9), 1471–1495.

Verbeke, A., & Kano, L. (2016). An internalization theory perspective on the global and regional strategies of multinational enterprises. *Journal of World Business*, *51*, 83–92.

CHAPTER 5

GOVERNANCE FOR SUSTAINABILITY: ALAIN VERBEKE'S CONTRIBUTION TO INTERNATIONAL BUSINESS SCHOLARSHIP AND PRACTICE

Liena Kano[a] and Luciano Ciravegna[b]

[a]*University of Calgary, Canada*
[b]*INCAE Business School, Costa Rica*

ABSTRACT

Alain Verbeke is one of the world's leading thinkers on international business (IB) and globalization, a renowned scholar and educator who contributes to creating a better global business environment by addressing some of today's most critical challenges. He was one of the first scholars to advance a theoretically rigorous and practically significant perspective on international corporate social responsibility (CSR). Verbeke's work on international CSR is particularly impactful because it is rooted in IB theory and based on a realistic set of assumptions about the behavior of managers, policymakers, and other market and nonmarket stakeholders. In this chapter, the authors apply theoretical principles central to Verbeke's research – most notably behavioral assumptions of bounded rationality and bounded reliability – to analyze businesses' and societies' pace of progress in relation to stated environmental, social, and governance (ESG) goals. The authors argue that bounded rationality and reliability challenges create misalignment between stated/imposed commitments toward

Walking the Talk? MNEs Transitioning Towards a Sustainable World
Progress in International Business Research, Volume 18, 71–85

Published under exclusive licence by Emerald Publishing Limited
ISSN: 1745-8862/doi:10.1108/S1745-886220240000018005

ESG performance, and economic actors' ability to deliver on these commitments. The authors discuss examples of such misalignment, focusing on tensions among stakeholders, between stakeholder organizations and firms, and within firms. The authors propose that to be relevant for policy and practice, the sustainability research should be based on realistic microfoundational assumptions.

Keywords: New internalization theory; bounded reliability; international governance; sustainability; long-term energy transition; corporate social responsibility; ESG

INTRODUCTION

Multinational enterprises (MNEs) are subject to intensifying stakeholder expectations regarding their social and environmental performance. Corporate citizenship standards are imposed on MNEs by regulators, consumers, investors, powerful nongovernmental organizations (NGOs), and employees, with environmental management playing an increasing role in firms' CSR efforts (Verbeke & Lee, 2021). Rising climate change concerns and sharpened societal attention to MNEs' governance practices have caused an upsurge of international business (IB) research on the topic. In recent years, IB scholars have viewed environmental sustainability as one of the grand challenges facing managers, policymakers, and researchers alike (Yu et al., 2023). This is evidenced by special issues of scholarly journals dedicated to CSR topics (e.g., a joint special issue of the *British Journal of Management* and the *Journal of International Business Studies* on long-term energy (LTE) transition, 2021), as well as a sheer number of published work on the subject. A quick Google Scholar search for "multinational enterprises" and "environmental performance" yielded over 8,500 hits, with almost half of the studies produced in the past five years.

In this sense, Alain Verbeke's work on sustainability in IB was at least two decades ahead of its time. Verbeke engaged in analyzing the relationship between MNE strategy and environmental regulations as early as in 1998 (Rugman & Verbeke, 1998a, 1998b). His early work on the topic established environmental regulations and associated stakeholder pressures as distinct forces influencing MNE strategies (Rugman & Verbeke, 2000), but also positioned MNEs as critical players in enacting sustainability strategies across national borders. Verbeke emphasized the agency and importance of MNEs, seeing them not as passive recipients of rules imposed by external actors, but as active enablers of economically, socially, and environmentally sustainable outcomes. This work has also laid out the foundation for developing a distinct research stream on international CSR: Verbeke was one of the first scholars to explore the complexities and challenges of cross-border CSR and to examine how cultural, institutional, geographic, and economic factors shape MNE's international CSR practices (Rugman & Verbeke, 1998a, 2000).

In his subsequent work, Verbeke continued to explore relationships between MNE strategies, expectations of a wide range of international stakeholders (employees, communities, governments, NGOs, competitors, etc.), and firms' sustainability performance, focusing on the need for *efficient firm-level governance* in order to achieve alignment between business objectives and broader societal and environmental interests (see, *inter alia*, Backman et al., 2017; Buysse & Verbeke, 2003; Van Balen et al., 2021; Verbeke & Tung, 2013; Verbeke et al., 2017; Yuan et al., 2011). What's remarkable about Verbeke' contribution to sustainability research, however, is not just its pioneering nature, but also its theoretical rigor combined with managerial and regulatory significance. To this end, his CSR research should be considered in the context of his broader scholarly oeuvre.

Prof. Verbeke's research spans the fields of IB, strategic management, transportation and logistics, and entrepreneurship. His body of work addresses contemporary challenges of business and offers nuanced perspectives on how MNEs can continue to create value globally while addressing these challenges, for example: how internationally operating firms can improve cooperation with nonmarket stakeholders; how businesses can improve their CSR performance; and how relationships between corporate head offices and subsidiaries of multinational firms can be streamlined for the benefit of both the firm and the communities where it operates. His work represents engaged scholarship, which focuses on developing research streams that are highly relevant to industry and policy and are aimed at long-term impact on the scholarly community and managerial practice.

Verbeke has been able to achieve this impact by developing comprehensive theoretical frameworks to explain, in a realistic fashion, managerial decisions and behavior that guide firm-level outcomes and thus predict global economic activity. He is the intellectual force behind the *New Internalization Theory* (NIT) (Narula & Verbeke, 2015; Rugman & Verbeke, 1992, 2004; Verbeke & Lee, 2021) – an extension on classic internalization theory (Buckley & Casson, 1976), with a greater focus on *ongoing* economizing and value-creation properties of the MNE's entire governance system (Hennart & Verbeke, 2022), and with an explicit incorporation of microfoundational assumptions that underlie managerial decision-making (Verbeke & Ciravegna, 2018). Theoretical concepts developed by Verbeke, such as, for example, *location-bound versus nonlocation-bound firm-specific advantages* (FSAs), *recombination*, and *bounded reliability* (Verbeke & Lee, 2021), are quickly becoming adopted by scholars all over the world as key terms in the study of business organizations. Due to their parsimony and theoretical soundness, these concepts have wide-ranging applicability across research fields and form essential components of NIT and, more broadly, the general theory of the firm. *Bounded reliability* in particular is a foundational construct that explains behaviour of economic actors (Kano & Verbeke, 2015; Verbeke & Greidanus, 2009). Bounded reliability accounts for a broad range of reasons for commitment failures and thus offers an actionable toolkit for managers to address breakdowns in governance, including large-scale crises related to firms' sustainability performance (Kano et al., 2023; Verbeke et al., 2019).

In this chapter, we will explore how bounded reliability (which we think represents one of Verbeke's most brilliant and impactful ideas) affects MNEs' ability to "Walk the Talk" when it comes to their ESG performance, and, more broadly, how microfoundations of IB theory explain the current global "state of affairs" in terms of ESG, including our collective progress toward climate change mitigation. We start by briefly reviewing NIT and its behavioral assumptions: bounded rationality and bounded reliability. We then apply these constructs to analyze the alignment (or the lack thereof) between firm-level commitments toward ESG performance and outcomes, between policy and enactment, and between stated objectives and feasibly achievable results. We conclude by highlighting Alain Verbeke's critical role in scholarly work on international CSR and, more broadly, IB scholarship and practice.

NIT AND ITS BEHAVIORAL ASSUMPTIONS

NIT focuses on firm-level governance choices for cross-border transactions and sustainability of these choices over time. The main feature of NIT, as compared to classic internalization theory (Buckley & Casson, 1976), is that NIT implies a *dynamic* consideration of the MNE as a complex structure (either integrated or disintegrated), where economizing properties of each governance arrangement are considered in the context of relevant phenomena affecting the MNE's operations, such as the structure of the MNE's value chain, network orchestration, regional strategy, stakeholder management, etc. (Kano & Verbeke, 2019).

When organizing cross-border transactions, MNE managers are faced with three sets of governance decisions: (1) whether transactions must be performed internally or externally; (2) how externalized transactions should be organized, in terms of partner and location choices, contractual design, relationship management, etc.; and (3) how internal transactions should be organized within the MNE, in terms of organizational structures, incentive and monitoring systems, communication routines, etc. An optimal mix of governance elements chosen should, on balance, allow the MNE to achieve superior management of transactions on three dimensions: (1) *reducing bounded rationality,* or the limits on managers' ability to process and act on important information (Simon, 1961); (2) *reducing bounded reliability*, or the limits on managers' ability to meet open-ended commitments (Kano & Verbeke, 2015); and (3) *creating and maintaining a supportive organizational context for value creation* (Grøgaard & Verbeke, 2012; Verbeke & Kenworthy, 2008). In NIT, "efficient governance" is defined as a set of structural and managerial governance tools that best facilitate the above economizing objectives. Efficient governance is an intermediate outcome toward subsequent value creation. MNEs that make comparatively inefficient governance choices will fail to create value, suffer competitive losses and will, in the long term, switch to more efficient governance forms.

Economizing is achieved by safeguarding against two common behavioral fallacies attributed to economic actors: bounded rationality and bounded reliability. We briefly review the two microfoundations below.

Bounded rationality reflects individuals' *scarcity of mind*, or limited capacity to process complex information and make optimal choices. In the context of international transactions, bounded rationality has four main sources: (a) necessary incompleteness of all information; (b) economic actors' limited capacity to process and interpret information; (c) multifacetedness of information, whereby different actors within the multinational network may select different facets of information as the basis of critical strategic decisions; and (d) divergence in judgment of identical information, whereby actors with different cultural backgrounds and experiences may interpret the same information differently (Simon, 1961, 1982, 1997; Verbeke & Yuan, 2005).

Despite its wide adoption in studies across various management research fields, including IB, the concept of bounded rationality does not tell the whole story when it comes to managerial behavior; bounded rationality accounts for cognitive biases and limitations of economic actors, but does not explain commitment failures related to imperfect effort. Alain Verbeke and colleagues' foundational work on the *bounded reliability* construct aims to fill this gap and develop a comprehensive set of microfoundations that explain a full range of failures in the realm of organizational goal pursuit, to facilitate accurate comparative analysis of various governance forms (Kano & Verbeke, 2015, 2019; Lumineau & Verbeke, 2016; Verbeke & Greidanus, 2009; Verbeke & Kenworthy, 2008).

As a complement to the bounded rationality construct, bounded reliability reflects *scarcity of effort* to make good on open-ended promises and assumes that managers are intendedly reliable, but only boundedly so (Kano & Verbeke, 2015; Verbeke et al., 2019). Bounded reliability extends Williamsonian assumption of opportunism, defined as "self-interest seeking with guile" (Williamson, 1981, p. 1545), by explaining instances of commitment nonfulfilment where no guile is present. Bounded reliability includes the following three dimensions:

(a) Opportunism as intentional deceit, which may occur ex ante (pre-commitment) or ex post (post-commitment), but in both cases is driven by a strong form of self-interest of the opportunistic party.
(b) Benevolent preference reversal, which means reverting commitments made in good faith. Benevolent preference reversal can take two forms. *First*, good faith reprioritization refers to situations whereby economic actors make ex ante commitments with benevolent intent, but postpone efforts to make good on original promises due to changing priorities over time, up until such a point where the commitment can no longer be fulfilled. *Second*, scaling back on overcommitment refers to managers' tendency to make benevolent yet unrealistic commitments that then need to be scaled back.
(c) Identity-based discordance, which means that individuals break commitments due to an attachment to a conflicting identity. Identity-based discordance can also take two forms: regression, which occurs when people abandon good-faith commitments in favor of ingrained undesirable behavior and routines, and divided engagement, which captures multi-actor situations with different sets of well-meaning actors identifying with conflicting goals and identities.

The need to economize on the full set of behavioral constraints described above will drive managerial choices on both structural governance design (e.g., multi-divisional vs functional organization, geographically dispersed vs localized organization, and a hierarchy vs a global value chain (GVC)) and strategic governance design (e.g., relative focus on formal coordination vs reliance on relational/social capital building, entrepreneurial orchestration). The best "real world" governance design refers to the one best equipped to meet bounded rationality and bounded reliability challenges.

Managing bounded rationality and bounded reliability challenges is a central concern also when it comes to the sustainability dimensions of the MNE, especially when comparing the MNEs' stated intent and formal ESG credentials to its actual environmental and social impact. The microfoundational lens presents an effective angle from which to analyze governance challenges related to MNEs' ESG performance, including compliance issues, LTE transition, and crises and breakdowns in MNEs' sustainability strategies, which may or may not be malevolently intended. In the realm of governance for sustainability, both actors imposing commitments (i.e., governments, regulators, NGOs, societal stakeholders) and actors enacting commitments (individual companies) are subject to bounded rationality and bounded reliability constraints (Verbeke & Hutzschenreuter, 2021; Verbeke & Lee, 2021). Failing to account for and safeguard against these constraints creates misalignment among stakeholder needs, and between MNE objectives and its governance system. Avoiding such misalignment through efficient governance is a focus of much of Verbeke's work on international sustainability.

In the next sections, we discuss how individual-level behavioral constraints affect MNEs' ability to create economic value while promoting sustainability objectives, focusing on two levels of tension: the tensions within stakeholder organizations and between firm-level actors and stakeholders (extra-MNE), and internal tensions within the firm (intra-MNE). Our analysis is inspired by Alain Verbeke's insightful writings on the subject and augmented by current examples of misaligned governance leading to damaging environmental, societal, and business consequences.

GOVERNANCE FOR SUSTAINABILITY: ALIGNING COMMITMENTS WITH OUTCOMES

Extra-MNE Tensions

MNEs are recognized for their unique ability to create value for a wide range of immediate stakeholders, including shareholders, partners, customers, and employees, while also creating positive benefits for the environment and the society, through disseminating innovation, creating jobs, and engaging in targeted ESG initiatives (Verbeke et al., 2022). However, ESG initiatives that MNEs undertake are often driven by intense pressure by governments and activist groups, rather than real needs of the business and the community. Unfortunately, societal stakeholder pressures to achieve certain ESG objectives may be detached from reality or driven by ideology rather than facts and science, which places unreasonable expectations on MNE managers. Further, regulators may not understand fully the nature of technological innovation and the size of capital investment required from MNEs in order to achieve compliance. More broadly, regulators and activists alike may not understand the full range of impacts of a particular policy and thus advocate for a course of action that creates negative externalities and hinders, rather than promotes, ESG progress.

This scenario is evident, *inter alia*, in the realm of LTE transition – the shift from production and consumption of carbon-emitting energy supplies such as fossil fuels toward renewable energy sources (Bass & Grøgaard, 2021), as part of the global fight against climate change. The shift requires massive capital expenditures and significant technological innovation from MNEs (Verbeke & Hutzschenreuter, 2021), who are pushed toward LTE transition by both market and nonmarket forces. On the market side, renewable energy markets are growing quite rapidly, spurred by the world-wide recognition of the urgent need to fight climate change; this growth has created considerable entrepreneurial opportunities for MNEs, stimulated new FSA development, and spurred a number of productive multi-layer collaborations among MNEs, technology startups, and research institutions (Bass & Grøgaard, 202; Verbeke et al., 2017).

Yet, by all official accounts, LTE transition is currently off-track, with most global climate pledges and national plans falling short of the projected path despite record global investment into energy transition technology in 2022 (International Renewable Energy Agency (IRENA), 2023). Bounded rationality and reliability of policymakers is, at least in part, the culprit.

Issue complexity and heterogeneity of stakeholder needs. Global and national actors designing LTE transition policies for companies and industries often do not have specialized expertise in the subject matter and do not comprehend fully the environmental implications of regulations that they are trying to develop and implement. The European Union's (EU) Renewable Energy Directive (RED) is a case in point. RED, a piece of European legislation tasked with funding LTE transition projects, classifies wood products as a renewable fuel. As a result of the classification, a significant portion of EU subsidies toward renewable energy (up to $13bn Euros in 2021) was directed to burning wood pellets, which, according to scientists, is an environmentally destructive practice. Wood burning releases more CO_2 per unit than fossil fuels, causes harmful pollution, and damages European forests (Rankin, 2022; Reynolds, 2022). In 2021, more than 500 scientists called on EU and world leaders to end wood burning subsidies. The EU has agreed to phase down the share of trees counted as renewable energy but has rejected calls for a complete phase-out of this energy source. On paper, EU is making significant progress toward LTE transition, with 22% of all energy consumed in the EU in 2021 coming from renewable sources (Reynolds, 2022). However, with the large proportion of the renewables coming from wood burning, the shift is not achieving tangible environmental benefits but rather is replacing one environmentally harmful fuel with another equally (or more) harmful one. This is an issue of bounded rationality, whereby policymakers fail to grasp the full complexity of the issue at hand, but can also be an expression of bounded reliability, with virtue signaling on the part of governments and regulators seeking to accelerate LTE transition. Divided engagement is also at play, whereby multiple macro-level agents (RED, the EU Parliament, country-level governments and committees, Greenpeace, industry associations, and NGOs, to name but a few) work toward the same ultimate goal of climate change mitigation but fail to establish a coordinated strategy that would take into account multiple, complex, and competing objectives and agendas.

Similar challenges could emerge with regard to the electrification of transport. The USA, China, and the EU have been heavily subsidizing electric vehicles (EVs) (Yang, 2023). The transition to EVs will cause a dramatic upsurge in the demand for electricity. The two largest economies of the world, the USA and China, rely on fossil fuels for most of the electric power they produce, with coal, the most polluting source, amounting to 63% of the electricity generated in China (EIA US Energy Information Administration, 2023). As a result, the transition to EVs may not have the intended impact on the environment, instead shifting the location of fossil fuel burning from the engine of cars to the power plants that produce electricity. In this example, LTE outcomes are hindered by the bounded rationality of policymakers due to the complexity of the problem, which encompasses multiple industries (EVs, power generation, infrastructure, and batteries) and government departments (transport and infrastructure, industry, environment, electrical grid, and power generation). Additionally, policymakers are boundedly reliable and may simply fail to comply with their LTE pledges, or they may assume that others will be in charge when the electric power generation problem truly escalates.

Disconnect between macro-level policy and firm-level reality. Policymakers imposing LTE transition incentives and penalties on firms and industries often do not understand firm-level reality. Policies are usually macro-level oriented and not designed with firm-level implications in mind (Verbeke & Hutzschenreuter, 2021). Neither regulators nor societal activists typically comprehend fully the nature of technology required to implement the transition, the size of capital investment on the MNE part, or the realistic time frame for LTE transition implementation. Moreover, regulatory frameworks affect firms heterogeneously, depending on firms' unique FSAs and experiences (Georgallis et al., 2021), yet regulators are typically not equipped to take firm heterogeneity into account when designing policy – in other words, "one size fits all" policy does not unambiguously promote LTE transition. This is a bounded rationality issue that results in unrealistic expectations imposed on firms. Rather than promoting and enabling investments in LTE, unrealistic policy requires costly compliance on behalf of MNEs and in fact diverts funds and managerial effort from actual climate change mitigation (Backman et al., 2017), resulting in negative consequences for both the firm and the environment. Bounded reliability on behalf of regulators may play a role when imposed commitments are politically motivated. Unrealistic policies (either ill-informed or motivated by self-interest) are met with resistance from MNEs; MNEs are in turn subjected to further pressure from nonmarket actors, and a vicious cycle of bounded reliability results (with no tangible innovation outcomes).

Negative societal externalities of ill-informed ESG pressures can be also observed in other realms, for example, when it comes to promoting a "full chain responsibility" approach to GVC governance. Full chain responsibility means that lead firms in GVCs must accept responsibility for ESG performance of all its partners in the value chain, including subcontractors and partners with whom they do not have direct contractual relationships (Narula, 2019). The push toward full chain responsibility was spurred by the Rana Plaza disaster in 2013, when a garment factory in Dhaka, Bangladesh, collapsed, killing over 1,100 workers. The incident raised public awareness of labor conditions in apparel value chain, and initially triggered concerted action by governments and NGOs to enforce labor standards in GVCs with operations in emerging economies (James et al., 2019). The contemporary meaning of full chain responsibility is broader and implies the lead firm's responsibility for a range of impacts, including environmental stewardship, social upgrading, and ethical sourcing (Wang & Li, 2019).

For lead MNEs, the pressure for full chain responsibility means engaging in extensive coordination and monitoring, which is difficult and costly, especially when suppliers are located in high-distance countries with opaque institutions (Ciravegna & Nieri, 2022). In an effort to enforce strict labor and environmental standards throughout the GVC, lead MNEs often adopt a "cascading compliance" (Narula, 2019, p. 1623) approach, which means that compliance responsibility is shifted down to the next tier of suppliers. This leaves Tier 1 and 2 suppliers with the responsibility to monitor and coordinate ESG performance of their own sub-tier contractors. Often, these suppliers find the monitoring burden too costly and risky and choose instead to internalize previously outsourced functions, which *de facto* exclude the smallest and most vulnerable actors from the GVC. Full chain compliance pressures can thus limit local enterprise growth and reduced employment prospects among community members who depend on MNEs to avoid abject poverty (Narula, 2019). That is, unrealistic demands on MNEs have a real potential to harm, rather than improve, community welfare.

Intra-MNE tensions

In the previous section, we have discussed scenarios where bounded rationality and reliability of powerful nonmarket stakeholders result in unfair and unrealistic pressures on MNEs. Implementing such imposed commitments may destroy firm value, and lead to MNEs naturally resisting compliance (Rugman & Verbeke, 1998a, 1998b). However, in some cases firms' interests align, at least partially, with those of nonmarket stakeholders (Verbeke et al., 2017). When that is the case, MNEs may be able to create "win–win" scenarios for the firm and societal stakeholders through what Verbeke and Lee (2021) termed "benevolent resource recombination" – that is, recombining FSAs across product and geographic space while reconciling global and local priorities.

Yet such benevolent resource recombination across geographic borders is inherently challenging. The pressures of competition and international media and NGO scrutiny force MNEs to seek "social license to operate" from multiple stakeholders, for example, by entering implicit or explicit social contracts in host countries, and by investing in headquarter-driven ESG projects to develop an international reputation for corporate citizenship (Verbeke et al., 2023; Verbeke & Lee, 2021). Delivering on these social contracts, however, is constrained by the challenges of bounded rationality and bounded reliability. These challenges arise from running a dispersed network of affiliates, depending on external actors for complementary resources, having to interpret and enact decisions of host country governments and stakeholders, and having to comply with both international and local rules and norms. As a result, MNEs may fail to meet their implicit and explicit international ESG commitments.

Decoupling of standards and implementation. Formal adoption of international standards and certifications presents an easily observable example of MNEs making explicit commitments that they are unable to keep. Firms use international certifications to signal socially and environmentally responsible behavior, but struggle to comply, because it is challenging to monitor geographically diversified organizations (Strike et al., 2006). Subsidiaries may diverge from the headquarter mandates and adopt a lax attitude about monitoring environmental and labor standards. For example, Walmart attempted to implement fair labor practices across its foreign locations, but the headquarters has not invested enough resources/developed requisite capabilities to manage compliance. As a results, subsidiaries in emerging economies continued to mistreat factory workers (Roberts & Bernstein, 2000; Strike et al., 2006). That is, MNEs can be intendedly socially/environmentally responsible, but only boundedly so, to the extent that they can effectively diffuse and monitor ESG practices across networks of affiliates (Nieri et al., 2023). What results is "decoupling" of standards from practice (Giuliani et al., 2017, p. 294), meaning that MNEs decouple their conduct aimed at signaling good credentials from their actual actions. Empirical evidence confirms that certifiable standards are not effective for governing MNEs' environmental conduct, due to decoupling of certification and implementation. Aravind and Christmann (2011) demonstrate that in the absence of high quality standard implementation, there is no difference in environmental performance between ISO-14001certified and noncertified manufacturing plants. Similarly, Giuliani et al. (2017) investigate whether coffee farms that have been granted private socio-environmental certification by global buyers display better ESG conduct than noncertified farms and find that certification has no impact on social conduct, and that the environmental impacts of certification are contingent on a number of firm- and macro-level conditions.

Such decoupling may be a result of bounded rationality and benevolent expressions of bounded reliability, whereby headquarters lack the requisite knowledge and capabilities to monitor and enforce compliance across the network. However, decoupling can also be opportunistic. Turning a blind eye to labor abuses and environmentally damaging operations in host countries can be a consequence of *institutional duality* (Kostova & Roth, 2002) – that is, MNE managers adapt their ESG practices to the institutional context where they operate. For example, managers engage in corrupt practices where such norms prevail or seek "pollution heavens" for highly emitting operations (Surroca et al., 2013).

Unfortunately, certifications have also been used to enable, rather than constrain, opportunistic behavior. MNEs have been shown to leverage international certifications to diverge attention from severe abuses of human rights, and to use their "good credentials" to interfere with investigations (Ciravegna & Nieri, 2022; Whiteman & Cooper, 2016). A further problem is related to the fact that NGOs in charge of monitoring and evaluating compliance with international standards often depend on financial funding from the very entities they are certifying, which creates a conflict of interest (Neilson, 2008).

To summarize, MNEs experience severe challenges when implementing stated ESG commitments across their networks; formal sustainability codes and certification schemes do not appear to be effective in addressing these challenges. The key scholarly, regulatory, and managerial question is thus not whether MNEs adhere to formal standards (either imposed or voluntarily adopted), but whether they have proper governance systems in place to economize against bounded rationality and reliability challenges and create value for the firm, the society, and the environment. In extreme cases, a firm's governance approach does not merely fail to implement ESG initiatives, but actually triggers significant negative consequences for a range of stakeholders. We refer to such instances as sustainability crises, which we discuss next.

Sustainability crises. A sustainability crisis is a severe disruption of prevailing routines and processes, which triggers redirection of resources and leads to significant negative consequences linked either to social/employment issues, health/safety/environment, or economic/financial issues, with likely spillovers among these three pillars of sustainability. Examples of large-scale sustainability crises have been plentiful over the past two decades and include the Enron collapse, accounting fraud at WorldCom, Merck's Vioxx recall, Mattel toy recall, Ford/Firestone controversy, BP Deepwater Horizon accident, and the Volkswagen emission scandal, to name but a few.

The court of public opinion is quick to attribute sustainability crises to MNE managers' intentional and calculative wrongdoing. While some individuals involved in the above scandals indeed embezzled funds, spread misinformation, engaged in unsafe practices, and violated employee rights with the sole purpose to maximize profit and increase their own welfare, the reality of the crises is infinitely more complex. Attributing sustainability crises exclusively to managerial opportunism is not only excessively pessimistic, but also ineffective in terms of uncovering real-world crisis trigger and designing potential solutions. We propose that sustainability crises may be triggered by well-intended yet dysfunctional actions, in addition to opportunistic behavior of parties involved.

For example, in a qualitative investigation of Volkswagen's emissions crisis, Kano et al. (2024) identify a range of situational context variables at the macro-, firm-, and micro-levels that facilitated the crisis, and argue that it is important to distinguish between intentional deceit and benevolent causes of governance breakdowns. The authors argue that opportunism was indeed a core feature of the crisis, supported by a corporate culture that institutionalized ethical lapses and did not tolerate dissent. However, the crisis was also spurred by multiple bounded rationality and reliability issues stemming from institutional and cultural differences between locations where the crisis unfolded, challenges in transferring knowledge across units, and tensions between the corporate headquarters and subsidiary employees. Governance for sustainability should therefore include safeguards against all expressions of bounded rationality and bounded reliability (beyond opportunism).

WALKING THE TALK IN IB RESEARCH, POLICY, AND PRACTICE

What does the above analysis mean for MNEs, regulators, and IB scholars? We have argued that broad macro-level policies and formal structures at the corporate level are powerless to impose sustainable action or to correct sustainability failure, in the absence of efficient international governance. Regulators must engage collaboratively with multiple stakeholders and especially with MNEs, to move away from "one size fits all" policies and to understand unique challenges and opportunities that firms face in achieving sustainability outcomes.

For managers, the key implication of our discussion is that achieving positive societal and environmental outcomes while creating value for the firm hinges on implementing *practical* mechanisms to improve the quality of international governance. *First*, proper organizational structures must be in place to facilitate information flows and enable headquarter control over units' ESG practices. *Second*, managerial governance tools must be deployed to economize on multiple expressions of bounded rationality and reliability. The latter include relational tools, orchestration routines, corporate diplomacy, and significant entrepreneurial effort on behalf of MNE managers.

To IB scholars, we would like to issue a call for action that has been central to Alain Verbeke's entire body of work: attention to individual motivations is paramount. We must extend our sustainability research agenda to study real causes of governance failures at the micro level, with an emphasis on developing actionable solutions. Many sustainability scholars have intuitively embraced opportunism as a way to explain corporations' failure to achieve sustainability objectives (Laufer, 2003). This assumption is readily picked up by the general public and reinforced by advocacy NGOs, who view MNEs' sustainability initiatives as greenwashing, corporate posturing, and political rent-seeking. However, this position is unhelpful in advancing either scholarly or practical sustainability agenda: it only tells us half of the story and therefore does not help MNEs develop real-world, impactful sustainability strategy. In order to be relevant for policy and practice, our sustainability research should be driven by scientific principles rather than ideology, and based on a complete, realistic set of microfoundational assumptions.

CONCLUSION

The objective of this chapter is to pay tribute to some of Alain Verbeke's insightful ideas, by applying them to the theme of the volume: "Walking the Talk – Transitioning Toward a Sustainable World." However, Prof. Verbeke's outstanding contribution to IB scholarship, as well as IB policy and practice, extends beyond corporate sustainability issues. Frameworks developed by Verbeke – most notably NIT – have been applied successfully to study a range of contemporary IB phenomena, including, *inter alia*, MNE regional strategy, emerging market multinationals, the future of work, global innovation, headquarter–subsidiary relationship, governance of GVCs, family governance, and digital globalization. The explanatory power of NIT lies in its parsimony, realistic micro-level assumptions, focus on firm-level governance, and elegant logic connecting various levels of analysis. The impact of Verbeke's scholarly work comes from deep engagement with theory and painstaking precision in its application, powerful intellect, remarkable erudition, and, importantly, his unwavering "globalism." For the duration of his scholarly career, Verbeke has been a vocal proponent and defender of IB.

In an era when researchers and the general public alike have developed an unfortunate fascination with "blaming and shaming corporate bad guys" (Wickert & Schaefer, 2011, p. 3), when the business world has been shaken by recurrent crises, and when de-globalization rhetoric is abound, Alain Verbeke has been consistently advancing an important message: intelligent and responsible globalization, supported by proper governance, is good for the world in the long run. This is what makes him one of today's most important thinkers on IB and globalization.

REFERENCES

Aravind, D., & Christmann, P. (2011). Decoupling of standard implementation from certification: Does quality of ISO 14001 implementation affect facilities' environmental performance? *Business Ethics Quarterly*, *21*(1), 73–102.

Backman, C. A., Verbeke, A., & Schulz, R. A. (2017). The drivers of corporate climate change strategies and public policy: A new resource-based view perspective. *Business & Society*, *56*(4), 545–575.

Bass, E., & Grøgaard, B. (2021). The long-term energy transition: Drivers, outcomes, and the role of the multinational enterprise. *Journal of International Business Studies*, *52*(5), 807–823.

Buckley, P. J., & Casson, M. C. (1976). *The future of the multinational enterprise*. Macmillan.

Buysse, K., & Verbeke, A. (2003). Proactive environmental strategies: A stakeholder management perspective. *Strategic Management Journal*, *24*(5), 453–470.

Ciravegna, L., & Nieri, F. (2022). Business and human rights: A configurational view of the antecedents of human rights infringements by emerging market firms. *Journal of Business Ethics*, *179*(2), 431–450.

EIA US Energy Information Administration. (2023). *What is U.S. electricity generation by energy source?* https://www.eia.gov/todayinenergy/detail.php?id=53959 and https://www.eia.gov/tools/faqs/faq.php?id=427&t=3

Georgallis, P., Albino-Pimentel, J., & Kondratenko, N. (2021). Jurisdiction shopping and foreign location choice: The role of market and nonmarket experience in the European solar energy industry. *Journal of International Business Studies*, *52*(5), 853–877.

Giuliani, E., Ciravegna, L., Vezzulli, A., & Kilian, B. (2017). Decoupling standards from practice: The impact of in-house certifications on coffee farms' environmental and social conduct. *World Development*, *96*, 294–314.

Grøgaard, B., & Verbeke, A. (2012). Twenty key hypotheses that make internalization theory the general theory of international strategic management. In A. Verbeke & H. Merchant (Eds.), *Handbook of research on international strategic management* (pp. 7–30). Edward Elgar.

Hennart, J.-F., & Verbeke, A. (2022). Actionable and enduring implications of Oliver Williamson's transaction cost theory. *Journal of International Business Studies*, *53*, 1557–1575.

International Renewable Energy Agency (IRENA). (2023). *World energy transitions outlook 2023*. Retrieved June 7, 2023, from https://mc-cd8320d4-36a1-40ac-83cc-3389-cdn-endpoint.azureedge.net/-/media/Files/IRENA/Agency/Publication/2023/Mar/IRENA_WETO_Preview_2023.pdf?rev=c4c2398e169a4243ad37cf67dc441fa8

James, P., Miles, L., Croucher, R., & Houssart, M. (2019). Regulating factory safety in the Bangladeshi garment industry. *Regulation and Governance*, *13*(3), 431–444.

Kano, L., & Verbeke, A. (2015). The three faces of bounded reliability: Alfred Chandler and the microfoundations of management theory. *California Management Review*, *58*(1), 97–122.

Kano, L., & Verbeke, A. (2019). Theories of the multinational firm: A microfoundational perspective. *Global Strategy Journal*, *9*(1), 117–147.

Kano, L., Verbeke, A., & Simoes, S. (2024). Governance failure and firm-level crises: The case of the Volkswagen emissions scandal. In A. Goerzen (Ed.), *Research handbook of international corporate social responsibility*. Edward Elgar.

Kostova, T., & Roth, K. (2002). Adoption of an organizational practice by subsidiaries of multinational corporations: Institutional and relational effects. *Academy of Management Journal, 45*(1), 215–233.

Laufer, W. S. (2003). Social accountability and corporate greenwashing. *Journal of Business Ethics, 43*, 253–261.

Lumineau, F., & Verbeke, A. (2016). Let's give opportunism the proper Back seat. *Academy of Management Review, 41*(4), 739–741.

Narula, R. (2019). Enforcing higher labor standards within developing country value chains: Consequences for MNEs and informal actors in a dual economy. *Journal of International Business Studies, 50*, 1622–1635.

Narula, R., & Verbeke, A. (2015). Making internalization theory good for practice: The essence of Alan Rugman's contributions to international business. *Journal of World Business, 50*(4), 612–622.

Neilson, J. (2008). Global private regulation and value-chain restructuring in Indonesian smallholder coffee systems. *World Development, 36*(9), 1607–1622.

Nieri, F., Rodriguez, P., & Ciravegna, L. (2023). Corporate misconduct in GVCs: Challenges and potential avenues for MNEs. *Journal of Industrial and Business Economics, 50*(1), 193–207

Rankin, J. (2022). EU limits subsidies for burning trees under renewable energy directive. *The Guardian*. Retrieved June 7, 2023, from https://www.theguardian.com/environment/2022/sep/14/eu-limits-subsidies-for-burning-trees-under-renewable-energy-directive

Reynolds, M. (2022). *Europeans are burning trees to keep warm. Wired*. Retrieved June 7, 2023, from https://www.wired.co.uk/article/eu-forests-energy-crisis

Roberts, D., & Bernstein, A. (2000, October 2). A life of fines and beating. *Business Week*, pp. 122–29.

Rugman, A. M., & Verbeke, A. (1992). A note on the transnational solution and transaction cost theory of multinational strategic management. *Journal of International Business Studies, 23*(4), 761–771.

Rugman, A. M., & Verbeke, A. (1998a). Corporate strategies and environmental regulations: An organizing framework. *Strategic Management Journal, 19*(4), 363–375.

Rugman, A. M., & Verbeke, A. (1998b). Corporate strategy and international environmental policy. *Journal of International Business Studies, 29*, 819–833.

Rugman, A. M., & Verbeke, A. (2000). Six cases of corporate strategic responses to environmental regulation. *European Management Journal, 18*(4), 377–385.

Rugman, A. M., & Verbeke, A. (2004). A perspective on regional and global strategies of multinational enterprises. *Journal of International Business Studies, 35*(1), 3–19.

Simon, H. A. (1961). *Administrative behaviour* (2nd ed.). Macmillan.

Simon, H. A. (1982). *Models of bounded rationality and other topics in economics*. Collected Papers (Vol. 2). MIT Press.

Simon, H. A. (1997). *Models of bounded rationality, volume 3: Empirically grounded economic reason*. MIT Press.

Surroca, J., Tribó, J. A., & Zahra, S. A. (2013). Stakeholder pressure on MNEs and the transfer of socially irresponsible practices to subsidiaries. *Academy of Management Journal, 56*(2), 549–572.

Strike, V. M., Gao, J., & Bansal, P. (2006). Being good while being bad: Social responsibility and the international diversification of US firms. *Journal of International Business Studies, 37*, 850–862.

Van Balen, M., Haezendonck, E., & Verbeke, A. (2021). Mitigating the environmental and social footprint of brownfields: The case for a peripheral CSR approach. *European Management Journal, 39*, 710–719.

Verbeke, A., & Ciravegna, L. (2018). International entrepreneurship research versus international business research: A false dichotomy? *Journal of International Business Studies, 49*, 387–394.

Verbeke, A., Ciravegna, L., Lopez, L. E., & Kundu, S. K. (2019). Five configurations of opportunism in international market entry. *Journal of Management Studies, 56*(7), 1287–1313.

Verbeke, A., & Greidanus, N. S. (2009). The end of the opportunism vs trust debate: Bounded reliability as a new envelope concept in research on MNE governance. *Journal of International Business Studies, 40*, 1471–1495.

Verbeke, A., & Hutzschenreuter, T. (2021). Imposing versus enacting commitments for the long-term energy transition: Perspectives from the firm. *British Journal of Management*, *32*(3), 569–578.

Verbeke, A., Kano, L., & Johnston, A. (2022). The full canvas: Exploring the bright and dark sides of international business strategy. In H. Merchant (Ed.), *The new frontiers of international business: Contributions to management science* (pp. 13–51). Springer.

Verbeke, A., & Kenworthy, T. (2008). Multidivisional versus metanational governance of the multinational enterprise. *Journal of International Business Studies*, *39*(2), 940–956.

Verbeke, A., & Lee, I. (2021). *International business strategy: Rethinking the foundations of global corporate success*. Cambridge University Press.

Verbeke, A., Osiyevskyy, O., & Backman, C. A. (2017). Strategic responses to imposed innovation projects: The case of carbon capture and storage in the Alberta oil sands industry. *Long Range Planning*, *50*(5), 684–698.

Verbeke, A., & Tung, V. (2013). The future of stakeholder management theory: A temporal perspective. *Journal of Business Ethics*, *112*, 529–543.

Verbeke, A., & Yuan, W. (2005). Subsidiary autonomous activities in multinational enterprises: A transaction cost perspective. *Management International Review*, *45*(2), 31–52.

Wang, S., & Li, D. (2019). Responding to public disclosure of corporate social irresponsibility in host countries: Information control and ownership control. *Journal of International Business Studies*, *50*, 1283–1309.

Whiteman, G., & Cooper, W. H. (2016). Decoupling rape. *Academy of Management Discoveries*, *2*(2), 115–154.

Wickert, C., & Schaefer, S. (2011). Revisiting corporate sustainability: Towards a critically-performative research agenda. In *Academy of Management Best Paper Proceedings*.

Williamson, O. E. (1981). The modern corporation: Origins, evolution, attributes. *Journal of Economic Literature*, *19*, 1537–1568.

Yang, Z. (2023, February). How did China come to dominate the world of electric cars? *MIT Technology Review*. https://www.technologyreview.com/2023/02/21/1068880/how-did-china-dominate-electric-cars-policy/

Yuan, W., Bao, Y., & Verbeke, A. (2011). Integrating CSR initiatives in business: An organizing framework. *Journal of Business Ethics*, *101*, 75–92.]

Yu, H., Bansal, P., & Arjaliès, D. L. (2023). International business is contributing to environmental crises. *Journal of International Business Studies*, *54*, 1151–1169. https://doi.org/10.1057/s41267-022-00590-y

PART II

WALKING THE TALK IN SUSTAINABILITY ISSUES

CHAPTER 6

THE SOCIETAL RESPONSIBILITIES OF CORPORATIONS IN THE 21ST CENTURY: FOUR TYPES OF CHALLENGES, FOUR TYPES OF STRATEGIES

Jacqueline Mees-Buss

University of New South Wales, Australia

ABSTRACT

An in-depth analysis of how senior managers in a large multinational corporation interpret their social and environmental responsibilities revealed that, notwithstanding formal corporate interpretations, discrepancies persisted in their interpretation of what was expected of them and how to implement it. Two fault lines emerged: (1) an instrumental versus a normative interpretation of corporate societal responsibilities, and (2) a focus on 'doing less/no harm' versus 'doing more good'. This chapter introduces a theoretical framework that combines these fault lines to form four quadrants that each represent a different set of challenges managers face as they commit to improving their organisation's impact on society. Rather than adjudicate between them, a holistic interpretation of corporate social responsibility (CSR) takes all four types into account. But the four types of challenges differ considerably in nature and thus in the strategic approach that is necessary to deal with them. In this chapter, each quadrant is discussed in detail. What characterises the issues in this quadrant, what mindset, and what strategy are necessary to address them? The chapter concludes with the observation that the framework, and the taxonomy of types of CSR

Walking the Talk? MNEs Transitioning Towards a Sustainable World
Progress in International Business Research, Volume 18, 89–106

ISSN: 1745-8862/doi:10.1108/S1745-886220240000018006

challenges that it brings to the fore, creates greater awareness of how industries are confronted with different sets of challenges and thus need different strategic approaches. A better understanding of these differences may lead to more support, in particular for those managers who work in industries that face a disproportionate share of one particular type of challenges, the 'nasty trade-offs'.

Keywords: Corporate social responsibility; CSR strategy; CSR taxonomy; SDGs; wicked problems; prisoners' dilemmas

INTRODUCTION

Capitalism, Porter and Kramer wrote in 2011, is under siege. Greater attention must be paid to the impact of commercial activity on the natural and social environment, or corporations will lose their legitimacy in society. But even though every self-respecting corporation in the 21st century seems to have pledged to make a positive contribution to society and has an agenda outlining their social and environmental commitments, defining their broader societal responsibilities remains a topic of debate. In varying degrees, companies commit (or feel obliged to commit) to concepts such as environmental sustainability, net-zero carbon, the circular economy, creating shared value (CSV), corporate responsiveness, corporate citizenship, the United Nations' sustainable development goals (SDGs), environmental-social-governance (ESG) impact, profit & people & planet, or simply a 'better future for all'. Amid all these concepts and acronyms, what exactly do they believe they are responsible for? What should be on their social agenda and how much is enough?[1] Also, in the academic literature, defining the societal responsibilities of corporations has been an ongoing topic of debate (Garriga & Melé, 2004; McWilliams et al., 2006; Wickert, 2021). 'Numerous definitions of corporate social responsibility (CSR) have been proposed and often no clear definition is given, making theoretical development and measurement difficult', McWilliams et al. (2006, p. 1) lamented. Since then, there has been a growing consensus that CSR[2] is not only an essentially contested concept (Okoye, 2009), but is also paradoxical in nature (Luo et al., 2020).

In this chapter, I offer a framework to help disentangle the ongoing debate on CSR. An analysis of 20 years of corporate reporting and 69 interviews with Unilever managers who have grappled with the challenge of interpreting their corporate societal responsibilities in the first two decennia of the 21st century, led to the conclusion that there are two fault lines that polarise their interpretations of CSR. Similar fault lines can be found in the academic literature (Mees-Buss, 2014). Rather than adjudicating between competing interpretations, I propose a framework that is more in line with contemporary interpretations of CSR: a framework that embraces the paradoxical nature of CSR by bringing these interpretations together. The framework exposes a taxonomy of four fundamentally different types of challenges managers face as they deal with the complex and often paradoxical demands of delivering shareholder, *and* multistakeholder, *and* environmental value: low-hanging fruit, nasty trade-offs, societal opportunities, and matters of conscience. These different types of challenges each demand a different strategic approach. I will argue that a better understanding and acknowledgement of these differences will contribute to a better understanding of what can and cannot be expected from corporations that sincerely want to make a positive contribution to society.

In the final section of the chapter, I will discuss a number of other issues the framework allows us to see, such as the need for a more balanced scorecard for social and environmental impact, the need to acknowledge the different mixes of challenges confronting different industries, and the opportunities for further theoretical development by focussing on issues within each quadrant, rather than theories that apply to all four quadrants.

COMPETING INTERPRETATIONS OF THE SOCIETAL RESPONSIBILITIES OF CORPORATIONS

Maybe the confusion about corporations' responsibilities towards society should not come as a surprise. Social and environmental issues such as climate change, persistent poverty, gender inequality, etc. are all 'wicked problems'. Wicked problems have many stakeholders, no easy solutions, they are complex, interrelated, systemic, their root-cause is unclear, and every attempt to improve them changes them or potentially leads to other problems in the system (Rittel & Webber, 1973). The wickedness of the problems implies that even what may look like a good 'solution' can still lead to unintended social or environmental damage.

Therefore, the question what role and responsibilities corporations have in addressing these wicked problems cannot easily be answered. Every answer is unavoidably politically and ideologically imbued. Davis (1960) called CSR a 'nebulous idea' (p. 70). Okoye (2009) concluded that CSR is an 'essentially contested concept' (Gallie, 1956); Sheehy (2015) described CSR as essentially dialectical; a concept that is marked by strong normative, ideological, and political differences of opinion (Scherer & Palazzo, 2007).

An in-depth case study of how Unilever, often presented as the poster child of CSR, collectively made sense of its broader societal responsibilities over time (2000–2020), revealed that the way the company constructed, and reconstructed CSR in public documents, in particular in its Annual Reports and Sustainability Reports, was ever changing, swayed by changes in sensemaking by the top management team; by isomorphic forces at an institutional level; and not in the last place, as a result of pressures from advocacy groups, NGOs, etc. (for analysis of the first decade, see Mees-Buss & Welch, 2014). In-depth interviews with individual managers operating in different parts of the globally dispersed organisation, revealed that, although Unilever may have always presented itself as united around a single definition of what it means to be a socially responsible corporation, there has always been diversity among its senior managers on what this meant. What transpired from the interviews, were two fault lines that divided the managers' opinions when they discussed their interpretation of the societal responsibilities of the corporation.

The first fault line centred around a fierce ideological controversy: should social and environmental responsibility be interpreted and implemented as a concept that entails doing good while simultaneously improving the profitability and competitiveness of the firm? Or should it be interpreted as a more normative concept that redefines the moral duties and responsibilities of the corporation as 'doing what is necessary for society' even where benefit to the bottom line cannot always be guaranteed (Mees-Buss & Welch, 2019)? Campaigns such as 'Dove's campaign for real beauty' (2003–2008) caused a lot of debate and controversy. The campaign addressed the wrong done to women's self-esteem by the beauty industry. Years of promoting unrealistic ideals of beauty, they claimed, had harmed women's self-esteem and had caused a global epidemic of eating disorders among women. Unilever managers were divided, some claiming this was a great example of the company taking its societal responsibilities seriously, others fiercely opposed. Unless the campaign could be linked to more profit for Unilever, they believed it was undermining the business. Over time, the brand's support for women's self-esteem remained, but the beauty industry was no longer accused as a co-creator of the problem.

This ideological fault line can also be found in the debate on the societal responsibilities of corporations in the academic literature since it took off in the 1950s. Howard Bowen (1953, p. xi) posed the question 'What responsibilities to society may businessmen reasonably be expected to assume?' Bowen made a plea for greater social consciousness among powerful businessmen. Not long after, Theodore Levitt (1958) published an article in the *Harvard Business Review* entitled 'The Dangers of Social Responsibility' in which he cautions against this idea, positing that 'government's job is not business, and business's job is not government' (p. 3). Joseph W. McGuire (1963) tried to reconcile the opposing views: 'The idea of social responsibilities supposes that the corporation has not only economic and legal obligations, but also certain responsibilities to society which extend beyond these obligations' (p. 144). But when Lockett et al. (2006) reviewed the CSR literature from its early beginnings till 2006, they observed that since the 1970s the instrumental non-normative interpretation started to dominate the field of CSR. The core research question in many of these studies is whether there is a business case for CSR: Does it pay to do good?

Prevailing interpretations such as CSV (Porter & Kramer, 2006, 2011) or the fortune at the bottom of the pyramid (BOP) (Prahalad & Hart, 2001) typically support the instrumental view, that companies have to look for win–win solutions based on the firm belief that shareholder return cannot and does not need to be compromised. They assume that there is indeed a business case for CSR. At the other side of the debate there are the interpretations that link CSR to virtue ethics and the normative belief that companies first and foremost have a duty towards society (e.g. Heugens et al., 2008). The most common terminology used for these two positions is the instrumental versus the normative approach to CSR (e.g. Jones & Wicks, 2018). I will use these terms for the framework as well.

Although less controversial, the other fault line that was dividing Unilever's managers' interpretation of CSR dealt with the question of which issues belonged on the CSR agenda and which did not. Should CSR activities focus primarily on eliminating all negative social and environmental impact from the company's own value chain, in other words, on 'righting the wrongs of the past', or should they also include other initiatives that have nothing to do with righting wrongs, but that do contribute to a better society? An example of righting wrongs would be Unilever's project to convert all their plastic laminate toothpaste tubes to recyclable toothpaste tubes by 2025. An example of a project that has nothing to do with righting their wrongs of the past, would be Unilever's 'H is for handwashing' campaign for Lifebuoy. Lifebuoy is a hygienic handwash soap. The brand collaborates with non-government organisations (NGOs) to develop media campaigns and school programmes that teach children to wash their hands regularly to prevent the spread of infectious diseases, a problem Unilever did not co-create, but can help to alleviate while simultaneously creating goodwill (and profit) for one of its key brands. While some would refer to the latter as simply good marketing, others refer to it as a great example of a company using its strengths to create social value. I will refer to these two positions as CSR with a focus on doing less harm versus CSR with a focus on doing more 'good'.

Analysing these debates and interpretations over time highlights that definitions of the societal responsibilities of the corporation are all built on a subjective interpretation of what exactly this responsibility entails. Each of them bears the risk of narrowing the CSR[3] agenda. As I will show in the next section, combining the different interpretations rather than adjudicating between them is more in line with emerging, contemporary interpretations of CSR as inherently paradoxical in nature. Combining the fault lines as 'both/and' conditions for responsible business conduct, leads to a comprehensive framework of four equally important but very different types of challenges, each very different in nature and demanding a different strategic approach. Rather than focus on only one or two, *all four* should be considered for the CSR agenda if the company wants to legitimise and shield its position as a socially responsible company in the 21st century.

THE FRAMEWORK

The different interpretations of what it means to be a socially responsible corporation inevitably prioritise different types of issues for the corporate social impact agenda while neglecting others. As Van Tulder and van Mil (2023) point out in their seminal work on the principles for sustainable business, societal trust in organisations continues to decline and corporate scandals continue to occur. To restore trust and legitimacy, corporations need to embrace a set of extended, sometimes seemingly contradictory 'fiduciary duties'. A set of duties that considers *all* stakeholders, including future generations (Van Tulder & van Mil, 2023, pp. 351–353); a set of duties that includes doing less harm *and* doing more good; and strives to identify win–win solutions *and* doing the right thing, even if there is no direct link to profit. A theoretical framework that combines rather than discriminates between these different types of interpretations, offers an overview of all the types of issues that may need to be considered for the corporate social impact agenda (see Table 6.1).

First, there is the dichotomy between either focussing on *doing less harm* – issues where the firm currently has a negative impact on society – and/or on *doing more good*: issues for which the company is not (co)responsible, but for which they have the resources and capabilities to make a positive contribution. Then, there is the ideological dichotomy between an *instrumental* approach to CSR – CSR as a contemporary way to compete, focussing on issues for which a win–win solution can be found – or a *normative* approach to CSR – an approach to CSR that is primarily based on ethical concerns, focussing on issues for which there may or may not be a win–win solution, but the company considers it its responsibility to address them anyway, either because they have co-created them or because they are in a unique position to do so.

As I will argue in what follows, the four types of issues all deserve a place on the agenda but all need their own strategic approach. Issues in each of these four quadrants represent different types of challenges. Addressing the areas where the company currently has a negative impact on society through profitable solutions represents the proverbial 'low-hanging fruit'. The issues must be on the agenda and present both an opportunity and a threat (in case the company fails to address them). 'Societal opportunities' offer the company the opportunity to gain profit *and* strengthen its reputation by making a positive contribution to society, not because they are part of the problem, but because they have the unique competencies to do so. Negative impact that is not easily addressed profitably presents the firm with what I call 'nasty trade-offs'. They are 'nasty' because they present the company with a difficult dilemma, often what is known as a prisoners' dilemma. The 4th quadrant, matters of conscience, present the company with a

Table 6.1. A Framework for the Typology of CSR Issues.

	Focus on doing less harm = addressing issues where the firm currently has a negative impact on society. Righting the wrongs of the past	*Focus on doing more good* = addressing issues where the firm has the resources and capabilities to make a positive contribution to societal issues
INSTRUMENTAL approach: CSR = win–win solutions	1. *Low-hanging fruit* Righting wrongs through win–win solutions	2. *Societal opportunities* Profitable ventures that successfully address societal issues
NORMATIVE approach: CSR = doing the right thing	3. *Nasty trade-offs* Righting wrongs even though there are no (obvious) win–win solutions	4. *Matters of conscience* Addressing societal issues without a direct link to profit

moral challenge. The issues here are those where the firm – given its competencies and/or assets – can make a significant contribution to societal issues (e.g. threats to public safety or health), but is unlikely to ever make a profit from the initiative. Although addressing the issues may contribute to the company's reputation, this is not guaranteed, and it is hard to predict if the financial benefit will outweigh the costs.

Utilising this framework to set the corporate social impact agenda not only cuts through the noise of disparate interpretations and recommendations about what should be on the agenda, but also provides structure to the strategic debate on how to address different types of issues. When to compete? When to collaborate? And when and why is lobbying for stricter industry regulation important? In the following section, I will discuss each quadrant in detail, discussing what type of challenges they present and what this implies for the strategic approach necessary to address them.

FOUR TYPES OF CHALLENGES

Quadrant 1: Low-hanging Fruit

Whether pressured by external threats, or as a result of an awakening of their conscience, many companies have in recent years taken a more critical look at their value chain, including all their internal and external stakeholders, to identify the areas where they could be accused of doing social or environmental harm. It takes brutal honesty, critical self-awareness, and open-mindedness to admit that some of the activities or habits that were developed long ago – often with good intentions and often with broad social acceptance – may no longer be acceptable. Henry Ford's commitment to the mass production of affordable cars for every household was legendary. Over time, however, the automobile's dependence on non-renewable fossil fuel, its contribution to CO_2 emissions, the social hazards of traffic accidents, and congested cities, have faced car manufactures with the challenge to respond to these wicked problems they co-created.

Rather than trying to ignore these issues, or brush them off as externalities, companies in the 21st century are expected to address them. Ideally, a profitable solution is found, thus turning these threats into opportunities. What overly optimistic instrumentalists may overlook, is the wicked nature of most social and environmental issues. The positive impact of the 'solutions' that are identified, may not last. Even the best solutions may create their own set of new problems over time. Electric cars reduce the negative impact of the automotive industry on the environment without jeopardising the profitability of the industry. In the short term, they certainly offer a win–win solution. This does not mean the net impact on the environment is now neutral. While electric cars have zero tailpipe emissions, the problem of minimising CO_2 emissions has been passed on to the industries responsible for electricity generation and the challenge now lies with them to move to zero emissions. Unilever's recyclable toothpaste tubes will be better than the non-recyclable tubes, but recycling is not without its own challenges.

In other words, profitable improvements are possible, but truly responsible companies stay alert and are aware that every solution may one day give rise to new issues.

Successful strategies in this quadrant move from identification to a learning cycle of innovation, experimentation, and feedback. Identification demands systems thinking; critical awareness of all the direct and indirect negative social and/or environmental impact the company is having along the total value chain. Already in 2008, Unilever started a major project to critically analyse and map the social and environmental impact of each of their 200+ brands, from procurement to production, distribution, usage, and ultimately disposal. They found, for example, that most of the water pollution in their home & personal care division came from usage of their products in consumers' homes, not from their factories. Finding products that deal with the issues identified, needs a strong and creative innovation team or access to external innovators to collaborate with. Reduced in-home water consumption became the briefing for Unilever's internal and external innovation teams. The result was a series of new products such as a dry spray to refresh clothes that are not really dirty in order to reduce wash loads, or a no-rinse hair conditioner that, the company claims, should shorten time in the shower by 99 seconds. None of these can yet claim global success, but they are good experiments. Experimentation is the best way to learn how the world, in all its complexity and unpredictability, responds to ideas. Constant feedback and open-mindedness will lead to better next-generation solutions (see Box 1).

Box 1.

Quadrant 1. Low-hanging fruit

Examples: hybrid and electric cars, vegetable-based meat, replacing plastic bags with cheaper and recyclable paper bags, cleaner more efficient factories, greater workplace safety, etc.

- Type of challenge:
  - Challenge to improve wicked problems.
- Mindset:
  - Brutal honesty, critical self-awareness, and open-mindedness.
  - Awareness that today's solutions may be the source of tomorrow's problems.
- Strategic tools:
  - Awareness of the company's negative impact; systems thinking.
  - Mapping tools to build understanding of the complexity of the issues.
  - Innovation; think outside the box.
  - Experiment. If possible, run multiple experiments.
  - Create good feedback loops to remain self-critical and monitor the potential side-effects of the 'solution'.

Because responses to wicked problems at best improve the problem but seldom come without their own set of problems, a robust strategy will never rely on just one experiment. KLM, the Dutch airline company, is planning to reduce their environmental impact through the replacement of fossil fuels by sustainable bio-jet fuel. A first passenger flight using sustainable synthetic kerosene took place in February 2021. A great innovation but currently still significantly more expensive than fossil fuel, so simultaneously the company is trying to reduce its fuel consumption by improving the aerodynamics of its aircrafts. A third experiment involves the replacement of some of its most frequent short-haul flights with high-speed electric trains that would further reduce the impact on the environment per passenger.

As mentioned above, companies successful in this quadrant need to remain vigilant and create good feedback loops to maintain self-critical and monitor the potential side-effects of their 'solutions'. High-speed electric trains, for example, are clean and fast but need tracks made with large amounts of concrete sleepers. The CO_2 emission from concrete production is significant and the main component of concrete is sand, a natural product that is rapidly becoming a scarce commodity. Today's solutions solve yesterday's problems but create the problems of tomorrow.

Quadrant 2: Societal Opportunities

Porter and Kramer (HBR 2011) suggested that to maintain legitimacy businesses need to reconnect their profit to societal progress; to develop products and services that address societal concerns, thus turning socially responsible behaviour into a profitable and contemporary way of doing business rather than a trade-off. Products and services that fall into this quadrant do not right the company's wrongs of the past. The company has no obligation to contribute to these societal needs and issues, but does so because they can and because it is profitable to do so. In order to contribute to societal issues through win–win solutions, many companies aligned their innovation strategies with the United Nations' SDGs. Examples include products such as Unilever's Pureit, a non-electric water purifier for households in areas where safe drinking water is scarce or unavailable; IKEA's air purifier Förnuftig ('clever' in Swedish) designed to clean the air in apartments in polluted cities; or services such as Grameen Bank (founded in Bangladesh in 1983) providing microfinance services and non-collateralised loans for the poor in order to alleviate poverty. Hindustan Unilever's 'Project Shakti' provides a different kind of win–win solution. Women in remote rural areas in India are provided with training, relevant IT skills, smart phones, etc. and become sales representatives for the company. As a result, Unilever now reaches an additional 165,000 remote villages, while more than 70,000 rural women were offered a job and education.

One could argue that also cause-related marketing belongs in this quadrant, for example, Tesco's 'Computers for Schools' programme or Lifebuoy's 'H is for Handwashing' campaign. In both cases, the big brand partners with schools and NGOs. The NGOs and schools gain money and profile, and the company gains goodwill (and ultimately profit) by contributing to a good cause.

As for the solutions in Quadrant 1, creating a better society is a wicked challenge. Companies can contribute to the betterment of problems; they cannot solve them. Where the types of challenges in this quadrant differ from the challenges in the 1st quadrant is that here, the starting point is to identify the company's core competencies. In the 1st quadrant, the starting point is the identification of environmental damage and social injustice done along the company's value chain. The strategic competency necessary to be successful in this quadrant is the ability to link core competencies, assets, and skills to the social and environmental challenges the world is facing today. Banks have the competencies, assets, and skills to provide financial services, and are thus best placed to help alleviate poverty by offering specially designed micro-loans to the poor. Universities, on the other hand, have competencies that can be used to contribute to global issues such as gender inequality or lack of education.

Experimentation remains crucial. And as before, corporations need to be careful not to overclaim their contribution to the mitigation of the problems, but simply strive to become net positive contributors to the SDGs, where necessary working together with NGOs, governments, and any other parties to maximise their impact on society (see Box 2).

Box 2.

Quadrant 2. Societal opportunities

Examples: Water and air cleaners, solar panels, micro-loans, most health and hygiene products, etc.

- Type of challenge:
  - Opportunity to contribute to wicked problems.
- Mindset:
  - Doing well by doing good.
  - Commitment to the SDGs, the environment, social justice.
- Strategic tools:
  - Awareness of key social and environmental issues challenging society; SDGs.
  - Awareness of the company's unique competencies, assets, and skills, and how they can be used to address societal needs.
  - Experiment. If possible, run multiple experiments.
  - Where relevant, collaborate with other businesses, NGOs, government departments etc. addressing the same societal needs.

Quadrant 3: Nasty Trade-offs

This is the most challenging quadrant. Issues that fall into this quadrant are often industry wide problems that present the company with a difficult trade-off. Righting the wrongs may be costly and for issues that fall into this quadrant, win–win solutions are unlikely to be found in the near future. Examples are industry wide dependence on toxic ingredients or non-renewable materials that cannot be replaced without risk or extra costs (e.g. in the tobacco industry or oil & gas industry), reliance on low-cost labour (e.g. in the fast fashion industry), broad acceptance of bribes to win tenders (e.g. in the global construction industry), international taxation habits that are legal, but increasingly regarded as unethical (e.g. in the global tech industry), etc.

Issues in this quadrant are not only wicked, but they also present the company with a dilemma. It is thereby important to distinguish between two types of dilemmas: moral dilemmas and prisoners' dilemmas. In a moral dilemma, it is clear what would be the right thing to do, but there is a price to be paid. Not doing the right thing is tempting because it may be easier, cheaper, or more convenient. Prisoners' dilemmas are dilemmas whereby doing the right thing could lead to significant losses for the good company while enriching their non-complying competitors, thereby often exacerbating the issue instead of improving it. The only way out of this dilemma is to make sure that the other players (competitors) also do the right thing, either through voluntary agreement or enforced regulation.

Examples of straightforward moral dilemmas abound. Volkswagen's decision to install a device to falsely record the emissions of its diesel cars facilitated early entry in the highly competitive US market, but once found out, it caused substantial reputational damage. BP's decision to delay regular safety checks on oil rigs to save time and money caused one of the largest oil spills in history and destroyed their carefully built image as a reliable and environmentally responsible company under the banner of 'BP Beyond Petroleum'. In both cases, the problem was not that the company did not know what would have been the right thing to do, but the company culture rewarded market success more than doing the right thing.

As in the 1st quadrant, the identification of these problems needs brutal honesty, critical self-awareness, and open-mindedness. But more is needed when there are no clear win–win solutions. Decision-makers also need a strong moral compass and an ethical corporate culture that supports socially responsible behaviour, to resist the temptation and choose the right, rather than the most profitable option. In 2018, Australia's Westpac Banking Corporation made a public statement refusing to fund Adani's Carmichael coal mining project. The bank emphasised that – in line with their strategy not to support non-sustainable ventures – they would not help to facilitate a new coal mine, despite strong pressure from the government to do so. By taking the courageous step to refuse, the bank stepped away from a $1bn concessional loan.

Prisoners' dilemmas add another complexity to this trade-off. Many large multinational companies have minimised their taxes by booking most of their profit in low-tax countries and tax havens. Although this practice may not be illegal, it does break the social contract and contributes to their negative public image. But if one corporation would stop this practice and would pay taxes where taxes are due, without the assurance that competitors will follow suit, the good corporation would put its future at risk. They are caught in a prisoners' dilemma.

The Prisoner's Dilemma

The prisoners' dilemma describes a particular type of situation whereby two (or more) players must decide if they will cooperate with the other or not (Axelrod, 1984). The best outcome would be if both parties do cooperate, or in the case of CSR, 'do the right thing'. But the temptation is to defect, because if one party defects, while the other party cooperates, the defector is likely to make significant gains, while the co-operator loses out. For example, if two competing businesses have agreed to stop paying bribes, being the sole defector could lead to significant gains; while being the sole co-operator (or adherent), could lead to significant losses. If both parties defect, the bad practice will continue. If both cooperate, the bad practice could be eradicated. In conclusion, when caught in a prisoners' dilemma, cooperating is the right thing to do but only if competitors can be trusted to cooperate as well. Blind trust is a large risk. Prisoners' dilemmas, therefore, are not solved by simply (recklessly) doing the right thing. Instead, there are two strategies to deal with a prisoners' dilemma: ensuring collaboration (building trust) or lobbying for regulation.

Corporate decision-makers in highly competitive markets will often find themselves in a prisoners' dilemma. Any industry practice that has a negative impact on society but that cannot be easily replaced without a significant risk to the company's market position as a result of a price increase and/or loss in quality, efficiency or profit, can place companies in this type of dilemma. An important strategic skill necessary to navigate these types of dilemmas is the ability to form constructive industry alliances. Many industries have formed roundtables where the societal issues they face as an industry are discussed, and joint solutions are developed. Responsible Health Initiative (RHI), for example, was founded by leading companies within the health industry to boost sustainability in the companies' value chains. In its vision statement, the RHI says it wants to harmonise industry standards, and to work together to reduce risk and increase visibility in the supply chain. Other examples include the Roundtable for Sustainable Palm Oil (RSPO), established to promote sustainable palm oil through global standards and multistakeholder governance; or the global network of chemical companies striving for sustainable development and support of the SDGs in the chemical industry.

Lobbying for regulation is another tool that can be used, either by individual companies or by industry groups, to manage the risks of the prisoners' dilemma. If, for example, all companies in the mining industry would be legally bound to net-zero carbon emissions by 2050, then all companies would have to divest their fossil fuel interests and pay the cost of transitioning to net-zero alternatives. In other words, the cooperators would not be at risk because defectors will be punished by law. Lobbying for the legislation of what is the right thing is a good strategy to deal with the prisoners' dilemma (see also De los Reyes et al., 2017 who refer to these strategies as 'CSV+').

It must be noted that political lobbying and industry cooperation come with their own set of moral challenges. Such cooperative relationships are necessary, but easily misused to gain competitive advantage, instead of regulating moral behaviour in the group. Here too, a strong corporate culture promoting ethical conduct is necessary to guarantee that all managers who act on behalf of the company know the difference between collaborating for the common good and collaborating to create self-advantage. Scorecards to measure a company's commitment to solving prisoners' dilemmas therefore should not measure individual company outcomes, but should measure their commitment to collaboration, their corporate values, their policy positions, and their lobbying activities.

Morally correct, but costly decisions also bear the risk of losing shareholders. According to the 2021 PwC survey of global investors,[4] only 19% of investors committed to ESG impact, said they were prepared to take a hit on their returns exceeding one percentage point in the pursuit of these ESG goals. Increasingly, companies seek shareholders' support before announcing decisions that positively impact society but have a negative impact on their bottom line. Unilever placed its 2020 climate transition action plan before its shareholders to seek a non-binding advisory vote on its emission-reduction targets and the plans to achieve them. In May 2021, Shell put its energy transition strategy to a vote of shareholders at its Annual General Meeting. Publicly owned companies need the support of corporate investors to prioritise the public good before profit.

The question that remains is: what to do when cooperation and regulation fail to bring all parties in line? When doing good does not lead to doing well, but favours the less scrupulous? Speaking up and creating media attention, rather than succumbing to behaviour that is no longer seen as responsible, has proven to be a successful strategy. By making their position public, companies may get sufficient public support for doing the right thing and inspire other players in the industry to follow their example (see Box 3).

Box 3.

Quadrant 3. Nasty trade-offs

Examples of agreements to collaborate: Roundtable for sustainable palm oil, UN social compact, industry collaboratives such as the Responsible Healthcare Initiative, and Together for Sustainability (TfS), a global network of chemical companies.

- Type of challenge:
  - Moral dilemmas and/or prisoners' dilemmas

- Mindset:
  - Brutal honesty, critical self-awareness, and open-mindedness PLUS a strong moral compass, collaboration, and a willingness to trust and be trustworthy.
- Strategic tools:
  - Awareness of the company's negative impact; systems thinking.
  - Mapping tools to build understanding of the complexity of the issues.
  - Awareness of different types of dilemmas (moral and prisoners' dilemmas)
  - Organisational culture: a culture that supports socially responsible behaviour.
  - Industry collaboration to 'clean up' the industry.
  - Lobby for government regulations that resolve the prisoners' dilemmas.
  - Seeking shareholders' support to do the right thing.

Quadrant 4: Matters of Conscience

Issues in this 4th quadrant are not the same as random acts of charity or philanthropy. Although philanthropy is good, it is not a responsibility; it is supererogatory. Voluntary giving out of generosity is generally considered to be a personal decision people make with their own resources, not a corporate act done with shareholder money. Instead, issues in this quadrant are societal issues for which the company – unlike other companies or individuals – has the *unique* competencies, skills, and/or assets to help. But there is no business model to support such action, nor is there a moral obligation to act because the company is not responsible for the problems. In these situations, a company is confronted with the moral question if they see it as their responsibility to respond and do something or not. In other words, challenges in this quadrant are moral challenges.

At the height of the COVID-19 pandemic, French luxury-goods conglomerate LVMH repurposed some of their perfume lines to produce hand sanitiser. The company had the know-how, the packaging, the production lines, and the people. The hand sanitiser was offered for free to the Assistance Publique-Hôpitaux de Paris, one of the largest hospital systems in Europe. Another classic example is Merck's development of a cure for river blindness. While working on a cure for parasites in livestock, Merck's scientists discovered a new compound, ivermectin, that could possibly also cure river blindness, a disease that plagued millions of people living in remote areas in the developing world. Experience had shown that to develop a drug suitable for human use would take an average of 12 years and would cost the company about $200 million dollars, but the people affected by this disease would never be able to pay for its cost. The company made the bold decision to develop the drug anyway and once manufactured, to distribute it to patients free of charge. This decision turned Merck into a moral exemplar for their industry and guaranteed their place in the ethics hall-of-fame, but it is hard to tell if the costs were ever recovered. Such decisions need an authentic sense of social responsibility and a good dose of moral courage.

Moral courage is needed because neither success nor fame are guaranteed. There is always the risk that the consequences of the good act are not as foreseen, and the social benefits fail to materialise. Secondly, the motives must be authentic. If a hidden agenda is discovered, the grandiose act of fake altruism will be seen as hypocrisy and the company's image can be damaged instead of strengthened. Strategic tools, necessary to navigate the challenges in this quadrant, include discernment and discursive power backed up by a strong corporate culture and moral code of conduct. Discernment is important because profit cannot be sacrificed on a regular basis: when is sacrificing profit the right thing to do? The ability to recognise the right cause, at the right time, will boost the company morale, and is likely to get broad stakeholder support. Pharmaceutical companies like Merck would not survive if they gave out all their medicine for free to people who could not afford to pay for it; lawyers can only spend a part of their time to offering free legal aid. The COVID-19 pandemic laid bare many of these types of dilemmas: when is it important for companies to use their competencies and assets for the common good, even if it does not pay? Discursive power (rhetoric) will be important to frame and guide the internal and external decision-making. All stakeholders will have to be convinced that this is the situation in which the normal rules of the game do not apply, and that the company instead must respond to its societal duty. A moral code of conduct can give guidance and support in these dire circumstances. Johnson & Johnson's famous 'Credo', for example, states that in the end, the wellbeing of patients comes before shareholder return. This clear statement of priorities offers a good moral guide to managers confronted with moral challenges. Last but not least, like in the case of strategic dilemmas, seeking shareholder consent may be wise when making these decisions, in particular those that could have significant impact on the bottom line (see Box 4).

Box 4.

Quadrant 4. Matters of conscience

Examples: Companies stepping up to offer support when disasters hit, lawfirms offering free legal aid to people who cannot afford it, pharmaceutical companies developing medicine to be offered for free, etc.

- Type of challenge:
  - Moral challenges: The company does not have to act, but a good company will.
- Mindset:
  - An authentic sense of social responsibility; moral courage.
- Tools:
  - Awareness of the company's unique competencies, assets, and skills, and how they can be used to address societal needs.
  - Discernment: When to act and forgo profit?
  - Discursive power to convince and inspire other stakeholders (including shareholders) that this is the right thing to do.

FURTHER IMPLICATIONS OF LOOKING AT CSR AS FOUR DIFFERENT TYPES OF CHALLENGES

The fact that there is no single minded 'correct' interpretation of CSR puts CSR scorecards into question. What do they actually measure? Scorecards that are used to determine which companies should be awarded for their 'ESG', 'Sustainability', or 'SDG' performance, may be built on a narrow interpretation of CSR. Attention should be paid to the assumptions underlying them. What do they measure as indicators of a success: How do they define the social and environmental responsibilities of the corporation? Attention should also be paid to what Kaplan and Norton (1996) called lag and lead performance indicators: lag indicators such as CO_2 emissions measure past performance, while lead indicators such as a strong culture of integrity, or active lobbying for industry regulation, help to predict the company's future performance.

Industries differ in how much attention they will need to pay to each quadrant. Companies that compete in essential industries yet face nasty trade-offs that threaten their existence such as oil and gas companies, face a very different set of challenges from companies that compete in industries that mostly face low-hanging fruit and societal opportunities such as the basic consumer goods companies. The oil and gas industry has been heavily criticised and some third parties have broken off ties not to be tainted by their negative reputation. But although this may be seen as socially correct behaviour, it may well have the opposite effect. Technical University Delft was confronted with this issue when the university issued its climate mission. For more than a decade, researchers at the university had been working together with oil giant Shell. Could they continue this collaboration and still be true to their climate mission? When asked if they would end the collaboration, their response, after much deliberation, was that this was 'the classic dilemma between purity and achieving nothing or getting your hands dirty to get things done' (Wassink, 2022). They chose the latter. Companies like Unilever or Patagonia have gained our highest praise for their sustainability efforts, but our future depends on companies that are deeply stuck in nasty trade-offs. Getting our hands dirty may be the best strategy.

A taxonomy of types of CSR challenges can also help to bring more clarity in the academic field. Acknowledging different types of issues means acknowledging different definitions, strategic approaches, mindsets, and tools necessary to deal with the range of challenges involved in creating environmental sustainability and social justice.

In conclusion, it may be easier for a manufacturer of dairy products, of hygienic soaps, or of primary education to be seen as a socially responsible company than it is, for example, for an oil producer or manufacturer of steel. But in each industry, the companies who will make a real contribution to the sustainability goals of the 21st century are the companies who do not use one, but use all of the above strategies: companies who are willing to use their innovation capacity to shift away from damaging practices of the past to find better solutions, who initiate the debate and lobby for regulation and collaboration to put an end to damaging industry practices, who are willing to use their unique strengths to better the

SDGs, and who accept the moral challenges and trade-offs when this is important for the greater good. They are the co-creators of a world of greater social justice and ecological sustainability, no matter how wicked this challenge may be.

NOTES

1. For example, despite good intentions and an ESG rating that places the Royal Dutch Shell group among the best in its industry (S&P Global, 2020), ClientEarth recently accused Shell's top management of mismanaging their environmental responsibilities (Gayle, 2022), not long after also the Dutch Government Department for Environmental Affairs accused the company of hazardous negligence and failing its human rights obligations (Boffey, 2021).
2. Rather than Porter and Kramer's (2006) very narrow definition of 'CSR' to juxtapose it to 'CSV', I follow Wickert (2021), Van Tulder and van Mil (2023), and others who use CSR as an umbrella term for the debate about the role of business in society.
3. Not all companies refer to this as a 'CSR'. A more popular term today is the 'sustainability' or 'social' agenda. To avoid the confusion around the term CSR, the OECD adopted a new term, 'responsible business conduct' (Van Tulder & van Mil, 2023, p. 351). As mentioned above, I follow those who rather avoid introducing another term and use 'CSR' as an umbrella concept covering various interpretations and definitions.
4. PwC Global Investor Survey 2021.

REFERENCES

Axelrod, R. (1984). *The evolution of cooperation*. Basic Books.

Boffey, D. (2021, May 26). Court orders Royal Dutch Shell to cut carbon emissions by 45% by 2030. *The Guardian*.

Bowen, H. R. (1953). *Social responsibilities of the businessman*. Harper & Row.

Camillus, J. C. (2008). Strategy as a wicked problem. *Harvard Business Review*, *86*(5), 98–101.

Davis, K. (1960). Can business afford to ignore social responsibilities? *California Management Review*, *2*(3), 70–76.

de los Reyes Jr, G., Scholz, M., & Smith, N. C. (2017). Beyond the "Win-Win" creating shared value requires ethical frameworks. *California Management Review*, *59*(2), 142–167.

Gallie, W. B. (1956). Art as an essentially contested concept. *The Philosophical Quarterly*, *6*(23), 97–114.

Gayle, D. (2022, March 15). Shell directors sued for 'failing to prepare the company for net zero'. *The Guardian*.

Garriga, E., & Melé, D. (2004). Corporate social responsibility theories: Mapping the territory. *Journal of Business Ethics*, *53*(1–2), 51–71.

Heugens, P. P. M. A. R., Kaptein, M., & van Oosterhout, J. (2008). Contracts to communities: A processual model of organizational virtue. *Journal of Management Studies*, *45*(1), 100–121.

Jones, T. M. & Wicks, N. C. (2018). Convergent stakeholder theory. In *Business Ethics and Strategy* (pp. 361–376). Routledge.

Kaplan, R. S., & Norton, D. P. (1996). *The balanced scorecard: Translating strategy into action*. Harvard Business Press.

Levitt, T. (1958). The dangers of social responsibility. *Harvard Business Review*, *36*(5), 41–50.

Lockett, A., Moon, J., & Visser, W. (2006). Corporate social responsibility in management research: Focus, nature, salience and sources of influence. *Journal of Management Studies*, *43*(1), 115–136.

Luo, B. N., Tang, Y., Chen, E. W., Li, S., & Luo, D. (2020). Corporate sustainability paradox management: A systematic review and future agenda. *Frontiers in Psychology*, *11*, 579272.

McGuire, J. W. (1963). *Business and society*. McGraw-Hill.

McWilliams, A., Siegel, D. S., & Wright, P. M. (2006). Corporate social responsibility: Strategic implications. *Journal of Management Studies*, *43*(1), 1–18.

Mees-Buss, A. E. J. (2014). *Multinationals in transition: Unilever's response to the demand for corporate social responsibility in a changing world* [Ph.d. thesis, University of Sydney, Australia]. ses.library.usyd.edu.au

Mees-Buss, J., & Welch, C. (2014). Taming a wicked problem? Unilever's interpretations of corporate social responsibility 2000–2012. In *International business and sustainable development* (Vol. 8, pp. 265–291). Emerald Group Publishing Limited.

Mees-Buss, J., & Welch, C. (2019). Managerial ideologies dividing the corporate elite: A process study of the rise and fall of a counter-ideology. *Organization Studies*, *40*(4), 563–592.

Okoye, A. (2009). Theorising corporate social responsibility as an essentially contested concept: Is a definition necessary? *Journal of Business Ethics*, *89*(4), 613–627.

Porter, M. E., & Kramer, M. R. (2006). The link between competitive advantage and corporate social responsibility. *Harvard Business Review*, *84*(12), 78–92.

Porter, M. E., & Kramer, M. R. (2011). Creating shared value: Redefining capitalism and the role of the corporation in society. *Harvard Business Review*, *89*(1–2), 62–77.

Prahalad, C. K., & Hart, S. L. (2001). The fortune at the bottom of the pyramid. *Strategy & Business*, *26*(1), 1–14.

Rittel, H. W., & Webber, M. M. (1973). Dilemmas in a general theory of planning. *Policy Sciences*, *4*(2), 155–169.

Sheehy, B. (2015). Defining CSR: Problems and solutions. *Journal of Business Ethics*, *131*(3), 625–648.

Scherer, A. G., & Palazzo, G. (2007). Toward a political conception of corporate responsibility: Business and society seen from a Habermasian perspective. *Academy of Management Review*, *32*(4), 1096–1120.

Van Tulder, R., & van Mil, E. (2023). *Principles of sustainable business: Frameworks for corporate action on the SDGs*. Taylor & Francis.

Wassink, J. (2022). *Can TU Delft still work with the oil industry?* https://www.delta.tudelft.nl/article/can-tu-delft-still-work-oil-industry#

Wickert, Ch. (2021). Corporate social responsibility research in the *Journal of Management Studies*: A shift from a business-centric to a society-centric focus. *Journal of Management Studies*, *58*(8), E1–E17.

CHAPTER 7

BEHIND THE TALK: TOWARDS A CONCEPTUAL AND ANALYTICAL FRAMEWORK OF CORPORATE HUMAN RIGHTS REPORTING

Stefan Zagelmeyer

The University of Manchester, UK

ABSTRACT

International business research is showing an increasing interest in the link between international business and human rights. Despite extensive coverage of corporate social responsibility (CSR) and sustainability, the analysis and discussion of why multinational corporations include human rights in corporate reporting is still in its early stages. This chapter develops an analytical framework on corporate human rights reporting, with special emphasis on international business. The conceptual part of the framework draws on legitimacy theory, stakeholder theory and signalling theory. The analytical part distinguishes between factors inside the corporation and the external environment of business organisations operating internationally.

Keywords: Corporate reporting; human rights; legitimacy; stakeholder; signalling

Walking the Talk? MNEs Transitioning Towards a Sustainable World
Progress in International Business Research, Volume 18, 107–125

Published under exclusive licence by Emerald Publishing Limited
ISSN: 1745-8862/doi:10.1108/S1745-886220240000018007

INTRODUCTION

In line with calls to address grand challenges (Buckley et al., 2017), international business research is showing increasing interest in the link between international business and human rights (Giuliani & Macchi, 2014; Wettstein et al., 2019). Recent work has analysed human rights infringements by multinational companies (Ciravegna & Nieri, 2021; Fiaschi et al., 2017; Villiers, 2023) and other forms of corporate social irresponsibility (Giuliani et al., 2014; Lange & Washburn, 2012) and their responses (Maher et al., 2021), emphasising the importance of corporate information control (Wang & Li, 2019). While there has been extensive research on CSR and sustainability reporting (Marano et al., 2017; Tschopp & Huefner, 2015), the analysis and discussion of why multinational companies include human rights in their corporate reporting activities is still in its early stages (Demir et al., 2022; Mehra & Blackwell, 2016; Zagelmeyer, 2019).

This chapter develops a conceptual and analytical framework on corporate human rights reporting which can be used to support the analysis of corporate responses to accusations of human rights infringements and socially irresponsible practices. The core of the conceptual part of the framework draws on legitimacy theory, stakeholder theory and signalling theory. The analytical part of the framework emphasises the role of internal factors and aspects in the external environment of multinational corporations, with a specific focus on institutional factors affecting the decision-making of corporations operating internationally.

This chapter makes several contributions to existing research. First, it links the phenomenon of corporate human rights disclosure to different theoretical approaches and combines elements from legitimacy theory, stakeholder theory and signalling theory to provide a theory-informed conceptualisation of the phenomenon. Second, drawing on the extensive literature on drivers and determinants of non-financial reporting, it integrates existing work and develops an analytical framework that can be used to inform empirical research. Ultimately, the chapter aims to foster a better understanding of the existing variation of corporate human rights reporting across companies and across countries.

The chapter is organised as follows: the next section introduces the context of corporate human rights reporting, reviews existing research and outlines the different theoretical perspectives, which provide a foundation for the analytical framework. The subsequent section draws on existing corporate non-financial disclosure research to identify potential determinants of human rights disclosure, distinguishing between drivers related to the internal and external environment of business organisations. The last section of the chapter summarises and discusses implications and provides ideas for future research.

CONTEXT, THEORETICAL PERSPECTIVES AND ANALYTICAL FRAMEWORK

Corporate Commitment to and Reporting on Human Rights

The endorsement of the United Nations Guiding Principles (UNGPs) on Business and Human Rights and its 'protect, respect and remedy' framework on human rights and transnational corporations and other business enterprises (OHCHR, 2011; Ruggie, 2013) has triggered a substantive shift in the outlook on CSR within multinational corporations.

Although not legally binding, the UNGPs create a form of multilevel and polycentric governance system by establishing a set of global standards that cover all business enterprises and all human rights in all UN member states (Ruggie, 2013). The UNGPs confirm that firms have a duty not to do harm while conducting their business and stipulate that business enterprises should explicitly state their commitment to human rights, engage human rights due diligence and provide access to remedy (OHCHR, 2011).

As far as the corporate responsibility to respect human rights is concerned, extensive research has covered issues related to human rights risk analysis (Salcito & Wielga, 2018) as well as human rights due diligence (McCorquodale & Nolan, 2021; Smit et al., 2020). Yet, the issue of corporate commitments to human rights through policies and communication has received only relatively little attention.

While corporate human rights reporting has existed for some (Hubers & Thijssens, 2020; Islam & Jain, 2013), the UNGPs and related public policy initiatives have further encouraged corporations to engage in human rights disclosure activities (Demir et al., 2022; Mehra & Blackwell, 2016). Zagelmeyer (2019) shows that corporations vary with respect to their human rights reporting strategy, ranging from integrated reporting, usually through an annual review, to multichannel reporting strategies that may involve CSR reports, sustainability reports, codes of conduct and/or stand-alone human rights statements or reports.

Theoretical Perspectives

The literature on non-financial reporting is dominated by three theoretical approaches: legitimacy theory, stakeholder theory and signalling theory (Deegan, 2002; Dmytriyev et al., 2021; Drover et al., 2018; Hetze, 2016; Islam, 2017). The following section will briefly outline these approaches and integrate them into an analytical framework.

Legitimacy Theory

'Legitimacy is a generalised perception or assumption that the actions of an entity are desirable, proper or appropriate within some socially constructed system of norms, values, believes and definitions' (Suchman, 1995, p. 574). Legitimacy theory argues that in order to obtain or maintain a social licence to operate (Deegan, 2002), corporations need to comply with values and norms that prevail in a specific institutional environment (Hahn & Kühnen, 2013). This is a necessary resource and prerequisite for the corporation's existence, acceptance and successful operation in that institutional environment (Suddaby et al., 2017).

Social reporting, including human rights reporting, seeks to influence the perception of the legitimacy of the organisation by internal and external stakeholders by communicating how the organisation secures legitimacy (Aray et al., 2021; Zheng et al., 2015). Failure to do so would threaten continued operation in a specific geography. One significant difference compared to CSR reporting is that human rights reporting is based on a list of universal rights, institutionalised and embodied in international human rights law.

Stakeholder Theory

The stakeholder approach theory (Donaldson & Preston, 1995) argues that the corporation is not only acting in the interests of shareholders, that is, the owners of the company, but also has to 'balance a multiplicity of stakeholder interests that can affect or be affected by the firm's actions' (Thorne et al., 2014, p. 689). This is a political process in which different stakeholders urge the corporation to act in ways that are conducive to the goals of the respective stakeholder. This involves a two-way process of stakeholders and the corporation interacting in pursuit of their respective goals (Dmytriyev et al., 2021).

From this angle, social and human rights reporting performs the function of communicating to the different stakeholder groups that the corporation is considering (and trying to meet) their diverse interests and expectations (Tan et al., 2016). In addition to one-way communication, one may argue that the corporation engages in a dialogue with its relevant stakeholders (Thorne et al., 2014).

With respect to human rights, which stakeholders are most relevant for the corporation? Stakeholders may include human rights holders, non-governmental organisations (NGOs), financial analysts or the media, among others. To address this question, it is useful to distinguish between different categories of stakeholders. First, one may distinguish internal stakeholders (the owners of the company, the workforce, or different actors from within the company, such as works councils) from those external to the organisation (NGOs, government, consumers or suppliers).

Second, one may distinguish between primary and secondary stakeholders (Clarkson, 1995). Secondary stakeholders are not engaged in transactions with the corporation and are not essential for its survival, whereas primary stakeholders are constituents on which the firm relies for its continuous survival and prosperity. Here one may distinguish between two dimensions. The legal dimension relates to the licence to operate, usually granted by governments in a particular geography. Corporations seek legitimacy (and/or a positive) reputation with the actor/agent is in the position to issue the social 'licence to operate'. The economic dimension relates to corporate performance and survival in the market: one critical aspect is reputation with customers in the context of industry competition, which forces companies to align with customers' expectations.

What is considered legitimate by the stakeholders in a particular social environment, and how does this affect the two-way interaction between stakeholders and the corporation? Companies have to first identify the relevant stakeholders, their interests and expectations and then devise strategies to address these. To add complexity, there may be variation in the definition, understanding, and perception of legitimacy across different stakeholders (e.g. trade unions vs financial analysts) and across different societies (e.g. governments in totalitarian vs liberal democratic political systems). Who are the relevant stakeholders in (say) Russia or China compared to Germany or the USA? Especially in pluralist societies, there may be the potential for conflicting interests with respect to a diverse body of stakeholders, while organisations considered legitimate in one society (e.g. trade unions and human rights NGOs) may not be legal and/or legitimate in another. The strategic importance of the different stakeholders to the corporation and the relative power of the respective stakeholder groups interact to influence corporate behaviour.

Legitimacy theory and the stakeholder approach may be used to explain the existence of multichannel social reporting, which may involve annual reports, codes of conduct/ethics, sustainability reports, CSR reports and human rights reports. Different target groups and different reporting needs (or requirements) are approached through the different reporting channels. In this respect, Thorne et al. (2014, p. 690) argue that different forms of reporting are sought to 'reduce the external costs or diminish pressure being imposed by external stakeholders', adding that using a specific reporting channel may be more effective for strategically influencing the perception of a specific stakeholder, even if this is not necessarily related to the respective corporate behaviour.

Signalling Theory

Based on the assumption of information asymmetries in the relationship of two parties, signalling theory suggests that credibility can be improved by one party sending signals (i.e. relevant information) to the other, thus improving decision-making and cooperation and avoiding adverse selection when it comes to the choice of corporation partners (Drover et al., 2018; Spence, 1974).

With respect to the potential manipulation of information in the social reporting process (Waniak-Michalak et al., 2018), Fifka and Loza Adaui (2015, p. 292) claim that stakeholders are almost certainly provided with trustworthy and correct information, as manipulated information would very likely be 'uncovered by NGOs or journalists through investigative practices'. This, however, assumes that NGOs and the media are independent and have the appropriate resources and access. Alternatively, stakeholders would press for correct information concerning the economic, social and environmental effects of the corporation's operations in a situation where it faces competitive pressures (Fifka & Loza Adaui, 2015).

As far as human rights reporting is concerned, this perspective suggests that companies should provide information on their human rights-related values, commitment and activities to the various stakeholders to indicate that the corporation considers their interests and meets their expectations, to create and/or maintain a positive image and reputation with the stakeholders to gain their approval, and thus the legitimacy required to receive and maintain the social licence to operate in the respective institutional setting (Morrison, 2014). Corporations may vary in their reporting channels and intensity of reporting, depending on their assessment of the interests of their respective stakeholders. This may also include hiding or only selectively disclosing human rights-related information (Hahn & Kühnen, 2013).

Analytical Framework 1 – Domestic Setting

The three approaches outlined above can be combined in an analytical framework which can be used to describe, explain, analyse and discuss corporate human rights disclosure activities embedded in a specific institutional field or setting. The analytical framework in Fig. 7.1 involves a hypothetical corporation (CORP1) operating in a domestic national environment (country 1) with a specific institutional setting. In this setting, CORP1 is interacting with two hypothetical internal stakeholders (SH1 and SH2), for example, a company-level works council, and the owners of the company. In addition, the corporation is interacting with two external stakeholders, SH3 and SH4, for example, a human rights NGO and a financial analyst.

Based on their interests, all four stakeholders have specific expectations for CORP1 with respect to human rights (i.e. EXP1 for stakeholder SH1 and EXP3 for stakeholder SH3). In line with signalling theory, CORP1 may send information signals containing information on the corporations' human rights-commitment to the stakeholders (e.g. S2 to SH2 and S4 to SH4), for example, in the form of a human rights report or a more general annual report.

As far as the signalling process is concerned, we consider the decisions about (i) whether or not to engage in human rights reporting; (ii) the extent, quality and quantity of human rights reporting; and (iii) the mix of specific reporting channels as a decision based on cost-benefit considerations. It is important to note that this cost-benefit analysis may also be applied to situations where the company is legally required to report on specific issues. While the existing literature does not explicitly or comprehensively analyse the elements of cost-benefit analysis related to corporate human rights disclosure, the literature on determinants of CSR and sustainability reporting provides valuable insights.

In their conceptual model, Kuo et al. (2016) consider the following elements in the analysis of social reporting: relevant target readers, motivations for CSR, barriers to CSR and goals of CSR. At the core of their model are the goals of CSR reporting,

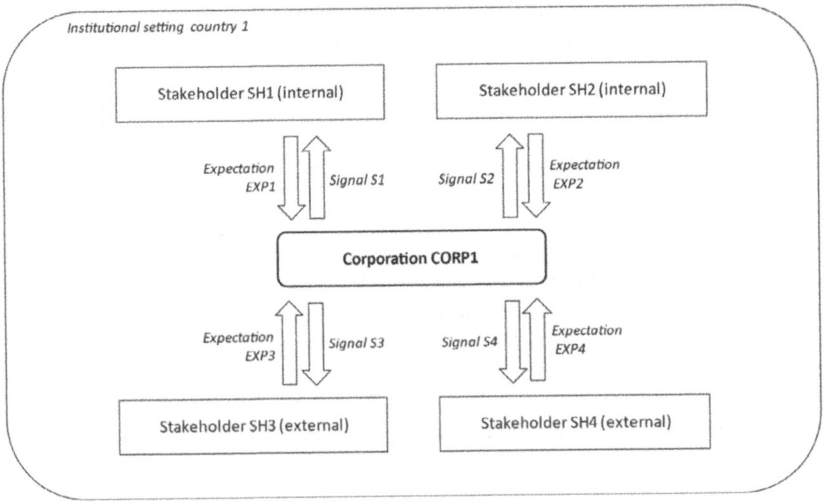

Fig. 7.1. Analytical Framework 1 – Domestic Setting.

Key: CORP = corporation, SH = stakeholder, EXP = expectation and S = signal.

which include economic and ethical goals. The economic goals relate to different aspects of the business management (e.g. strategy development, marketing, human resource management, organisational aspects and controlling), which are intended to increase business performance and profitability. Ethical goals include compliance issues (e.g. legal requirements) and voluntary social reporting items (e.g. awareness of social and environmental issues in the community; Kuo et al., 2016). This conceptual model can be adapted to analyse corporate human rights disclosure.

Previous analysis has highlighted the importance of distinguishing between different actors and stakeholders, and the pressures on corporations to obtain legitimacy in the eyes of internal and external stakeholders, for example, though signalling. The distinction between internal and external aspects is elaborated further by the inside-out approach versus outside-in approach distinction (Dienes et al., 2016). The inside-out approach involves companies in which social reporting is based on a deliberate decision by key decision-makers, driven by internal factors (e.g. particular human rights-related values to which the company subscribes). This perspective is compatible with the resource-based view in strategic management (Barney, 1991), suggesting that human rights-related values, policies and practices can be incorporated as an integral element of the business model to gain competitive advantage. In contrast, the outside-in approach involves companies that respond to external shareholder expectations by engaging in corporate human rights reporting activities to gain legitimacy with the respective stakeholders. This perspective has an institutionalist underpinning and is compatible with stakeholder and legitimacy theory.

DRIVERS OF CORPORATE HUMAN RIGHTS DISCLOSURE

The framework introduced above sets out the main actors and their relationships within a particular institutional context, which is very much in line with actor-centred institutionalism (Scharpf, 1997; Zagelmeyer, 2019), and emphasises either corporate values or stakeholder expectations as drivers of corporate human rights disclosure. This section will build on the conceptual part of the framework and further elaborate on the specific factors and determinants that influence corporations' decisions to engage in human rights reporting, distinguishing drivers (i) operating from within the corporation from and (ii) factors in the external environment.

Internal Drivers

Internal drivers operate from within the organisation, pressuring management to engage in human rights reporting activities. The subsequent sections will briefly introduce and describe these drivers and the respective lines of reasoning.

Business Model

There has been a long and controversial debate in business ethics on the existence of corporate moral development, that is, whether business organisations of business organisations as a distinct and genuine social entity can develop morally, or whether moral development in the context of business only occurs at the level of the individual members within the organisation (Sridhar & Camburn, 1993; Wilmot, 2001). While a detailed discussion of this topic is beyond the scope of this chapter, corporations can embed certain human rights related values in their business philosophy, which – if implemented – then affect the choice and design of the business model and corporate-/business-level strategies.

Levels of moral development have traditionally been analysed and discussed with respect to individual people (Kohlberg & Hersh, 2009). Managers in leadership positions may vary in terms of their personal moral development, which may affect their decisions on corporate human rights disclosure. Involvement in a human rights agenda may relate to the personal values of a person and/or their political position, or may be used to create a particular image or reputation to legitimise one's position and actions. An important additional consideration is whether the level of moral development and its consequences are attached to a person or a social role (e.g. manager or family member).

Capital Structure and Ownership Structure

With respect to the link between different measures of capital structure and sustainability, companies with an increased capital demand have a greater interest in providing information on sustainability to capital providers (Dienes et al., 2016). The information provided by companies through social reporting can provide useful information to internal and external investors. Tan et al. (2016) argue that investors take information about social reporting into account when making investment decisions. However, with respect to empirical evidence, Dienes et al. (2016, p. 172) conclude that 'inconsistent results make it impossible to draw any reliable conclusion as to the tendency of the relationship'. Also, the evidence in relation to the link between ownership concentration and sustainability is mixed and inconclusive.

A related topic concerns public and state ownership, or the phenomenon of being closely linked to the public sector or the government, an issue that has recently gained prominence in international business research (Mariotti & Marzano, 2019). Here, the line of reasoning is different. If a corporation is firmly embedded in the public sector, there may be hierarchical reporting relationships with institutions and/or actors in the political system, which in turn may have specific expectations with respect to corporate human rights reporting.

Corporate Governance

Another set of factors relates to corporate governance. There are considerable empirical studies on the link between characteristics of the corporate governance systems on the one hand and social reporting on the other, such as the size of the board of directors, the proportion of independent or female board members, the number of board meetings, the number of committees and/or the existence of a sustainability or CSR committee (Dienes et al., 2016; Ramón-Llorens et al., 2019). Furthermore, specific forms of representation in corporate governance, such as involving board-level employee representatives or government representatives in corporate decision-making, may support human rights reporting, assuming that the respective actors have a human rights agenda. In addition, particular groups of owners or their agents, for example, banks or insurance companies, may also affect whether (and how) human rights become a topic for reporting. Yet, empirical studies here have not yet yielded consistent results (Dienes et al., 2016, p. 170).

Additional Organisational Aspects

Other internal stakeholders influence whether and how companies engage in human rights reporting. Central to this are the organisational units and the individuals involved in the respective decision-making processes linked to corporate human rights reporting activities.

Habek and Wolniak (2015) highlight the relevance of those that prepare human rights reports, who often appear to be from the organisational unit responsible for CSR, sustainability and/or compliance issues. While business and corporate strategy will certainly have an impact on human rights reporting, there are internal coordination and communication processes in which different internal stakeholders engage with their own agendas, highlighting the political dimension of the respective decision-making processes. While financial reporting has traditionally been the domain of the accounting department, the accounting literature suggests its involvement in non-financial and social reporting has also become increasingly relevant.

In this context, the corporate general counsel, sometimes called chief legal officer, or corporate counsel, is of particular relevance (Ruggie, 2013). Partnering with executive management and the board, the general counsel is the corporation's main attorney and source of legal advice, covering governance, corporate and institutional legal matters. In addition, concerning human rights issues, it acts as a technical expert and counsellor on legal matters and leads and coordinates particular human rights-related initiatives and activities (Sherman, 2021).

A further relevant internal stakeholder is the marketing department, which is usually responsible for external public relations, which may influence social reporting in the process of managing corporate identity and the corporation's brand image and value. Social reporting may affect the public perception and reputation of the company, its brands, products or services in the respective markets through creating positive images. Accusations of human rights abuses and related scandals can have an adverse effect on the achievement of the organisation's goals, for example, through the responses of the consumers or the financial markets.

The human resource management department may also have a stake in human rights reporting, as labour rights issues are very closely linked to human rights (Compa & Diamond, 2003). Labour-related issues such as modern slavery, human trafficking, child labour, labour and trade union rights and international labour standards are especially relevant in business organisations that operate internationally. Various recent pieces of legislation, for example, the modern slavery act in the UK, or the implementation of human rights-related national action plans in the European Union, oblige companies to investigate and document potential human rights issues – most of which are labour-related – within their supply chains.

Firm Size
The review of empirical studies by Dienes et al. (2016) finds a positive relationship between firm's size and CSR disclosure activities, potentially resulting from internal and external factors. With respect to internal factors, larger firms will find it easier to cope with the administrative costs of developing the respective reports (Gallego-Álvarez & Quina-Custodio, 2016). The smaller a company, the less economical it will find social reporting from a cost-benefit point of view (Dienes et al., 2016). Concerning external factors, companies depending on the capital markets may need to comply with specific reporting requirements and are closely monitored by analysts, financial journalists and the interested public. Greater financing needs may also add pressure to provide more information to financial analysts and investors to reduce information asymmetries. Public visibility and attention may increase with the size of a corporation (Tan et al., 2016). Yet, empirical research on the link between voluntary disclosure and capital market exposure is inconclusive (Gallego-Álvarez & Quina-Custodio, 2016).

Financial Performance
There is a long and controversial discussion on the link between social reporting and financial performance. There are different lines of reasoning. Profitable companies may be more willing to assume the additional costs of producing and publishing a sustainability report (Gamerschlag et al., 2011). Furthermore, if social reporting is perceived by key decision-makers and actors to improve organisational and/or performance, this may create pressure to engage in respective social reporting activities. A further reason could be 'greenwashing', which refers to company concealing socially and environmentally irresponsible corporate practices or showcasing routine operating procedures as sustainability performance, using social reporting as a marketing tactic. Yet, based on a review of the relevant literature, Dienes et al. (2016, p. 169) conclude that 'an influence of profitability on CSR disclosure cannot be confirmed' and that profitability can be affected by the social reporting both positively and negatively.

External Drivers

External stakeholders may exert pressure on companies to publish or withhold information on human rights-related aspects of their operations. In other words, corporations are expected to signal specific information requested by stakeholders to legitimise their existence and operation in a particular social context.

Financial Markets, Consumers and NGOs
As mentioned, financial analysts and the financial markets play an important role, as they may be involved in the provision of capital for the company. Financial analysts engage in risk assessments, which also may cover human rights risks, affecting their recommendations to financial markets.

Pressure may also come from the respective product markets, for example, if consumers respond to reports on the negative impact of a corporation's activities. The extent of this pressure depends on the relevance and importance of the relevant market to the company, and how far consumers are able to organise collectively, for example, in consumer protection associations.

In addition to the government as a public stakeholder, NGOs may exert pressure for corporations to engage in human rights reporting activities (McVey, 2022). The effectiveness of these organisations depends on their resources, their campaigns and their relation to and impact on government activities, media and public opinion.

Governance Systems and Regulation

The UNGPs establish the nucleus for a polycentric business and human rights governance system, which includes elements of both private and public governance (Zagelmeyer, 2019). This governance system involves legally binding government regulations and requirements to collective self-regulation, such as voluntary participation in reporting initiatives. In terms of different levels of regulation, there may be agreements, recommendations or initiatives at international, national or regional levels. At the international level, international business and human rights governance includes the global reporting initiative, the United Nations Global Compact, the World Trade Organisation guidelines and Organization for Economic Cooperation and Development (OECD) guidelines. In addition, country-level regulation is expected to have a significant impact on social reporting. In recent years, several countries have issued legislation requiring companies to publish non-financial reports. Other countries have gone further and require companies to monitor and report the human rights situation of other companies involved in their supply chain, for example, the UK (Islam & Van Staden, 2021; Mai et al., 2022).

Industry-specific Actors and Institutions

There are several reasons why corporate human rights reporting practices vary across industries. There may be sectoral arrangements supporting or requiring specific types of reporting, organised and implemented through industry associations or other institutionalised industry-wide organisations at both national and international levels. Examples include banking, resource extraction and textile manufacturing.

Young and Marais (2012) show that specific industries are under high coercive and normative pressure to engage in social reporting. Sensitive (or so-called 'sin') industries (Grougiou et al., 2016) have greater potential to be environmentally and/or socially harmful and may be under increased public scrutiny and thus pressure to engage with external stakeholders. Corporations in such industries can be expected to be more transparent about their environmental and social strategies and spend more resources in these activities in response to the relatively higher political cost, greater regulatory burdens and the likelihood of environmental and social incidents. Transparency in social behaviour as communicated through social reporting seeks to build and/or restore public confidence in the corporation, which in turn will increase prices and trading volume (Tan et al., 2016). However, Tan et al. (2016) ask whether there are variations in terms of the existence, magnitude and effectiveness of social pressure across different countries.

The Media

The media is another external stakeholder and source of soft but effective pressure for responsible corporate behaviour through influencing public opinion (Amengual et al., 2022). The extent to which the media can perform this function depends to a large degree on its independence and available resources (Wang & Li, 2019). Media visibility can be measured, for example, by the number of hits on journals and magazines, newspaper rankings or other measures of media presence. Dienes et al. (2016) report that all studies included in their review show a significant positive association between CSR disclosure and media visibility of corporations, because companies affected more intensively by media visibility may be more responsive to stakeholder pressure to justify their activities.

INTERNATIONAL ASPECTS

While the analytical framework developed was designed to describe, analyse and discuss the corporate human rights reporting activities of companies operating in the domestic environment, the following extends the analysis to multinational corporations operating internationally. From the perspective of the corporation, moving into international operation adds complexity. While the core ideas of the analytical framework in Fig. 7.1 and the arguments presented in the previous section in relation to internal and external pressures on companies to engage in corporate human rights reporting still apply, multinational corporations must consider additional factors such as variation in markets and in formal and informal institutions across the nations in which it operates.

This section will first extend the analytical framework developed above to multinational corporations operating internationally. It will then elaborate on additional factors that may affect corporate human rights reporting activities. Notably, those international and comparative aspects may modify both internal and external drivers of corporate human rights disclosure.

Analytical Framework 2 – International Setting

Combining elements from stakeholder theory, legitimacy theory and signalling theory, Fig. 7.2 extends the analytical framework to describe, explain, analyse and discuss corporate human rights reporting activities developed above for a corporation operating in a domestic environment to a multinational corporation operating in an international business environment.

At the centre of the revised framework is the multinational corporation CORP1, which operates in two or more countries. In Fig. 7.2, we assume two subsidiaries SUB1 and SUB2 in two different host countries, country 2 and country 3, and headquarters in the country of origin, country 1, each with a country-specific institutional environment and setting. When moving from domestic to international operations, CORP1 faces additional stakeholders in the two host countries, namely SH5 and SH6 in country 2 and SH7 and SH8 in country 3. These stakeholders (SH5-8) may have specific and varying human rights-related expectations (EXP5-8, respectively), and the MNC responds through signals S5-8, respectively.

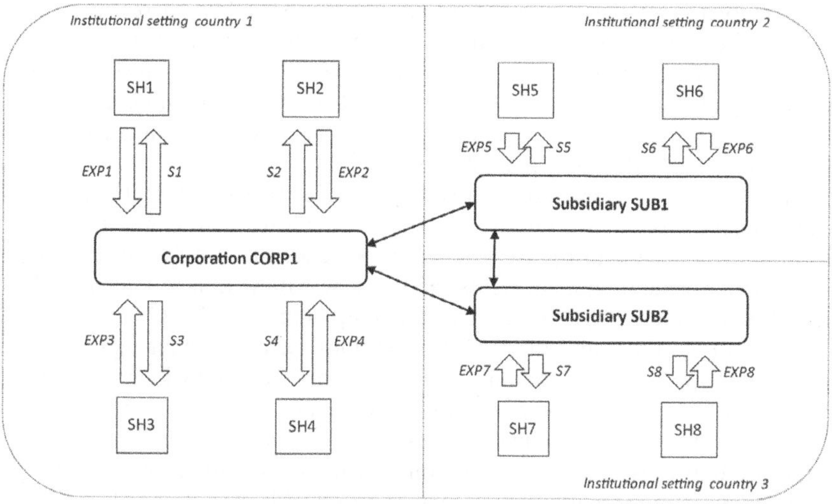

Fig. 7.2. Analytical Framework 2 – International Setting.
Key: CORP = corporation, SUB = subsidiary, SH = stakeholder, EXP = expectation and S = signal.

There may also be additional stakeholders at international levels, for example, organisations such as the United Nations or the International Labour Organization (ILO), which could be easily included in a revised version of the framework.

For simplicity, Fig. 7.2 only includes dyadic relationships between either the HQ or a subsidiary on the one hand, and the respective set of stakeholders within the respective country on the other. In real life, it may well be that headquarters engage in signalling and legitimacy related relationships with stakeholders across borders (e.g. between the headquarters in country 1 and stakeholder SH5). Furthermore, the model assumes that the headquarters and the subsidiaries share the same position on the corporation's human rights-commitment, while in fact there may be differences here between the headquarters and subsidiaries. Thus, multinational corporations operating in an international business environment may face more complex sets of stakeholders with specific (and potentially conflicting) legitimacy-related expectations as well as additional organisational contingencies. This may lead the corporation to engage in more complex signalling strategies concerning human rights disclosure.

International Aspects

Internationalisation processes lead to geographical expansion of the corporation's environment, in which the corporation seeks legitimacy for its activities vis-á-vis different public spheres and actors and an increased and more diverse number of stakeholders (Kolk & Fortanier, 2013). In this process, multinational enterprises furthermore face the challenges of liability of foreignness and being an easy target for interest groups due to their size and visibility. These aspects ultimately affect the corporation's human rights reporting activities. Yet, legitimacy-related stakeholder pressures may be country-specific and vary across countries, which may challenge headquarters with respect to the design and coordination of reporting strategies to meet the different expectations of various stakeholders (Corciolani et al., 2022; Sun et al., 2021).

At the international level, there is a complex framework of international agreements, regulations and institutions that seek to establish a governance framework for business and human rights, which may influence the stakeholder's expectations about corporate human rights reporting activities (Lim & Tsutsui, 2012). In addition to the UNGP, these include, among others, the UNGC, the OECD Guidelines for MNEs and the ILO. Conventions include human rights-related stipulations and recommendations to influence corporate behaviour and to harmonise and increase the comparability of corporate activities (Fortanier et al., 2011). Relevant regulations and pressures may also exist at regional level, such as in relation to the discussion and adoption of mandatory CSR reporting (Fifka & Loza Adaui, 2015) or mandatory human rights due diligence (Deva, 2023).

Taking the institutionalist perspective, corporations in a competitive environment will seek to meet the expectations of their stakeholders and to comply with the respective standards in order to gain legitimacy. In this process, Fortanier et al. (2011) argue that isomorphic pressures would push corporations to structure their activities in different institutional fields but operating in multiple fields would potentially be associated with pressures pulling in different directions. This may lead to the disclosure of different human rights reporting content to different stakeholders in different countries, and maybe even to differences in reporting between the headquarters and subsidiaries. The implications of cross-country variation in formal and informal institutions for corporate human rights reporting will be discussed in greater detail in the next section.

Comparative Aspects

Companies operating internationally may face different expectations in different countries. Stakeholders may vary in their interests, and there may be differences concerning the relevance and importance of the internal and external determinants of corporate human rights reporting, for example, with respect to governments, investors and consumers. It would be interesting to see whether owners from or based in countries with a positive human rights record have a distinctively different impact on corporate human rights disclosure, as compared to corporations based in countries associated with extensive human rights violations. These pressures may be different at the headquarters compared to subsidiaries, but the pressures may also vary across subsidiaries.

Recent literature increasingly emphasises the role of cultural characteristics and national cultural context in social and human rights-related policies and practices as well as human rights reporting. Using Hofstede's (2001) national culture value dimensions approach, Garcia-Sanchez et al. (2016, p. 15) find that organisations from countries 'characterized by higher long-term orientation, higher cultural values of collectivist, feminist and uncertainty avoidance and a lower power distance index, are more sensitive to publishing corporate social responsibility reports'. The country score on masculinity is associated with greater propensity to engage in social reporting, as a more competitive culture forces companies to disseminate information to stakeholders. The lower the score of uncertainty avoidance, the more easily members of a particular society will deal with uncertainty and there will be less demand for formal regulation, which may in turn lead stakeholders to ask for more specific information. Power distance, linked to the vertical stratification of a society, is associated with different levels of power status and low expectations in terms of quality and quantity of information, thus affecting human rights reporting negatively. Finally, actors in a society characterised by long-term orientation are thought more likely to save and invest, less likely to behave opportunistically, and thus more interested in valid, long-term information (Garcia-Sanchez et al., 2016).

The formal institutional context includes legal, regulatory and professional structures that may affect corporations' human rights reporting. Drawing on the varieties of capitalism approach (Hall & Soskice, 2001), Matten and Moon (2008) analyse variation in the emergence and diffusion of implicit and explicit CSR policies and practices across countries. While implicit CSR consists of values, norms and rules integral to the socio-economic system, explicit CSR requires that corporations engage voluntarily in socially responsible activities. Different socio-economic systems are associated with the two CSR categories. Liberal market economies tend to be associated with explicit CSR (i.e. companies explicitly decide, state and disseminate information on their position with respect to CSR), while, coordinated market economies tend to be associated with implicit CSR systems, which embed the respective values and norms in national institutions. As a consequence, corporations face less pressure to publish CSR-related information (Matten & Moon, 2008).

Lim and Tsutsui (2012) take this up, arguing that corporations operating in coordinated market economies may face greater pressure to engage in social reporting compared to those in liberal market economies. On the other hand, referring to the distinction between explicit and implicit CSR, they argue that the more liberal a system is, the more likely corporations are to engage in explicit social reporting. Following Jackson and Apostolakou (2010), liberal marker economies are linked with explicit forms of CSR and social reporting, as this information can meet stakeholder demands in the absence of government regulation. The opposite is true for coordinated market economies, as CSR is embedded in the socio-economic system and leaves less scope and less need for individual corporate social reporting activities, emphasising implicit CSR (Gjølberg, 2009).

CONCLUSIONS

This chapter develops a conceptual and analytical framework on corporate human rights disclosure, which combines elements of legitimacy theory, stakeholder theory and signalling theory. It considers organisational and institutional aspects affecting corporate human rights reporting and can be applied to both domestic and multinational corporations. While the potential internal and external determinants as well as international aspects were introduced separately, it is important to note that these may be interrelated.

The chapter opens up several avenues for further research. Existing research shows variation with respect to the incidence and intensity of corporate human rights reporting and various disclosure channels across companies and countries (Zagelmeyer, 2019). As previous research on social reporting does not distinguish between different disclosure channels, the analytical framework could inform future empirical research on the determinants of the incidence and intensity of different channels across companies and countries. Furthermore, future research is encouraged that analyses internal and external determinants of corporate human rights disclosure using multilevel data. This may be helpful for the analysis of different configurations of stakeholders that influence corporate human rights reporting activities.

Knorringa and Nadvi (2016) suggest that characteristic differences in responsible behaviour of firms may be constituted by different social contracts within their institutional environments. In this respect, future research could examine how the requirements and complexity regarding legitimacy requirements change when corporations expand internationally and thus encounter different social contract spaces with differentiated institutional settings.

REFERENCES

Amengual, M., Mota, R., & Rustler, A. (2022). The 'court of public opinion': Public perceptions of business involvement in human rights violations. *Journal of Business Ethics*, *185*, 49–74. http://doi.org/10.1007/s10551-022-05147-5

Aray, Y., Dikova, D., Garanina, T., & Veselova, A. (2021). The hunt for international legitimacy: Examining the relationship between internationalization, state ownership, location and CSR reporting of Russian firms. *International Business Review*, *30*(5), 101858. http://doi.org/10.1016/j.ibusrev.2021.101858

Barney, J. (1991). Firm resources and sustained competitive advantage. *Journal of Management*, *17*(1), 99–120.

Buckley, P. J., Doh, J. P., & Benischke, M. H. (2017). Towards a renaissance in international business research? Big questions, grand challenges, and the future of IB scholarship. *Journal of International Business Studies*, *48*(9), 1045–1064. http://doi.org/10.1057/s41267-017-0102-z

Ciravegna, L., & Nieri, F. (2021). Business and human rights: A configurational view of the antecedents of human rights infringements by emerging market firms. *Journal of Business Ethics*, *179*, 431–450. http://doi.org/10.1007/s10551-021-04861-w

Clarkson, M. B. E. (1995). A stakeholder framework for analyzing and evaluating corporate social performance. *Academy of Management Review*, *20*(1), 92–117.

Compa, L. A., & Diamond, S. F. (Eds.). (2003). *Human rights, labor rights, and international trade*. University of Pennsylvania Press.

Corciolani, M., Giuliani, E., Humphreys, A., Nieri, F., Tuan, A., & Zajac, E. J. (2022). Lost and found in translation: How firms use anisomorphism to manage the institutional complexity of CSR. *Journal of Management Studies*. http://doi.org/10.1111/joms.12877

Deegan, C. (2002). The legitimising effect of social and environmental disclosures – A theoretical foundation. *Accounting, Auditing & Accountability Journal*, *15*(3), 282–311.

Demir, M., Min, M. K., & Coppola, L. D. (2022). Discrepancies in reporting on human rights: A materiality perspective. *Thunderbird International Business Review*, *64*(2), 169–178. http://doi.org/10.1002/tie.22255

Deva, S. (2023). Mandatory human rights due diligence laws in Europe: A mirage for rightsholders? *Leiden Journal of International Law*, *36*(2), 1–26.

Dienes, D., Sassen, R., & Fischer, J. (2016). What are the drivers of sustainability reporting? A systematic review. *Sustainability Accounting Management and Policy Journal*, *7*(2), 154–189. http://doi.org/10.1108/sampj-08-2014-0050

Dmytriyev, S. D., Freeman, R. E., & Hörisch, J. (2021). The relationship between stakeholder theory and corporate social responsibility: Differences, similarities, and implications for social issues in management. *Journal of Management Studies*, *58*(6), 1441–1470. http://doi.org/10.1111/joms.12684

Donaldson, T., & Preston, L. E. (1995). The stakeholder theory of the corporation: Concepts, evidence and implications. *Academy of Management Review*, *20*(1), 65–91. http://doi.org/10.3138/9781442673496-011

Drover, W., Wood, M. S., & Corbett, A. C. (2018). Toward a cognitive view of signalling theory: Individual attention and signal set interpretation. *Journal of Management Studies*, *55*(2), 209–231.

Fiaschi, D., Giuliani, E., & Nieri, F. (2017). Overcoming the liability of origin by doing no-harm: Emerging country firms' social irresponsibility as they go global. *Journal of World Business*, *52*(4), 546–563. http://dx.doi.org/10.1016/j.jwb.2016.09.001

Fifka, M., & Loza Adaui, C. R. (2015). Corporate social responsibility (CSR) reporting—Administrative burden or competitive advantage? In L. O'Riordan, P. Zmuda, & S. Heinemann (Eds.), *New perspectives on corporate social responsibility – Locating the missing link* (pp. 285–300). Springer Gabler.

Fortanier, F., Kolk, A., & Pinkse, J. (2011). Harmonization in CSR reporting MNEs and global CSR standards. *Management International Review*, *51*(5), 665–696. http://doi.org/10.1007/s11575-011-0089-9

Gallego-Álvarez, I., & Quina-Custodio, I. A. (2016). Disclosure of corporate social responsibility information and explanatory factors. *Online Information Review*, *40*(2), 218–238. http://doi.org/10.1108/OIR-04-2015-0116

Gamerschlag, R., Moller, K., & Verbeeten, F. (2011). Determinants of voluntary CSR disclosure: Empirical evidence from Germany. *Review of Managerial Science*, *5*(2–3), 233–262. http://doi.org/10.1007/s11846-010-0052-3

Garcia-Sanchez, I. M., Cuadrado-Ballesteros, B., & Frias-Aceituno, J. V. (2016). Impact of the institutional macro context on the voluntary disclosure of CSR information. *Long Range Planning*, *49*(1), 15–35. http://doi.org/10.1016/j.lrp.2015.02.004

Giuliani, E., & Macchi, C. (2014). Multinational corporations' economic and human rights impacts on developing countries: A review and research agenda. *Cambridge Journal of Economics*, *38*(2), 479–517. http://doi.org/10.1093/cje/bet060

Giuliani, E., Macchi, C., & Fiaschi, D. (2014). The social irresponsibility of international business: A novel conceptualization. In R. van Tulder, A. Verbeke, & R. Strange (Eds.), *International business and sustainable development* (pp. 141–171). Emerald Group Publishing Limited.

Gjølberg, M. (2009). Measuring the immeasurable? *Scandinavian Journal of Management*, *25*(1), 10–22. http://doi.org/1016/j.scaman.2008.10.003

Grougiou, V., Dedoulis, E., & Leventis, S. (2016). Corporate social responsibility reporting and organizational stigma: The case of "sin" industries. *Journal of Business Research*, *69*(2), 905–914. http://doi.org/1016/j.jbusres.2015.06.041

Habek, P., & Wolniak, R. (2015). Factors influencing the development of CSR reporting practices: Experts' versus preparers' points of view. *Inzinerine Ekonomika – Engineering Economics*, *26*(5), 560–570. http://doi.org/10.5755/j01.ee.26.5.7690

Hahn, R., & Kühnen, M. (2013). Determinants of sustainability reporting: A review of results, trends, theory, and opportunities in an expanding field of research. *Journal of Cleaner Production*, *59*, 5–21. http://doi.org/10.1016/j.jclepro.2013.07.005

Hall, P. A., & Soskice, D. (2001). An introduction to the varieties of capitalism. In P. A. Hall & D. Soskice (Eds.), *Varieties of capitalism: The institutional foundations of comparative advantage* (pp. 1–70). Oxford University Press.

Hetze, K. (2016). Effects on the (CSR) reputation: CSR reporting discussed in the light of signalling and stakeholder perception theories. *Corporate Reputation Review*, *19*(3), 281–296. http://doi.org/10.1057/s41299-016-0002-3

Hofstede, G. H. (2001). *Culture's consequences: Comparing values, behaviors, institutions, and organizations across nations* (2nd ed.). Sage.

Hubers, F., & Thijssens, T. (2020). Human rights reporting under increasing institutional pressure. *Maandblad Voor Accountancy en Bedrijfseconomie*, *94*(7/8), 303–312. http://doi.org/10.5117/mab.94.50425

Islam, M. A. (2017). CSR reporting and legitimacy theory: Some thoughts on future research agenda. In M. Aluchna & S. O. Idowu (Eds.), *Dynamics of corporate social responsibility: A critical approach to theory and practice* (pp. 323–339). Springer: Cham.

Islam, M. A., & Jain, A. (2013). Workplace human rights reporting: A study of Australian garment and retail companies. *Australian Accounting Review*, *23*(2), 102–116. http://doi.org/10.1111/auar.12009

Islam, M. A., & Van Staden, C. J. (2021). Modern slavery disclosure regulation and global supply chains: Insights from stakeholder narratives on the UK Modern Slavery Act. *Journal of Business Ethics*, *180*, 455–479 http://doi.org/10.1007/s10551-021-04878-1

Jackson, G., & Apostolakou, A. (2010). Corporate social responsibility in Western Europe: An institutional mirror or substitute? *Journal of Business Ethics*, *94*(3), 371–394. http://doi.org/10.1007/s10551-009-0269-8

Knorringa, P., & Nadvi, K. (2016). Rising power clusters and the challenges of local and global standards. *Journal of Business Ethics*, *133*, 55–72. https://doi.org/10.1007/s10551-014-2374-6

Kohlberg, L., & Hersh, R. H. (2009). Moral development: A review of the theory. *Theory Into Practice*, *16*(2), 53–59. http://doi.org/10.1080/00405847709542675

Kolk, A., & Fortanier, F. (2013). Internationalization and environmental disclosure: The role of home and host institutions. *Multinational Business Review*, *21*(1), 87–114. http://doi.org/10.1108/15253831311309500

Kuo, T. C., Kremer, G. E. O., Phuong, N. T., & Hsu, C. W. (2016). Motivations and barriers for corporate social responsibility reporting: Evidence from the airline industry. *Journal of Air Transport Management*, *57*, 184–195. http://doi.org/10.1016/j.jairtraman.2016.08.003

Lange, D., & Washburn, N. T. (2012). Understanding attributions of corporate social irresponsibility. *Academy of Management Review*, *37*(2), 300–326. http://doi.org/10.5465/amr.2010.0522

Lim, A., & Tsutsui, K. (2012). Globalization and commitment in corporate social responsibility: Cross-national analyses of institutional and political-economy effects. *American Sociological Review*, *77*(1), 69–98. http://doi.org/10.1177/0003122411432701

Maher, R., Neumann, M., & Lykke, M. S. (2021). Extracting legitimacy: An analysis of corporate responses to accusations of human rights abuses. *Journal of Business Ethics*, *176*(1), 609–628. http://doi.org/10.1007/s10551-020-04678-z

Mai, N., Vourvachis, P., & Grubnic, S. (2022). The impact of the UK's Modern Slavery Act (2015) on the disclosure of FTSE 100 companies. *The British Accounting Review*, *55*(3), 101115. http://doi.org/10.1016/j.bar.2022.101115

Marano, V., Tashman, P., & Kostova, T. (2017). Escaping the iron cage: Liabilities of origin and CSR reporting of emerging market multinational enterprises. *Journal of International Business Studies*, *48*(3), 386–408. http://doi.org/10.1057/jibs.2016.17

Mariotti, S., & Marzano, R. (2019). Varieties of capitalism and the internationalization of state-owned enterprises. *Journal of International Business Studies*, *50*(5), 669–691. http://doi.org/10.1057/s41267-018-00208-2

Matten, D., & Moon, J. (2008). "Implicit" and "explicit" CSR: A conceptual framework for a comparative understanding of corporate social responsibility. *The Academy of Management Review*, *33*(2), 404–424. http://doi.org/10.2307/20159405

McCorquodale, R., & Nolan, J. (2021). The effectiveness of human rights due diligence for preventing business human rights abuses. *Netherlands International Law Review*, *68*(3), 455–478. http://doi.org/10.1007/s40802-021-00201-x

McVey, M. (2022). Untangling the authority of external experts in the corporate implementation of the UN Guiding Principles on Business and Human Rights. *Journal of Human Rights*, *21*(5), 620–638. http://doi.org/10.1080/14754835.2022.2105646

Mehra, A., & Blackwell, S. (2016). The rise of non-financial disclosure: Reporting on respect for human rights. In D. Baumann-Pauly & J. Nolan (Eds.), *Business and human rights* (pp. 276–284). Routledge.

Morrison, J. (2014). *The social license – How to keep your organization legitimate*. Palgrave Macmillan.

OHCHR. (2011). *Guiding principles on business and human rights – Implementing the United Nations "protect, respect and remedy" framework*. Office of the UN High Commissioner for Human Rights.

Ramón-Llorens, M. C., García-Meca, E., & Pucheta-Martínez, M. C. (2019). The role of human and social board capital in driving CSR reporting. *Long Range Planning*, *52*(6), 101846. http://doi.org/10.1016/j.lrp.2018.08.001

Ruggie, J. G. (2013). *Just business: Multinational corporations and human rights*. W. W. Norton & Company.

Salcito, K., & Wielga, M. (2018). Corporate human rights risk assessment: Aligning what is measured and managed. *Nordic Journal of Human Rights*, *36*(4), 411–435. http://doi.org/10.1080/18918131.2018.1547529

Scharpf, F. W. (1997). *Games real actors play: Actor-centered institutionalism in policy research*. Westview Press.

Sherman, J. F. (2021). The corporate general counsel who respects human rights. *Legal Ethics*, *24*(1), 49–72. http://doi.org/10.1080/1460728x.2021.1979731

Smit, L., Holly, G., McCorquodale, R., & Neely, S. (2020). Human rights due diligence in global supply chains: Evidence of corporate practices to inform a legal standard. *International Journal of Human Rights*, *25*(6), 945–073.

Spence, M. (1974). *Market signaling: Informational transfer in hiring and related screening processes*. Harvard University Press.

Sridhar, B. S., & Camburn, A. (1993). Stages of moral development of corporations. *Journal of Business Ethics*, *12*(9), 727–739.

Suchman, M. (1995). Managing legitimacy: Strategic and instituitonal approaches. *Academy of Management Review*, *20*(3), 571–610.

Suddaby, R., Bitiktine, A., & Haack, P. (2017). Legitimacy. *Academy of Management Annals*, *11*(1), 451–478.

Sun, P., Doh, J. P., Rajwani, T., & Siegel, D. (2021). Navigating cross-border institutional complexity: A review and assessment of multinational nonmarket strategy research. *Journal of International Business Studies*, *52*(9), 1818–1853. http://doi.org/10.1057/s41267-021-00438-x

Tan, A., Benni, D., & Liani, W. (2016). Determinants of corporate social responsibility disclosure and investor reaction. *International Journal of Economics and Financial Issues*, *6*(4), 11–17.

Thorne, L., Mahoney, L. S., & Manetti, G. (2014). Motivations for issuing standalone CSR reports: A survey of Canadian firms. *Accounting, Auditing & Accountability Journal*, *27*(4), 686–714.

Tschopp, D., & Huefner, R. (2015). Comparing the evolution of CSR reporting to that of financial reporting. *Journal of Business Ethics*, *127*(3), 565–577. http://doi.org/10.1007/s10551-014-2054-6

Villiers, C. (2023). A game of cat and mouse: Human rights protection and the problem of corporate law and power. *Leiden Journal of International Law*, *36*(2), 1–24. http://doi.org/10.1017/s0922156522000632

Wang, S. L., & Li, D. (2019). Responding to public disclosure of corporate social irresponsibility in host countries: Information control and ownership control. *Journal of International Business Studies*, *50*(8), 1283–1309. http://doi.org/10.1057/s41267-019-00224-w

Waniak-Michalak, H., Sapkauskiene, A., & Leitoniene, S. (2018). Do companies manipulate CSR information to retain legitimacy? *Inzinerine Ekonomika-Engineering Economics*, *29*(3), 352–360. http://doi.org/10.5755/j01.ee.29.3.18946

Wettstein, F., Giuliani, E., Santangelo, G. D., & Stahl, G. K. (2019). International business and human rights: A research agenda. *Journal of World Business*, *54*(1), 54–65.

Wilmot, S. (2001). Corporate moral responsibility: What can we infer from our understanding of organisations? *Journal of Business Ethics*, *30*(2), 161–169.

Young, S., & Marais, M. (2012). A multi-level perspective of CSR reporting: The implications of national institutions and industry risk characteristics. *Corporate Governance: An International Review*, *20*(5), 432–450. http://doi.org/10.1111/j.1467-8683.2012.00926.x

Zagelmeyer, S. (2019). MNEs and human rights: From responsibilities to rights to governance. *Critical Perspectives on International Business*, *16*(2), 117–135. http://doi.org/10.1108/cpoib-07-2018-0063

Zheng, Q., Luo, Y., & Maksimov, V. (2015). Achieving legitimacy through corporate social responsibility: The case of emerging economy firms. *Journal of World Business*, *50*(3), 389–403. http://doi.org/10.1016/j.jwb.2014.05.001

CHAPTER 8

THE EUROPEAN UNION'S CROSS-BORDER ADJUSTMENT MECHANISM AND ITS IMPLICATIONS FOR INTERNATIONAL BUSINESS

Viviana Pilato and Ari Van Assche

HEC Montréal, Canada

ABSTRACT

Carbon leakage – where multinational enterprises (MNEs) transfer carbon-intensive production activities to countries with laxer emissions constraints for cost purposes – is one of the main mechanisms through which international business (IB) contributes to climate change. This chapter discusses a new policy initiative called the Carbon Border Adjustment Mechanism (CBAM) that the European Union (EU) introduced in May 2023 to fight carbon leakage. The authors analyze the logic of CBAM and discuss how it will likely influence IB both in industries that are directly targeted by CBAM and related industries that will face spillover effects.

Keywords: Carbon Border Adjustment Mechanism; international business; carbon emissions; foreign direct investment; carbon leakage

Walking the Talk? MNEs Transitioning Towards a Sustainable World
Progress in International Business Research, Volume 18, 127–140

Published under exclusive licence by Emerald Publishing Limited
ISSN: 1745-8862/doi:10.1108/S1745-886220240000018008

1. INTRODUCTION

Climate change is a global public goods problem that is notoriously difficult to address without policy collaboration among countries (Khan & Munira, 2021). An often-heard concern in this regard is that companies can easily circumvent more stringent climate measures in a country by moving carbon-intensive production activities to a region with laxer stringent measures, leading to a phenomenon that is often labeled "carbon leakage" (Assogbavi & Dées, 2023; Böhringer et al., 2018; Grubb et al., 2022; Jakob, 2021; Mehling et al., 2019).

The issue of carbon leakage has become an especially heated topic in countries that have adopted environmental policies such as carbon pricing to curb greenhouse gas emissions (GHG) (Grøgaard et al., 2022). Carbon pricing policies can boost carbon leakage through two mechanisms. First, it provides a competitive advantage to foreign products relative to domestic goods because it increases production costs in carbon pricing implementing countries (Cosbey et al., 2019). Second, reducing carbon emissions with a carbon pricing policy means also reducing fossil fuels' demand in countries under carbon pricing regulation, incentivizing fossil fuel consumption through local global fuel prices in non-regulated countries and, therefore, their carbon emissions too (Cosbey et al., 2019; Jakob, 2021). Without appropriate countermeasures, there is therefore, the apprehension that carbon leakage can undermine the effectiveness of national policies aimed at limiting carbon emissions and can more generally decrease public support for climate policy.

The EU's decision in May 2023 to implement a CBAM is, in this context, a noteworthy policy initiative that attempts to disincentivize carbon leakage in the wake of Europe's decision to reach carbon neutrality by 2050. Complementing its current Emissions Trading System (ETS), CBAM imposes a tariff on the carbon emitted during the production of carbon-intensive goods that are imported into the EU: iron & steel, cement, fertilizers, hydrogen, aluminum, and electricity generation. The aim of this policy is to reduce the desire of companies to move carbon-intensive activities to non-EU countries with laxer environmental regulations.

In this chapter, we aim to explain this novel mechanism and discuss the potential implications for IB. We have organized the chapter into five sections. In Section 2, we will use recent updates to the OECD's Trade in Value Added (TiVA) data to analyze the geographic and industrial origin of foreign CO_2 emissions embodied in EU final demand. In Section 3, we will discuss the specifics of CBAM and how it is distinct from previous climate change policies. In Section 4, we will then analyze how CBAM may influence IB. We provide a set of concluding comments in Section 5.

2. CARBON EMISSIONS EMBODIED IN THE EU's TRADE

Recent updates to the OECD's TiVA database allow us to uncover interesting insights into the problem of carbon leakage for the EU. The database combines the OECD Inter-Country Input-Output dataset with statistics on CO_2 emissions

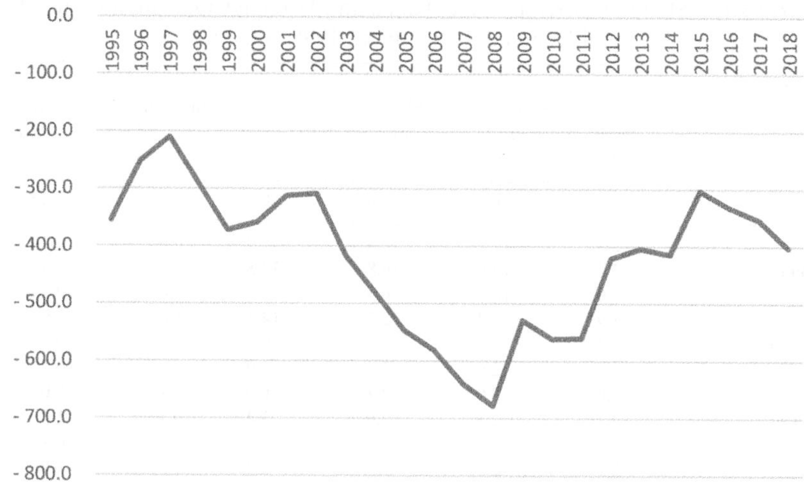

Fig. 8.1. Carbon Dioxide Emissions Embodied in the EU's Net Trade, 1995–2018. *Source*: Own elaboration, based on TiVA database in OECD website available at https://www.oecd.org/sti/ind/measuring-trade-in-value-added.htm.

from fuel combustion and other industry statistics to provide estimates of the amount and geographic distribution across economies of the CO_2 emissions that are embodied in international trade (Yamano & Guilhoto, 2020). In this section, we use the database to analyze where the foreign CO_2 emissions embodied in EU final demand originate from and which industries are primarily responsible.

Fig. 8.1 demonstrates that the EU, throughout the timeframe of the dataset (1995–2018), has always been a net importer of embodied carbon. In other words, the EU generally releases fewer carbon emissions for the production of their total exported goods and services than the amount generated by their trading partners for producing their total imported goods and services. The amount nonetheless fluctuates over time. The EU's net imports grew from 1995, reaching a peak in 2008, and have been gradually declining since.

Table 8.1 digs further into the origin of foreign CO_2 emissions that are embodied in EU final demands. As it is shown in columns 4 and 5 of Table 8.1, some of the world's largest economies are key contributors to foreign CO_2 emissions embodied in EU final demand. In 2018, China alone was responsible for almost 25% of foreign CO_2 emissions in EU final demand, which was more than doubling of the share compared to 1995. Russia and the USA rounded up the top three with 14% and 8% of foreign CO_2 emissions in EU final demand, respectively. While this finding should not be entirely surprising since large economies are generally responsible for larger shares of foreign value added embodied in EU final demand (see columns 2 and 3 in Table 8.1), it is noteworthy that emerging markets such as China, Russia, and India have a significantly larger foreign share in terms of CO_2 emissions than in terms of value added, while the opposite is the case for developed countries such as the USA, the UK, and Switzerland.

Table 8.1. Share of Foreign Value Added and Foreign CO_2 Emissions in EU Final Demand, by Country, 1998 and 2018.

Country	Share of Foreign Value Added in EU Final Demand		Share of Foreign CO_2 Emissions in EU Final Demand		Ratio of Foreign CO_2 Emissions Versus Foreign Value Added in EU Final Demand	
	1998	2018	1998	2018	1998	2018
USA	22.3	17.2	12.8	7.7	0.6	0.4
China	**3.3**	**13.6**	**10.8**	**24.5**	**3.3**	**1.8**
UK	18.0	10.6	6.5	3.3	0.4	0.3
Russia	**3.9**	**6.9**	**22.7**	**14.1**	**5.8**	**2.0**
Switzerland	6.4	5.0	1.1	0.7	0.2	0.1
Japan	8.5	2.9	2.6	2.0	0.3	0.7
India	**1.0**	**3.0**	**2.1**	**6.0**	**2.1**	**2.0**
Turkey	**2.7**	**2.8**	**2.0**	**3.4**	**0.7**	**1.2**
Norway	2.5	2.8	1.7	1.1	0.7	0.4
Korea	**2.1**	**2.2**	**3.1**	**2.2**	**1.5**	**1.0**
Saudi Arabia	1.2	1.6	0.4	1.2	0.3	0.8
Brazil	1.7	1.5	1.0	1.2	0.6	0.8
Canada	**1.7**	**1.4**	**1.8**	**1.5**	**1.1**	**1.1**
Singapore	0.9	1.3	0.8	1.4	0.9	1.1
Kazakhstan	**0.3**	**1.1**	**1.2**	**1.8**	**4.0**	**1.6**
Thailand	**1.0**	**1.0**	**1.4**	**1.4**	**1.4**	**1.4**
Taiwan	**1.8**	**1.0**	**1.4**	**1.5**	**0.8**	**1.5**
Rest of the world	20.7	24.1	26.6	25.0	1.3	1.0
Total	100.0	100.0	100.0	100.0		

Source: Own elaboration, based on TiVA database in OECD website available at https://www.oecd.org/sti/ind/measuring-trade-in-value-added.htm.

Columns 6 and 7 of Table 8.1 demonstrate this latter trend more clearly by showing the ratio of foreign CO_2 emissions versus foreign value added in EU final demand for the years 1995 and 2018. It clearly illustrates the stylized fact that the EU generally imports goods and services with a higher CO_2 emission intensity relative to value added from emerging markets. In contrast, imports into the EU from developed countries generally have a lower CO_2 emission intensity relative to value added.

Table 8.2 finally demonstrates the contribution of different industries to the foreign CO_2 emissions embodied in EU final demand. Column 5 of Table 8.2 shows that in 2018 three industries were responsible for more than half of foreign CO_2 emissions embodied in EU final demand: agriculture, hunting, forestry (29%); electricity, gas, steam, air conditioning supply (16%); Basic metals (8%). Column 7 of Table 8.2 shows that these three industries are also the sectors with the highest CO_2 emissions relative to value added, suggesting that carbon leakage may well be the largest in these industries. Given these numbers, it is not surprising that the EU's CBAM primarily targets the six industries iron & steel, cement, fertilizers, aluminum, hydrogen, and electricity generation.

Table 8.2. Share of Foreign Value Added and Foreign CO_2 Emissions in EU Final Demand, by Industry, 1998 and 2018.

Industry	Share of Foreign Value Added in EU Final Demand		Share of Foreign CO_2 Emissions in EU Final Demand		Ratio of Foreign CO_2 Emissions and Foreign Value Added in EU Final Demand	
	1998	2018	1998	2018	1998	2018
Wholesale and retail trade	14.1	12.8	1.7	1.4	0.1	0.1
Mining and quarrying energy	7.9	12.4	5.9	5.9	0.7	0.5
Financial and insurance activities	6.6	6.9	0.4	0.4	0.1	0.1
Professional, scientific, and technical activities	5.7	6.7	0.3	0.5	0.1	0.1
Administrative and support services	3.6	5.0	0.4	0.7	0.1	0.1
Computer, electronic, and optical products	6.0	3.7	0.7	0.4	0.1	0.1
Agriculture, hunting, and forestry	**3.9**	**3.6**	**29.1**	**29.1**	**7.4**	**8.1**
IT and other information services	1.9	3.5	0.1	0.2	0.1	0.1
Land transport and transport via pipelines	**3.6**	**3.3**	**5.1**	**5.0**	**1.4**	**1.5**
Chemical and chemical products	**3.1**	**2.6**	**3.9**	**3.1**	**1.3**	**1.2**
Textiles, wearing apparel, and leather	2.6	2.4	1.2	0.6	0.4	0.2
Real estate activities	2.2	2.2	0.1	0.1	0.0	0.0
Publishing, audiovisual, and broadcasting	2.0	2.2	0.1	0.1	0.1	0.1
Basic metals	**2.3**	**2.2**	**5.7**	**7.8**	**2.5**	**3.6**
Coke and refined petroleum products	**1.5**	**2.0**	**4.0**	**3.9**	**2.8**	**1.9**
Food products, beverages, and tobacco	2.0	1.8	0.7	0.6	0.3	0.3
Machinery and equipment n.e.c.	2.8	1.8	0.4	0.2	0.1	0.1
Motor vehicles, trailers, and semi-trailers	2.3	1.6	0.2	0.2	0.1	0.1
Warehousing and transportation support activities	1.4	1.6	0.3	0.6	0.2	0.4
Pharmaceuticals, medicinal and chemical products	1.2	1.5	0.6	0.8	0.5	0.6
Electricity, gas, steam, and air conditioning supply	**1.6**	**1.5**	**12.7**	**15.6**	**7.9**	**10.7**
Manufacturing n.e.c.	**1.5**	**1.5**	**1.7**	**2.4**	**1.1**	**1.6**
Other sectors	**20.4**	**17.1**	**24.8**	**20.6**	**1.2**	**1.2**
Total	**100.0**	**100.0**	**100.0**	**100.0**		

Source: Own elaboration, based on TiVA database in OECD website available at https://www.oecd.org/sti/ind/measuring-trade-in-value-added.htm.
Note: Bold indicates the industries where the ratio of foreign CO_2 emissions and foreign value added in EU final demand is larger than 1 in 2018.

3. CARBON BORDER ADJUSTMENT MECHANISM

In order to realize its 2019 European Green Deal (European Commission, 2019), the European Commission in July 2021 unveiled a bundle of proposals under the name of the "Fit for 55" package (European Commission, 2021a) to decrease net GHG emissions by at least 55% by 2030 in comparison to 1990. CBAM featured among these proposals (European Commission, 2021b) with the aim of phasing out the EU's ETS gradually from 2026 and reducing carbon leakage (Assogbavi & Dées, 2023; Grubb et al., 2022).

A carbon border adjustment is "a trade measure designed to level the playing field between domestic producers facing costly climate policy and foreign producers with no or little constraint on their GHG emissions" (Ismer & Neuhoff, 2007; Monjon & Quirion, 2011, p. 1958). Specifically, it is "a charge on the carbon content of imported products that is intended to ensure treatment equivalent to domestic carbon pricing, potentially combined with rebates for the carbon content of exports" (Keen et al., 2022, p. 210). Therefore, a carbon border adjustment can take the form of (1) a mere carbon tax, or (2) it can require importers to buy allowances from an allowance pool or an ETS (Keen et al., 2022; Mehling et al., 2019). The latter considers two options: the cap-and-trade system, as the EU ETS system, where the ETS sets a cap on the amount of GHG emissions that companies that are covered by the system can emit, and companies receive or buy emission allowances, which they can also trade ("Emissions Cap and Allowances", n.d.; "EU Emissions Trading System (EU ETS)", n.d.; European Commission, 2015); and the output-based pricing system (Fischer & Fox, 2012; Monjon & Quirion, 2011), which is similar to the ETS system, but while in the ETS system, the cap is reduced every year (European Commission, 2022); in the output-based pricing system, the amount of emissions can increase if there is an increase in the production ("Nova Scotia Transitions to New Carbon Pricing System", 2023).

While the EU ETS system put a cap on the amount of emissions allowed, creating uncertainty on the price traded in the ETS trading system (Dechezleprêtre et al., 2023), CBAM aims to set a "fair price" on the amount of embedded carbon emissions produced in the fabrication of highly carbon content goods that are imported in the EU ("Carbon Border Adjustment Mechanism", n.d.). CBAM will allow equalizing the competition within manufacturers subject to EU ETS and foreign producers not covered by a similar system (Morgado Simões et al., 2023).

In a similar way to the EU ETS, also CBAM will use certificates for embedded carbon emissions of goods imported in the EU (European Commission, 2023). Starting in 2026, when CBAM will be fully in force, EU importers will be required to purchase CBAM certificates equal to the carbon price paid if goods would have been manufactured with the EU's carbon pricing system, therefore equal to the weekly EU carbon price (European Commission, 2023; Morgado Simões et al., 2023). The price of the CBAM certificates depends on "the weekly average auction price of the EU ETS allowances expresse d in €/tonne of CO_2 equivalents emitted" (European Commission, 2023, p. 7). In addition, there is the possibility of a deduction from CBAM if the non-EU manufacturers have already purchased a carbon price related to the embedded carbon emissions emitted for the imported goods' production in another country (European Commission, 2023). This deduction allows products not to undertake a double price, and this, in addition to the price of the CBAM certificates at the weekly EU carbon price and the recognition of an inferior CBAM payment for firms exporting in the EU that have already undertaken decarbonizing efforts, allows CBAM to respect the "no less favourable treatment" for imported products (European Commission, 2023). CBAM has been designed to be compliant with the World Trade Organization (WTO) rules, and it finally became a regulation on May 10, 2023; and the CBAM Regulation will be in force starting from 2026, with a transitional period from October 1, 2023 to December 31, 2025. The transitional period will allow for introducing CBAM gradually and for applying it to imports of carbon-intensive goods at higher risk of carbon leakage in six emission-intensive industries: electricity, iron & steel, cement, aluminum, fertilizers, and hydrogen (European Commission, 2023). In the remainder of this chapter, we will call them CBAM industries. Importers from these sectors will have to report the direct embedded GHG emissions and also the indirect emissions for a subgroup of products covered (such as fertilizers and cement). During the transitional period, only direct embedded GHG emissions (scope 1) will be used for the CBAM charge, while the inclusion also of indirect emissions (scope 2 and scope 3) and the possibility of extending CBAM to other goods will be re-evaluated in the future (European Commission, 2023).

CBAM applies to goods imported in the EU from any non-EU countries, with the exemption of the European Economic Area (EEA) and Switzerland because those countries are already covered by the EU ETS system or by an ETS linked to the EU ETS (European Commission, 2023; Morgado Simões et al., 2023).

CBAM will progressively replace the EU ETS system. The EU ETS system works as a cap-and-trade, meaning that it puts a cap on the maximum amount of GHG emissions, and importers must buy allowances to import their goods, and there are some free allowances given in order to reduce carbon leakage (Dechezleprêtre et al., 2023; European Commission, 2023). This system has been implemented in different phases since 2005, but notwithstanding its effort, several challenges still remain. While the cap-and-trade system allows to decide the maximum amount of emissions permitted, there is no limit for the EU carbon price, which unfortunately has raised quite drastically from 2020 to 2023, reaching now 100€per tonne of CO_2 (Morgado Simões et al., 2023). Moreover, each year, the amount of free allowances is reduced (European Commission, 2023), increasing the chances of carbon leakage from importers and therefore requiring a new mechanism, such as CBAM.

As CBAM works as well with certificates, during the transitional phase, importers covered by the CBAM regulation are required to disclose their embedded carbon emissions, but also the nitrous oxide (N_2O) and the perfluorocarbons in some cases (for some fertilizers and aluminum goods), on a quarterly basis (European Commission, 2023). The CBAM report should include certain information, among which the actual amount of embedded emissions, the total amount of indirect emissions, the quantity amount of goods, and the type

(Morgado Simões et al., 2023). The importers need only to submit the CBAM report without paying the CBAM financial adjustment during the transitional period (European Commission, 2023), allowing CBAM to be gradually adopted and to phase out the ETS system from 2026 (Morgado Simões et al., 2023). CBAM financial adjustment will be in effect in 2026, and from 2026 to 2034, CBAM will be enforced to "the proportion of emissions that does not benefit from free allowances under the EU ETS" (European Commission, 2023, p. 2) to allow the complete transition between the two systems. Starting on October 1, 2023, the CBAM report will be submitted through the CBAM Transitional Registry (European Commission, 2023).

While narrowing the implementation of CBAM only to six industries – iron & steel, cement, fertilizers, aluminum, hydrogen, and electricity generation – was considered prudent to remain in compliance with WTO rules and to reduce the administrative complexity of implementing the border adjustment mechanism (Cosbey et al., 2019), CBAM aims to reduce carbon leakage (Korpar et al., 2022; Ladly, 2012; Mehling et al., 2019), but also incentivize third countries to implement carbon pricing policies (Cosbey et al., 2019; Keen et al., 2022; Ladly, 2012; Mehling et al., 2019).

4. IMPLICATIONS FOR IB

In the previous section, we have described CBAM as a policy framework that works in parallel with the EU ETS by applying a carbon price to imported goods that is equivalent to the carbon price applied to goods manufactured in the EU, through the use of CBAM certificates. The mechanism will initially apply only to six CBAM industries: iron & steel, fertilizer, aluminum, electricity production, hydrogen, and cement sectors. In the future, the EU has planning to expand CBAM to other goods, such as cars and organic chemicals.

In this section, we focus on the various ways in which the implementation of CBAM will influence IB, which we organize by types of industries.

4.1. IB in CBAM Industries

CBAM increases the competitiveness of EU-based production activities (domestically- or foreign-owned) in CBAM industries relative to foreign-based production activities in unregulated regions outside of the EU (Keen et al., 2022; Ladly, 2012). This is because CBAM requires foreign-based firms in CBAM industries to pay a carbon tax in proportion to the emissions embedded in their imports. The competitive advantage that CBAM generates is especially large relative to emerging economies such as China and Russia, which have a comparative advantage in CBAM products.

The increased competitiveness of EU-based production activities affects IB in CBAM industries in several ways (see Table 8.3). First, it reduces EU imports and dampens EU outward foreign direct investment (FDI) in CBAM industries as consumers shift away from the relatively more expensive foreign-based

Table 8.3. Impact of CBAM on EU Trade and FDI.

Industry	Effect of CBAM
CBAM industry	
Exports	?
Imports	↓
Outward FDI	↓
Inward FDI	↑
Downstream sector of CBAM industry	
Exports	↓
Imports	↑
Outward FDI	↑
Inward FDI	↓

Source: Own elaboration.

products toward goods produced in the EU (Dechezleprêtre et al., 2023; Mehling et al., 2019). Second, CBAM increases EU inward FDI in CBAM industries as firms attempt to "tariff jump" over the newly imposed carbon taxes to improve their access to consumers in the EU common market. Third, it has an unclear impact on EU exports in CBAM industries as the increase in competitiveness of EU-based production activities only applies if the raw materials are sold in the EU common market.

In line with its goal, the combined effects of CBAM on IB reduce carbon leakage (Grubb et al., 2022; Keen et al., 2022; Korpar et al., 2022) from the EU to unregulated non-EU regions in CBAM industries. Indeed, the reduction in imports and increase in net FDI into the EU implies that companies are reacting to the policy framework by moving carbon-intensive production activities away from non-regulated countries (especially emerging markets) to the EU (Jakob, 2021).

Finally, the exemption of the EEA and Switzerland from paying carbon tariffs might incentivize companies to use these countries as transshipment corridors. In this case, non-EU operating in unregulated companies could export their CBAM goods to EEA or Switzerland without having to pay a carbon tariff and then re-export it to the EU where CBAM would not apply (Kortum & Weisbach, 2017). There are currently few rules in place to avoid such transshipments.

4.2. IB in Downstream Sectors from CBAM Industries

Discussions on CBAM often ignore that the policy framework has opposing implications for the competitiveness of EU-based production activities in downstream sectors, such as manufacturing, that heavily rely on CBAM inputs. That is, CBAM reduces the competitiveness of EU-based production activities in the sectors directly downstream from CBAM industries, incentivizing companies to move production activities in these downstream industries outside of the EU to avoid having to pay carbon taxes.

To see this, it is important to understand that CBAM only increases the prices of CBAM raw materials when they are directly imported into the EU and face a carbon tax. Carbon taxes do not apply to the EU imports of downstream industries that have relied on non-EU CBAM raw materials for their production. This is important because it implies that downstream sectors outside of the EU that intensively use CBAM raw materials gain competitiveness relative to their EU counterpart in ways that can substantially alter IB in these sectors. It may increase EU imports and EU outward FDI in downstream sectors from CBAM industries (see Table 8.3). It may also reduce EU inward FDI and exports in these same sectors.

Overall, these mechanisms may counter the EU's efforts to tackle carbon leakage as it provides an opportunity for companies to circumvent the carbon tax through the offshoring of downstream activities, generating significant global environmental impacts. It is for this reason that many experts call for CBAM to be wider in scope than it currently is, including many of the downstream sectors that intensively rely on inputs from CBAM industries (Keen et al., 2022; Mehling et al., 2019).

4.3. IB in Other Sectors

It is finally important to consider the indirect impact of CBAM on IB in other sectors. In our discussion above, we have not taken into account that a variety of non-EU countries may take actions in support of or against CBAM. We, in this section, identify a number of ways how this may happen.

First, emerging markets, such as China or Russia, which have a significant competitive advantage in CBAM industries, may well retaliate to the policy framework due to the negative effects that they generate on local production and employment (Cosbey et al., 2019; Erixon, 2021). Indeed, it is not unimaginable that these countries may impose tariffs on EU exports in downstream sectors from CBAM industries to reduce the hurt that CBAM imposes. At a time when the EU is attempting to revive its manufacturing activities, this type of retaliation could create further rifts between Western and Eastern EU countries as the latter has developed a significant comparative advantage in manufacturing activities.

Second, many other emerging economies are already calling out against CBAM due to its lack of consideration of the level of economic development of its trading partners when imposing carbon taxes. As we have seen, CBAM will have a disproportionately negative impact on developing countries that rely on trade to foster economic development (Zhong & Pei, 2022). For these countries, it is hard to stomach that the EU's climate change actions occur on the back of developing countries. These tensions will likely spark greater opposition to better global coordination of climate policies between the EU and emerging economies (Zhong & Pei, 2022).

Third, there is a general concern about the WTO compatibility of CBAM despite efforts by the EU to counter these critiques (Jakob, 2021; Ladly, 2012; Mehling et al., 2019). Indeed, CBAM may well be inconsistent with the WTO's rule of most favored nations, which requires that any advantage granted to the imported products of one WTO member must be accorded immediately and unconditionally to like products originating from all other WTO members (Cosbey et al., 2019).

Fourth, several developed countries have paid careful attention to CBAM to investigate if it is useful for them to adopt a similar policy framework. The formation of a "climate club" (Perdana & Vielle, 2023), a region of key importers surrounded by a carbon price at their borders, could, in this respect, further accelerate global climate mitigation (Perdana & Vielle, 2023) but also speed up decoupling between different geopolitical regions of the world and increase carbon pricing policies' credibility at international level (Keen et al., 2022)

Finally, prior studies suggest a possible combination of pricing policies with other policies, such as non-pricing instruments (Grubb et al., 2022), to strengthen the efforts against climate change (Khan & Johansson, 2022; Sumner et al., 2011), and as a complement to carbon pricing policies (Asian Development Bank, 2023; Black et al., 2022; Grubb et al., 2022; Santos, 2022).

Performance standards and technology standards are some non-pricing measures. A performance standard puts a threshold on the amount of emissions, like the case of the Ontario Emissions Performance Standards Program (Asian Development Bank, 2023). A technology standard requires that organizations should use a specific technology, as fuel switching (Asian Development Bank, 2023). Other standards include product standards (Grubb et al., 2022) which refer to particular specifications of the product, and process standards, which concern the process through which goods are created (Asian Development Bank, 2023). In this case, renewable portfolio standards (RPS) require that the production of electric power should come from renewable sources in a specific percentage (Black et al., 2022). Other non-pricing measures include labeling, the requirement for certain products to display their energy performance, and information campaigns to sensibilize and inform consumers but also companies, such as policies to educate about the construction of green buildings (Asian Development Bank, 2023). Finally, also financial instruments, such as the green bonds or other policies related to sustainable finance, are part of non-pricing instruments (Asian Development Bank, 2023).

5. CONCLUSION

This chapter has used a combination of data analysis and theoretical reflection to investigate the logic of the EU's recent CBAM policy and what its likely implications will be on European trade and FDI. We have pointed out that CBAM is in many ways unique since it is one of the first major efforts to reduce carbon leakage and associated negative externalities (Montiel et al., 2021). The main finding of our analysis is that CBAM will likely increase the EU's competitiveness and reduce carbon leakage in the six CBAM industries where carbon tariffs will apply to EU imports. At the same time, and discussed less in the literature, it will likely decrease the EU's competitiveness and increase carbon leakage in the downstream sectors that intensively use CBAM raw materials in their production. Finally, we have pointed out that significant uncertainty remains relating to the reactions of non-EU countries to CBAM. The policy hurts many emerging markets, which may react by putting in place retaliatory policies, but it may convince other developed countries to join forces and implement a similar type of environmental CBAM. In a world that is already rife with geopolitical tensions, these forces cannot be ignored.

Some challenges related to the implementation of CBAM still remain. For example, the calculation of embedded carbon emission is still difficult to achieve due to sometimes complex global value chains (GVCs), but also due to the difficulty of calculating indirect emissions. Future research could compare the CBAM effect on direct and indirect emissions and evaluate which one is more effective in reducing carbon leakage. Future studies could also look at the reconfiguration of GVCs for downstream sectors.

The design of CBAM is not easy, and different components should be taken into account, such as which countries, which products, and which industries but also how emissions are calculated and how the CBAM revenues are spent (Cosbey et al., 2019; Jakob, 2021; Mehling et al., 2019). Future research could look further into understanding if there could be different "stepwise implementation" strategies for carbon pricing (Khan & Johansson, 2022) or if the broader product scope and industries, in addition to expanding countries under CBAM regulation, could strengthen the CBAM effect (Keen et al., 2022).

A next step could also be to look at the effect of CBAM on economic performance in downstream and upstream CBAM sectors to understand which sectors have a higher impact on economic performance. In response to the effect of developing and emerging countries, future research could look into exempting mechanisms from CBAM in those countries to increase their competitiveness (Keen et al., 2022).

REFERENCES

Asian Development Bank. (2023). *Promoting climate action through non pricing policy measures* [Paper presentation]. Presented at the 2nd Sustainable Finance Working Group meeting. Asian Development Bank. https://g20sfwg.org/wp-content/uploads/2023/03/Promoting-Climate-Action-Through-Non-Pricing-Policy-Measures-ADB.pdf

Assogbavi, K. K. E., & Dées, S. (2023). Environmental policy and the CO_2 emissions embodied in international trade. *Environmental and Resource Economics*, *84*(2), 507–527. https://doi.org/10.1007/s10640-022-00734-6

Black, S., Minnett, D., Parry, I., Roaf, J., & Zhunussova, K. (2022). A framework for comparing climate mitigation policies across countries. *IMF Working Papers*, *2022*(254), 1. https://doi.org/10.5089/9798400228285.001

Böhringer, C., Carbone, J. C., & Rutherford, T. F. (2018). Embodied carbon tariffs. *The Scandinavian Journal of Economics*, *120*(1), 183–210. https://doi.org/10.1111/sjoe.12211

Carbon Border Adjustment Mechanism. Retrieved July 18, 2023, from https://taxation-customs.ec.europa.eu/carbon-border-adjustment-mechanism_en

Cosbey, A., Droege, S., Fischer, C., & Munnings, C. (2019). Developing guidance for implementing border carbon adjustments: Lessons, cautions, and research needs from the literature. *Review of Environmental Economics and Policy*, *13*(1), 3–22. https://doi.org/10.1093/reep/rey020

Dechezleprêtre, A., Nachtigall, D., & Venmans, F. (2023). The joint impact of the European Union Emissions Trading System on carbon emissions and economic performance. *Journal of Environmental Economics and Management*, *118*, 102758. https://doi.org/10.1016/j.jeem.2022.102758

Emissions Cap and Allowances. Retrieved July 17, 2023, from https://climate.ec.europa.eu/eu-action/eu-emissions-trading-system-eu-ets/emissions-cap-and-allowances_en

Erixon, F. (2021). *The good, the bad and the ugly: Taking stock of Europe's new trade policy strategy* [Policy Brief]. ECIPE, European Centre for International Political Economy. https://ecipe.org/publications/europes-new-trade-policy-strategy/

EU Emissions Trading System (EU ETS). Retrieved July 17, 2023, from https://climate.ec.europa.eu/eu-action/eu-emissions-trading-system-eu-ets_en

European Commission. (2015). *EU ETS handbook*. Retrieved from https://climate.ec.europa.eu/system/files/2017-03/ets_handbook_en.pdf

European Commission. (2019). The European Green Deal. Communication from the Commission. Retrieved from https://eur-lex.europa.eu/resource.html?uri=cellar:b828d165-1c22-11ea-8c1f-01aa75ed71a1.0002.02/DOC_1&format=PDF

European Commission. (2021a). *'Fit for 55': Delivering the EU's 2030 climate target on the way to climate neutrality*. Communication from the Commission. Retrieved from: https://eur-lex.europa.eu/legal-content/EN/TXT/HTML/?uri=CELEX%3A52021DC0550#:~:text=The%20Fit%20for%2055%20Package%20aims%20to%20deliver%20the%20EU's,drive%20stronger%20overall%20emissions%20reductions

European Commission. (2021b). *Proposal for a regulation of the European Parliament and of the Council establishing a carbon border adjustment mechanism*. Retrieved from https://eur-lex.europa.eu/resource.html?uri=cellar:a95a4441-e558-11eb-a1a5-01aa75ed71a1.0001.02/DOC_1&format=PDF

European Commission. (2022). *Report on the functioning of the European carbon market in 2021 pursuant to Articles 10(5) and 21(2) of Directive 2003/87/EC* (as amended by Directive 2009/29/EC and Directive (EU) 2018/410). Retrieved from https://eur-lex.europa.eu/legal-content/EN/TXT/PDF/?uri=CELEX:52022DC0516

European Commission. (2023). *Questions and answers: Carbon Border Adjustment Mechanism (CBAM)*. European Commission, Memo. https://taxation-customs.ec.europa.eu/system/files/2023-07/20230714%20Q%26A%20CBAM_0.pdf

Fischer, C., & Fox, A. K. (2012). Comparing policies to combat emissions leakage: Border carbon adjustments versus rebates. *Journal of Environmental Economics and Management*, *64*(2), 199–216. https://doi.org/10.1016/j.jeem.2012.01.005

Grøgaard, B., Sartor, M. A., & Rademaker, L. (2022). What merits greater scholarly attention in international business? *Journal of International Business Studies*, *53*, 1508–1518. https://doi.org/10.1057/s41267-022-00539-1

Grubb, M., Jordan, N. D., Hertwich, E., Neuhoff, K., Das, K., Bandyopadhyay, K. R., Van Asselt, H., Sato, M., Wang, R., Pizer, W. A., & Oh, H. (2022). Carbon leakage, consumption, and trade. *Annual Review of Environment and Resources*, *47*(1). https://doi.org/10.1146/annurev-environ-120820-053625

Ismer, R., & Neuhoff, K. (2007). Border tax adjustment: A feasible way to support stringent emission trading. *European Journal of Law and Economics*, *24*(2), 137–164. https://doi.org/10.1007/s10657-007-9032-8

Jakob, M. (2021). Climate policy and international trade – A critical appraisal of the literature. *Energy Policy*, *156*, 112399. https://doi.org/10.1016/j.enpol.2021.112399

Keen, M., Parry, I., & Roaf, J. (2022). Border carbon adjustments: Rationale, design and impact. *Fiscal Studies*, *43*(3), 209–234. https://doi.org/10.1111/1475-5890.12307

Khan, J., & Johansson, B. (2022). Adoption, implementation and design of carbon pricing policy instruments. *Energy Strategy Reviews*, *40*, 100801. https://doi.org/10.1016/j.esr.2022.100801

Khan, M. R., & Munira, S. (2021). Climate change adaptation as a global public good: Implications for financing. *Climatic Change*, *167*(3), 50. https://doi.org/10.1007/s10584-021-03195-w

Korpar, N., Larch, M., & Stöllinger, R. (2022). The European carbon border adjustment mechanism: A small step in the right direction. *International Economics and Economic Policy*, *20*, 95–138. https://doi.org/10.1007/s10368-022-00550-9

Kortum, S., & Weisbach, D. (2017). The design of border adjustments for carbon prices. *National Tax Journal*, *70*(2), 421–446. https://doi.org/10.17310/ntj.2017.2.07

Ladly, S. D. (2012). Border carbon adjustments, WTO-law and the principle of common but differentiated responsibilities. *International Environmental Agreements: Politics, Law and Economics*, *12*(1), 63–84. https://doi.org/10.1007/s10784-011-9153-y

Mehling, M. A., Van Asselt, H., Das, K., Droege, S., & Verkuijl, C. (2019). Designing border carbon adjustments for enhanced climate action. *American Journal of International Law, 113*(3), 433–481. https://doi.org/10.1017/ajil.2019.22

Monjon, S., & Quirion, P. (2011). Addressing leakage in the EU ETS: Border adjustment or output-based allocation? *Ecological Economics, 70*(11), 1957–1971. https://doi.org/10.1016/j.ecolecon.2011.04.020

Montiel, I., Cuervo-Cazurra, A., Park, J., Antolín-López, R., & Husted, B. W. (2021). Implementing the United Nations' sustainable development goals in international business. *Journal of International Business Studies, 52*(5), 999–1030. https://doi.org/10.1057/s41267-021-00445-y

Morgado Simões, H., Titievskaia, J., & Dobreva, A. (2023). *EU carbon border adjustment mechanism: Implications for climate and competitiveness* [Briefing EU Legislation in Progress]. European Union. https://www.europarl.europa.eu/thinktank/en/document/EPRS_BRI(2022)698889

Nova Scotia transitions to new carbon pricing system. (2023, February 10). Retrieved July 17, 2023, from https://icapcarbonaction.com/en/news/nova-scotia-transitions-new-carbon-pricing-system

Perdana, S., & Vielle, M. (2023). Carbon border adjustment mechanism in the transition to net-zero emissions: Collective implementation and distributional impacts. *Environmental Economics and Policy Studies, 25*, 299–329. https://doi.org/10.1007/s10018-023-00361-5

Santos, G. (2022). Climate change policy and carbon pricing. *Energy Policy, 168*, 112985. https://doi.org/10.1016/j.enpol.2022.112985

Sumner, J., Bird, L., & Dobos, H. (2011). Carbon taxes: A review of experience and policy design considerations. *Climate Policy, 11*(2), 922–943. https://doi.org/10.3763/cpol.2010.0093

Yamano, N., & Guilhoto, J. (2020). *CO_2 emissions embodied in international trade and domestic final demand: Methodology and results using the OECD Inter-Country Input–Output database*. [Working Paper, 2020/11]. OECD Science, Technology and Industry. https://doi.org/10.1787/8f2963b8-en

Zhong, J., & Pei, J. (2022). Beggar thy neighbor? On the competitiveness and welfare impacts of the EU's proposed carbon border adjustment mechanism. *Energy Policy, 162*, 112802. https://doi.org/10.1016/j.enpol.2022.112802

CHAPTER 9

HOW CAN LARGE INTERNATIONAL FINANCIAL FIRMS 'GO GREEN' AND 'WALK THE TALK'?

John Holland

The Adam Smith Business School, University of Glasgow, UK

ABSTRACT

How can large international financial firms go green in authentic ways? What enhances 'Net Zero action'? Changes in global banks, fund managers, and insurance firms are at the heart of green finance. External change pressures – combined with problematic firm predispositions – exacerbate barriers to change and promote scepticism about authentic Net Zero change. Field research reveals main elements, connections, and interactions of this question by considering financial firms as complex socio-technical systems (Mitleton-Kelly, 2003). An interdisciplinary/holistic narrative approach (De Bakker et al., 2019) is adopted to design a conceptual framework that can support a green 'behavioural theory of the financial firm' (green BTFF). The BTFF presents an international version (Peng, 2001) of the resource-based view (RBV) of the firm (Barney, 1991; Hart, 1995; Teece et al., 1997).

The approach of this chapter is aimed at closing knowledge gaps and realign values in financial markets and society. By raising awareness about organised hypocrisy and facades (Brunsson, 1993; Cho et al., 2015; Schoeneborn et al., 2020) in financial firms the chapter aims at overcoming the gap between

Walking the Talk? MNEs Transitioning Towards a Sustainable World
Progress in International Business Research, Volume 18, 141–165

Published under exclusive licence by Emerald Publishing Limited
ISSN: 1745-8862/doi:10.1108/S1745-886220240000018009

'talking' and 'walking' in the financial sector. The chapter defines testable firm-level hypotheses for 'Green Finance' (Poterba, 2021) as well as – by implication – tests for 'greenwashing'.

Keywords: International financial firms; climate change; complexity, knowledge; hypocrisy

1. INTRODUCTION – CHANGE, MOTIVATION, GAPS, AND PROBLEM

The research question of this chapter is: how can large international financial firms go green in authentic ways? Climate change and global and transnational demands for Net Zero change have had serious impact on the global financial system and financial firms – including banks such NatWest/RBS, Lloyds, and Wells Fargo, fund managers such Schroders, Baillie Gifford, and Blackrock, and insurance firms such as Aviva. Significant physical and transitional risks arise from climate change (IPCC, 2021), which create intensive world-wide institutional and civil society pressures for a change to green finance (EBA, 2021; GFANZ, 2021; UNEP FI, 2022). Glasgow COP26 (2021) noted that going 'Net Zero' involves ongoing reduction of greenhouse gas (GHG) emissions and ensuring that remaining emissions are balanced by implementing methods to absorb carbon dioxide from the atmosphere.

The motivation for the chapter originates in the central role attributed to financial firms in Net Zero change (EBA, 2021; GFANZ, 2021; UNEP FI, 2022). Green financial firms leverage the financial means to protect and restore nature, biodiversity, and human living systems (Dasgupta, 2021), thus building resilience into the global financial and socio-economic systems. By making financial firms and their Net Zero changes comprehensible, visible, and accountable to a wider community of stakeholders, science can contribute to climate change, but cannot remove problems such as organised hypocrisy and facades (Brunsson, 1993).

Changes in climate (IPCC, 2021) and global social systems (Dashwood, 2020, chapter 8; Falkner, 2021) create major problems for *financial firms* to organise knowledge, identify values, and define purpose. Financial firms face problems of understanding and responding to historic, current, and expected trends in climate change. They contribute to organised hypocrisy and facades by firms. Financial firms have formed global sector alliances or coalitions such as GFANZ (2021) and PRB (2019) to respond to problems. These 'communities of practice' (Lave & Wenger, 1991) have ameliorated knowledge and lowered value gaps in sectors such as banking, insurance, and fund management.

Nevertheless, practitioners still face knowledge and value problems. The rate and contested nature of change, combined with problematic internal dispositions makes it difficult to develop a holistic view in large, multifaceted international financial firms, with global operations (Larsen, 2017; Mathews, 2021). Larsen (2021) argued that bankers may not have the 'head space' or cognitive capabilities to deal with this complexity. They may struggle to deal with climate changes as well as nature-based risks and corporate social responsibility (CSR) issues. Cases occur where firms employ 'organised hypocrisy' and 'facades' to create ambiguity, and to deflect and absorb criticism about growing gaps between Net Zero change 'Talk and Walk' (Cho et al., 2015; Schoeneborn et al., 2020).

Knowledge gaps arise in practice because of problems of learning and adaption during rapid and complex change (Larsen, 2021). Knowledge gaps arise in academia because of the fragmented nature of literature, theory, and hypothesis testing concerning financial firms and their change processes. These arise in part because conceptual connections between finance theory and other social and management sciences remains very limited (Holland, 2019). Value gaps exist in both practice and academic domains because of the central role of traditional finance theory in explaining and influencing finance phenomena and decision making. This tradition concentrates on increases in financial value and 'crowds out' prosocial behaviour (Bénabou & Tirole, 2006) concerning Net Zero aims. There is very limited research done by traditional finance academics on matters of climate change (Diaz-Rainey et al., 2017; Hong & Scheinkman, 2020). These gaps play a role in financial firm use of 'organised hypocrisy' and 'facades'.

Overcoming the extant fragmented, partial approach requires a holistic response which in this chapter boils down to a 'behavioural theory of the financial firm' or BTFF (Holland, 2016, 2019) to also include issues of Net Zero oriented change in financial firms. Efforts to develop such an approach can be witnessed with international financial firms such as NatWest (Rose, 2020) and Schroders (Harrison, 2020). They try to adapt their traditional financial purpose, function, philosophy, and values, to change financial decision activities, products, and financial intermediation (Buckle & Beccali, 2011), to reflect climate change risks and sustainability aims (GFANZ, 2021). This involves a complex change process in financial firm intangibles such as organisation, culture, and knowledge. It involves major changes in the impact of the non-financial context in the decision making in financial firms to achieve Net Zero aims. The Green BTFF – as developed in this chapter – is used to explore how this is done and can be done.

2. RESEARCH METHODS

This chapter uses 'field-based stories' or 'empirical narratives' (Golden-Biddle & Locke, 2007) to explain how financial firms can 'go green'. Two sources of data are used. *Firstly*, public sources are used to develop detailed cases of change in six major international financial firms in the period 2010–2022. These include three banks (NatWest/RBS, Lloyds, and Wells Fargo), and three fund managers (Schroders, Baillie Gifford, and Blackrock). *Secondly*, the author attended practitioner conferences to gauge change across financial firms. The practitioner conferences are at the heart of the practitioner and policy debate about change in green finance.

They included initiatives by the City of London Corporation and partners. These events included, Green Finance Summit (2017, 2018, 2019), Green Horizon Summit (2020, 2021), and Finance for Impact Summit (2022). They included various

'Net Zero' Conferences (2021, 2022, 2023) hosted by Business Green, and events by CGFI (2019) and City Week (2020). They included events organised by the Ethical Finance Summit (GEFI) (2017, 2018, 2019, 2020, 2021, 2022, 2023). The Green Finance Institute (2021) provided videos of all major finance presentations and discussions at Glasgow COP26 (2021). The data were primarily based on firms committed to Net Zero aims (GFANZ, 2021).

Presentation of the chapter at academic conferences such as EIBA, BAFA, CSEAR, EIASM, BFWG, CSRCOM, and WFC, provided many theoretical ideas and literature from the academic community. The change narrative covers periods of rapid change, and major problems and crisis from 2010 to 2023 with an emphasis on climate change. The COVID-19 pandemic (2020–2022) showed how financial firms could rapidly re-organise their 'soft' socio-technical infrastructures, 'hard' technology, and working conditions in positive self-reinforcing cycles to deliver 'hard' financial services and products during major unanticipated changes. The 'financialisation' period before the global financial crisis (GFC) showed how shareholder wealth values became dominant in use of intangibles in financial firms (Holland, 2019). These show the potential for change in green intangibles and values in the years ahead. The case firms and industry conferences revealed such change for Net Zero aims.

Variation existed within case financial firms in term of specialist financial activity. However, they all shared similar financial intermediation processes. The conventional purpose and function of financial firms such as NatWest, Schroders, and Aviva, is to intermediate sources of funds (deposits, savings, and premiums) to create new uses of funds (loans, equity investment, and insurance contracts). The firm transforms risk, liquidity, size, and maturity of funds to provide financial services, products, and functions required in the real economy and society (Buckle & Beccali, 2011; Lewis & Davies, 1987). Financial firm's function is to deliver specialist financial services within their 'philosophy'. The firm's *conventional* values have focussed on how to create financial value for shareholders from these activities.

Despite the variation in specialist financial activities in cases such as NatWest, Aviva, and Schroders, the financial firms made similar green oriented changes to the 'soft' and 'hard' context to influence specialist financial decision making and common financial intermediation processes (or 'financial machine'). This multi-case design created opportunities for identifying common themes and differences across the cases (Yin, 2018). Five of the UK-based international financial firms (NatWest/RBS, Aviva, Lloyds, Schroders, and Baillie Gifford) had already been extensively interviewed by the author, on three separate occasions, during 1993 to 2017 and provided longer term context to the change process.

The grounded theory approach (Strauss & Corbin, 1998) breaks down the complexity of the green change problem in financial firm and network system into four comprehensible metaphors (Morgan, 1997) or themes to develop a big picture of 'what is going on' (Kay & King, 2020). Green change themes were manifested as green oriented changes: at the top of the firm ('head'); to socio-technical context ('house'); to interactions, processes (in 'community'); and to financial decisions (in 'machine'). It includes consequences and feedback. The use of metaphors reflects Morgan's (1997) approach to understanding complexity and change in organisations. The 'empirical narrative' was interpreted using relevant interdisciplinary literature (De Bakker et al., 2019) to develop an equivalent 'theoretical narrative' and form a green BTFF.

The green BTFF was developed from previous *finance oriented* BTFF (Holland, 2016, 2017, chapter 8, part 3, 2018, chapter 5, 2019). The green BTFF constitutes an exercise in 'theoretical sensitivity' whereby new work allows the author to return to the original financial firm phenomena and data with a new perspective (Strauss & Corbin, 1998). The full details of the green BTFF are outlined in Holland (2021) and are briefly summarised in this chapter.

3. A GREEN BEHAVIOURAL THEORY OF THE FINANCIAL FIRM

The green BTFF is based on a combination of insights from an 'empirical narrative' interpreted within a 'theoretical narrative' (Golden-Biddle & Locke, 2007). The narratives employs four metaphors (Morgan, 1997) – 'head', 'house', 'community', and 'machine' – to explore relevant dimensions of Net Zero change. Financial firms are seeking a subtle combination of, 'what we (believe we) are for, and what we are (becoming), affects what gets done' and 'what gets measured gets done'.

3.1. The Change Narrative – Brief Outline

The full details of the green BTFF are outlined in Holland (2021) and are briefly summarised below. *The first part of the* 'change narrative' involves top teams ('head') 'looking out' to learn (Pedler et al., 1997) about external climate change and 'looking in' and learning how to change the firm. *The second part* involves changes to the 'house' where top teams make green oriented strategic changes to purpose, 'socio-technical' infrastructure in the firm 'house', and its contextual influencing role on the use of financial resources. *The third part* concerns how firm social and knowledge resources are mobilised, how changes are enabled and driven by organisation processes. The latter involve green purposeful interactions at individual, team, firm, and network levels in the financial firm 'community'. The *fourth part* highlights how changes to the non-financial context and working conditions, supports Net Zero oriented, financial decisions, and products, in the financial 'machine'. The latter concern creation of new green oriented financial assets and liabilities (and derivatives of) in each firm's specialist domain such as banking, fund management, and insurance. This 'change narrative' is fully explored in Section 4.

A green BTFF can provide knowledge, information, and power to internal and external stakeholders to help them overcome *in part* the top team, core shareholders, and middle management tendencies to dysfunctional forms of organised hypocrisy and use of facades (Brunsson, 1993; Cho et al., 2015).

3.2. A Holistic Theory Interpretation

The change narrative is interpreted within overlapping and overarching theory frames including, systems theory (Mumford, 2000) and international theory of the firm (Bourdieu, 1990; Cyert & March, 1963; Peng, 2001). Financial firms and their stakeholder networks, operating in financial markets and changing environment, form complex systems made up of purpose, core elements and connections, and dynamic processes (Barile et al., 2018; Holland, 2014; Mitleton-Kelly, 2003). The financial firm's socio-technical system (Mumford, 2000) consists of top teams and purpose ('head'), contextual resources ('house'), interactions and conditions (in 'community' of employees and others), and a financial 'machine' of connected financial decisions. The emphasis is on 'doing the right thing' in financial decisions as defined by an explicit green oriented purpose such as in GFANZ (2021). This integrated system of non-financial and financial resources is the basis for transforming the inputs, through business activities, into outputs and outcomes in a financial 'machine' that aims to fulfil the organisation's strategic purposes (IIRC, 2013; VRF (Value Reporting Foundation), 2021) including Net Zero aims.

In terms of an international version (Peng, 2001) of the RBV of the firm (Barney, 1991; Hart, 1995; Teece et al., 1997), the large international case financial firm such as NatWest/RBS, Lloyds, Schroders, and Baillie Gifford develop unique firm specific competitive advantages, within sector alliance creation of shared co-operation advantages. This includes the orientation of 'head', 'house', 'community', and 'machine' in firms to complementary and integrated Net Zero and financial aims. Firm specific advantages and sector alliance advantages, each with dynamic capabilities (Teece, 2007), shape their competitiveness and legitimacy in global and national markets. The change in practice complements Peng's (2001) view that in the academic world 'a broad, expanding, and cumulative knowledge base is emerging to connect IB (International Business) and strategy research through the RBV'.

Cyert and March's (1963) 'behavioural theory of the firm' can additionally be used to argue that clearly structured and goal directed organisational processes are means for uncertainty avoidance and conflict resolution in the firm. Organisational processes such as top down, bottom up, and other *multi-dimensional social interactions* need to be used to develop, allocate, mobilise, and exploit context-based resources (social, knowledge, and technology) and financial resources for Net Zero aims.

In Bourdieu's (1990) terms, green change is about the exercise of power in the financial firm and external networks. Top teams, middle management, and front-line employees exercise and experience power pressures to change. Increasingly, the existential threat of climate change has forced powerful agents in large international financial firms to use their 'habitus' and 'capitals' to collaborate, share their power and co-operate in new forms of shared green oriented practice. These contribute to green oriented adaptations to habitus and capital of middle management, back office and front office in the internal financial firm field or 'house'. Top teams use these to influence likeminded individuals working together in specialist decision teams in the firm's hierarchy, or with customers to develop green oriented group habitus and capitals in shared social fields. These are critical to how they conduct *practice* (as strategic change and everyday activities) and interact with each other and with their shared fields.

4. HOW FINANCIAL FIRMS CAN GO GREEN – THE CHANGE NARRATIVE

The following subsections includes insights from four major parts of the integrated empirical change narrative concerning 'head (Section 4.1)', 'house' (Section 4.2), 'community' (Section 4.3), and 'machine' (Section 4.4). Each metaphor part of the empirical change narrative is supported by a brief theoretical interpretation from specialist literature and the holistic view from Section 3.

4.1. Strategic Changes in Team Strategic Thinking – and Top Teams or the 'Head'

The *first part of the* strategic 'change narrative' explores how changes at *the 'head' of the firm* are *strategic* means to respond to potential and actual harm caused by problems of top team ideology, ignorance, and indifference. These are means to oppose top team organised *hypocrisy* and facades, about Net Zero (Brunsson, 1993; Cho et al., 2015).

The 'change narrative' starts with how external climate change pressures are driving learning at board and executive team levels in 'learning organisations' (Pedler et al., 1997). Top teams in case firms such as NatWest (Rose, 2020) and Schroders (Harrison, 2020), and members of their peer groups such as specialist financial sector (Net Zero) alliances (GFANZ, 2021) are making sector wide coordinated changes as well as firm specific changes to firm purpose, function, and planning. This involves top teams ('head') 'looking out' (Pedler et al., 1997) to learn about climate change pressures, current and expected. Top teams actively engage with and learn from other top teams (peer groups), other elites in the financial community, and stakeholders in wider society. This involves individuals and teams collaborating in peer group alliances such as GFANZ (2021) within and across specialist finance sectors to understand change and develop the 'rules of the game' (Bourdieu, 1990).

A key ingredient in 'change at the top' involves leaders in leading financial firms adopting a science evidence base for metrics, decisions, and engagement. Leaders such Rose at NatWest and Harrison at Schroders (Aden et al., 2022) have ensured that their firms became signatories to the Science Based Targets initiative (SBTi) in 2020. Financial firms cooperate and iterate towards best Net Zero practice and outcomes before regulators such as SEC take control (Aden et al., 2022).

This collective use of science-based information is a basis to learn about uncertainty and complexity, induced by climate change physical risk, and about transition risks from associated political and regulatory changes. Top teams in case firms use this to identify new strategic opportunities and threats (Porter, 1985). They are learning how to develop adaptive strategic plans and strategic choices (Mirfenderesk & Corkill, 2009), to develop resilient firms capable of managing the new physical, political, and economic risks, and delivering core financial functions and performance in the economy. They seek 'complementary' financial value and Net Zero aims by focussing on economic activities that involve increasing use of non-hydrocarbon energy sources, and do not harm people and their social and economic organisation. The expectation and belief are these activities would not involve loss in growth opportunities and profits.

The existential threat of climate change means that top teams go beyond a competitive outlook. They work together in alliances (GFANZ, 2021) to create co-operative advantages with other international financial firms and with client firms in the real economy. They pursue both sustainable co-operative and competitive advantages in global finance, to deliver green oriented financial services and products, and create superior financial performance that supports shared climate change aims. In the RBV of the firm (Barney, 1991; Hart, 1995; Peng, 2001; Teece et al., 1997), the large international case financial firm develops a unique sustainable competitive advantage or SCA within global sector alliance creation of shared co-operation advantages.

A priority area is for top teams ('head') to change the 'head' and its associated SCA. This involves change in governance, composition, and capabilities of boards and executive teams, as well as behaviour or 'tone from the top'. These create top team capabilities to deal with climate change issues. In firms such as NatWest and Schroders (Rose, 2020; Harrison, 2020), this informs strategic choices, about their green oriented purpose and orientation of firms, and about performance metrics for achieving Net Zero outcomes.

Learning by top teams (Pedler et al., 1997) and integrated thinking (VRF (Value Reporting Foundation) (2021) are intended to iteratively lead to new knowledge, capabilities, and advantages in all teams, with top teams as major initiators of change. Thus, changes in and by top teams are significant drivers of coordinated changes to contexts, mechanisms, and technology – throughout the financial firm – to achieve sustainability and financial aims. These are key *parts* of the means to promote awareness amongst employee and external stakeholders about Net Zero change and its impact on financial decisions and transactions.

4.2. Changing Context as Adapted Infrastructure and Mechanisms – 'House' Change

Schroders and NatWest use climate change to stimulate thinking about structural change and behavioural change in firms (Harrison, 2020; Rose, 2020). Top teams use the green changes in their learning, understanding, and strategic thinking capabilities to 'look in' and make strategic changes to purpose, and planning, structures, and resources. This involves changes to 'enabling infrastructure' or 'soft', 'socio-technical' contexts (Mitleton-Kelly, 2003; Mumford, 2000) in the existing financial firm 'house', and its (contextual) influencing role on decision behaviour and use of financial resources.

The empirical narrative about socio-technical infrastructure ('house') change concerns three clusters of strategic change. The first cluster includes changes to: firm wide social and organisation structure, culture as 'organising' means, and to knowledge resources; at both macro and micro levels. The second cluster concerns green changes to control and influence mechanisms. These include communications and storytelling, top team behaviour, training, incentives ('soft' and 'hard'), explicit risk appetite, formal control systems, as well as changes in culture as a means of controlling and influencing. The third cluster involves green changes to technology and its impact on knowledge, social, and financial process. These changes are strategically matched to changes in the external environment (Teece et al., 1997) concerning climate change and Net Zero aims, and the need to develop internal resilience in the face of uncertainty. These green oriented intangibles (internal and external) are critical to directing green oriented behaviour in employees, to create information in the financial firm, and make green financial transactions possible. These choices help create a positive change 'atmosphere' in working conditions and their influence on employee behaviour and in interactions with customers.

New *technology*, adapted for Green oriented change, is a key factor in creating new socio-technical systems and working conditions (Mumford, 2000). Change in technology capabilities and uses of technology disrupts established activities; changes the nature of work and working patterns in the firm and teams; and alters power conditions within the firm and externally with customers in market social and economic networks. COVID-19 provides an example of this. Change in technology such as Schroder's development of Sustainex (Howard, 2019) is expected to have a similar impact. Software in 'Sustainex', and staff capabilities have been upgraded to reflect SBTi (2022) insights. The technology focusses on green oriented data processing to enhance information and metrics about costs and benefits of Net Zero changes. This decision support system provides information to decision makers at all levels.

The theoretical narrative for the above explores how case firms 'look in' and learn (Pedler et al., 1997) how to promote green oriented organisational change in the socio-technical infrastructure in the three clusters throughout the firm. Internal learning about change concerns how to adapt the socio-technical infrastructure ('house') and 'community' fields, how to mobilise control mechanisms in a *top-down way*, and how to exploit technology. These are the basis to influence working conditions in the firm 'community' and influence everyday behaviour and financial decisions in the financial 'machine' by individuals and teams in the firm. Learning by 'looking in' also occurs in the case financial firms during everyday actions (learning 'by doing' and 'by trial and error'), by imitation and during periods of reflection and discussion (reflexivity) (Pedler et al., 1997). Learning occurs through knowledge transfer mechanisms such as recruiting experienced managers, and attending green finance, Net Zero, and ethical finance conferences. Thus, the firms develop a green orientation to their 'socio-cultural and technical' elements (Mitleton-Kelly, 2003) or soft technology and measurement parts of the firm complex system.

Organisational literature and theory sources such as Burnes (2004), Whelan-Berry et al. (2003), Heyden et al. (2017), Lewin (1947), and Cyert and March (1963) further develop the organisational change narrative. The green changes are the basis to change and shape behaviour and actions. From the perspective of strategic choice and organisational change literature (Child, 1972), top teams in financial firms develop strategic responses to climate change pressures and to adjust the firm strategic position, organisation, operational capabilities, and behaviour. They do this to survive in new markets, and in socio-political and physical change circumstances.

The green oriented organisational control and influence mechanisms, such as culture, formal controls, and incentives, form additional ways of organising and integrating firms and enabling behaviour and actions (Beusch et al., 2016; Gond et al., 2012; Moon et al., 2011). Culture in a firm such as Schroders (Harrison, 2019, 2020) is an important learnt and evolved property of organisation life (Schein, 2004), and creating a Net Zero orientation to culture plays a significant role in influencing behaviour and change.

Financial firms develop and exploit these different but complementary ways of green oriented 'organising' in financial firms. Gond et al. (2012) define this as social organisation (social integration), shared capabilities (technical integration), and ways of thinking (cognitive integration). The combined and connected presence of these forms of organising support working conditions or enabling conditions (Mitleton-Kelly (2003), to 'allow' green oriented behaviour, enable Net Zero oriented information production and financial decisions.

In Bourdieu's (1986, 1990) terms, greening of the 'house' of infrastructure, control mechanisms, and technology reflect 'people' and 'community' learning activities. They arise from learnt changes to top team, middle management, and frontline employee 'habitus', 'capitals', and power in internal 'fields'.

Thus, all non-financial and financial parts of this complex system are the focus of change. The full set of integrated and green oriented socio-technical infrastructure factors embed green change (Lueg & Radlach, 2016) in financial firms. These integrated social and knowledge factors influence and control mechanisms and technology, prevent the dominance of financial values, and avoid the 'tragedy of the horizon' (Carney, 2015). They create a shared consciousness in all employees to influence financial decisions at firm, portfolio, and transaction levels to be Net Zero oriented at the point of decision making.

The above illustrates how top teams and employees change their views of non-financial resources as the prerequisite to change use of 'hard' financial resources to achieve hard performance targets, both financial and Net Zero green aims. This reflects Ortiz-de-Mandjona and Bansal's (2015) argument and evidence that, 'firms that adopt responsible social and environmental practices, relative to a carefully matched control group, have lower financial volatility, higher sales growth, and higher chances of survival over a 15-year period'.

4.3. Using 'People' Interactions – To Mobilise Mechanisms in Context and Influence Conditions in 'Community'

The third part of the strategic empirical 'change narrative' uses the 'community' metaphor to explore core dynamics in change. This concerns changes in lived experience, green awareness, and behaviour of individuals and teams in the 'community'. This involves many green oriented interactions, activities, and experiences by individuals and teams working in increasingly green oriented 'communities' in the firm and external networks. At the same time, this involves mobilisation and use of contextual resources and 'social forces' from a green oriented 'head' and 'house' such as 'tone from the top', and culture. This is not a hierarchy of top-down change alone (Heyden et al., 2017; Levy, 2014; Whelan-Berry et al., 2003). It forms a Net

Zero purposeful 'wave of change' over time (Hirsch, 2016; Morgan, 2013). All parts of firms and external networks recognise green purpose and interact, iterate, and adapt, as one complex system. It involves 'working with the grain', matching change to existing values and green norms in firms (Schein, 2004), society (Falkner, 2021), and principles of responsible behaviour (e.g. PRB, 2019).

Contextual influences or 'social forces' from *'head' and 'house'* change working and decision conditions in the 'community'. These collectively focus employee minds on how Net Zero aims can be incorporated with financial aims and behaviour at the point of financial decisions. Similar engagement with customers reflects all these changes. Changes in the 'community' are enabled and driven by goal driven organisation processes (Cyert & March, 1963). The organisational processes involve Net Zero oriented purposeful interactions and awareness activities at individual, team, firm community, and network levels.

This combined use of context and process seeks to change the working conditions and 'lived experience' of individual employees and teams in the firm 'community' and of stakeholders in external communities. Top teams seek to directly change the values, mindsets, and behaviour of employees throughout the firm community to become Net Zero oriented in financial decisions. This in turn is expected to improve individual and team performance, customer relations, and firm performance. In this regard, Esteban-Sanchez et al. (2017) found that during and after the financial crisis (2005–2010), banks with better employee relations and corporate governance had better corporate financial performance. During the crisis, better relations between employees and the community were valued positively by investors, which, in turn, increases corporate financial performance.

The interactions concern top-down processes such as setting a green oriented 'tone from the top', and promoting green oriented culture (Schein, 2004) in the firm 'community'. Rose (2020) at NatWest, and Harrison (2020) at Schroders have established climate change as the strategic priority of the board. This requires interactions throughout the firm community to involve individual and teams, interacting and debating with each other about green change to achieve Net Zero aims.

This occurs in multidimensional ways (through bottom up, lateral, and network interactions), within green oriented organisational structure and network contexts. Top teams, middle managers, and frontline team leaders mobilise formal and informal meetings, control mechanisms, and technology to promote these firm wide interactions.

More specifically, this approach uses direct and explicit activities for the promotion of Net Zero or green awareness. It focusses on how the proactive use of specialised firm wide awareness activities and interactions between employees, ensure all major finance decisions in the firm 'community' are 'Net Zero aware'. These include using green *changes* in top team behaviour to create a 'tone from the top' *and use of other direct and specialised means* to promote 'Net Zero awareness activities' amongst all employees. The *specialised* means include promotion of Net Zero purpose, training, and use of incentives and control systems. Net Zero awareness activities between individuals and teams also involve engagement, learning, use of good practice, promotion of green culture, and self-monitoring.

For example, leaders Rose (2020) at NatWest, and Harrison (2020) at Schroders, having made prior green changes to 'head' and 'house', also promote green oriented changes to purpose, training, incentives, targets, metrics, action timetables, and control systems. They mobilise formal and informal meetings, control mechanisms, and technology to promote these firm wide interactions about complementary Net Zero and financial aims.

Holland (2021) explored in more detail how the case financial firms used a range of *green* awareness activities as a fruitful standpoint to understand the wider interactions and complex changes in the 'community' and in 'head' and 'house' contexts. Awareness activities illustrate changes in the 'lived experience' and behaviour of employees, customers, and stakeholders during change. *Specialised and focussed* awareness activities include continuous engagement, managed learning, use of good practice, use of culture, and self-monitoring. These promoted interactions *improve* change experiences of employees, customers, and stakeholders.

The shared sense of purpose, multidimensional social interactions, and awareness activities are the primary means to mobilise the socio-technical infrastructure and firm wide controlling and influencing mechanisms to create desirable green oriented *working conditions* in the firm 'community'. The conditions include emotional enabling conditions such as trust, desire to co-operate, sense of ease of communications, and buy-in to the top-down green change story. The empirical narrative shows how case firms create working conditions in the firm 'community' to promote desirable decision conditions based on understanding of climate change risk, and ability to create green information about fund users and suppliers. It shows how they control their own behaviour and behavioural biases (Statman, 1999) relative to sustainability and financial aims. Change to working and decision conditions in firm 'community' are made to improve each decision team's focus on risk and opportunities relative to Net Zero (GFANZ, 2021) and financial aims. They are used to change economic advantage (Barney, 1991) and behaviour and create information for financial decisions.

Thus, the above 'community' metaphor (Morgan, 1997) provides insights into the 'lived experience' or 'life' of employees, customers, and stakeholders involved in green change in behaviour in internal and external 'communities' or social organisation in financial firms. It highlights the main 'head and house' factors at play and their connections in the green change process in the 'community' or social field. It concerns how they generate social forces to change behaviour and financial decision practice in the firm and with customers. Thus, new socio-technical structure and organisational process *shape* changes in decision behaviour in teams. Purpose led mechanisms (for control and influence) and purpose led 'people' interactions (multidimensional), and awareness activities, *drive* changes in the financial firm 'community'. Collectively, they influence 'ongoing engagement' with employees, customers, and stakeholders, as well as the delivery of new green products.

4.4. Changing Economic Conditions and Financial Decisions in a Financial 'Machine'

The fourth part of the change narrative discusses Net Zero change themes identified in the financial 'machine'. The change themes concern changes to financial decisions, especially their structure, processes, and outcomes.

Morgan (1997, p. 6) notes that when 'managers think of organisations as machines they tend to manage and design them as machines made up of interlocking parts that each play a clearly defined role in the functioning of the whole'. The financial firm 'machine' metaphor assumes the firm is a *formal* organisation (Silverman, 1970, p. 2) made up of highly structured and connected financial decisions in an integrated firm hierarchy. The 'machine' structure also consists of connected decision routines, decision conditions, and decision processes, used in all these hierarchy levels.

Within hierarchy, the 'machine' concerns purposeful decisions about transactions by frontline teams; specialist financial portfolios by middle management; and firm wide financial decisions by top teams (capital structure, mix of portfolios and products, etc.). This includes green oriented financial strategy, capital allocation, and overall financial risk management of the whole firm by top teams. It concerns Net Zero oriented capital allocation and risk management of specialist financial portfolios (assets and liabilities) by middle management teams. It includes specialist green oriented transaction decisions by frontline teams interacting with customers. All of these are supported by back office teams for information, risk, contracting, and transaction services.

The hierarchy of integrated financial decisions in case international financial firms is the collective basis to financially intermediate or transform risk, return, maturity, and size of input capital into desired capital outputs for customers and economy (Scholtens & van Wensveen, 2003). This integrated 'machine' is designed to deliver these finance functions for the economy. However, the GFC in 2007–2008 has shown (Holland, 2010) how the 'machine' metaphor has major problems, especially when finance value calculation, rationality, and logic, based in part, on finance theory, ignores the complexity of organisational life. Morgan (1997, p. 5) argues that the 'Machine metaphor is incomplete, ignores human aspects, is biased, and elevates rationality'.

This chapter recognises this issue by using extended metaphors of 'head', 'house', and 'community' to outline how organisational and human dimensions interact with the financial 'machine' in a complex system (Mumford, 2000). The Net Zero oriented and integrated non-financial social system elements (in 'head' and 'house') influence the social structure, purpose, values, and 'atmosphere' of *working conditions* in the internal 'community' and financial 'machine'. They influence and support decision conditions as adapted cognitive and emotional states in individual employees and their teams. They influence the hierarchy of financial decisions in the firm (firm wide, portfolio, and single), as well as decision routines used in all these hierarchy levels. These are new ways to deliver new Net Zero oriented financial decision activities and products.

In the financial 'machine', these 'social finance' resources are means to overcome, in part, the problems of 'head space' (Larsen, 2021) or cognitive capabilities to deal with complexity arising from climate change. They are a social organisation response (Gond et al., 2012) to ameliorate 'information overload' and limits cognitive abilities of individuals in teams (Lejarraga & Pindard-Lejarraga, 2020). They influence Net Zero oriented conditions (attitudes, mindset, emotions, behaviour, bias, and motivation) amongst individual employees and teams during their connected financial decisions and when transacting with customers. The green contexts and working conditions influence cognitive and emotional conditions in individual employees, their teams, and customers.

In Net Zero 'believer' case firms, the 'social finance' resources in 'head, house, and community' are used to modify employee behaviour with customers when setting terms and expectations of potential transactions during the negotiation, selling, and transaction processes. The case firms use SBTi (2022) sources and their own Net Zero oriented information systems (e.g. Sustainex at Schroders) to highlight Net Zero and financial metrics and wider information relevant to frontline employees and customers. This is a basis to reduce employee and customer *availability bias*. They also highlighted information on the history of gains and losses to reduce *anchoring bias* and ensure that employee and customer *loss aversion* (Tversky & Kahneman, 1974) is based on relevant information.

In addition, the green oriented financial firms such as NatWest (Rose, 2020) use their social resources and their deep understanding and experience of customers to improve customer financial literacy, Net Zero change literacy, and behavioural biases. They use this to create credible stories about the complementary Net Zero and financial value dimensions of the transactions they are selling. In Net Zero 'believer' firms, these actions are a basis to improve the quality of customers' financial and personal decisions. The aim is to increase trust, confidence, and desire of customers with to engage in Net Zero oriented transactions.

As a result, customers, especially in retail finance, have their green Net Zero, financial value rationality, cognitive and emotional conditions, and behaviour including biases, in part constructed and adapted for them by financial firms. They are constructed through social structures, knowledge, information environments, transacting environment, and selling behaviour, all designed by financial firms.

Large international financial firms jointly negotiate these conditions with large corporate customers. Green oriented customers are then expected to act within this negotiated context according to their behavioural finance biases, trust in financial firms, and their own views of Net Zero and financial value rationality. In transactions between large financial firms and with large corporates, often in wholesale market settings, economic rationality was more dominant than retail settings but still influenced by green change values from wider society and financial firm alliances (e.g. GFANZ, 2021).

It should be noted that even in the case of Net Zero 'believer' firms, scepticism is still required about the top teams and the above Net Zero change process with employees, and with both retail and wholesale customers. Events such as COVID-19 and the Russia–Ukraine war indicate a drift away from Glasgow COP26 (2021) aims. Given the complexity of internal organisation, activities and processes, and of external stakeholder demands, these 'believer' top teams are likely to use this situation to employ a form of managed 'organised hypocrisy' and 'facades' to create the time and space to execute the desired Net Zero changes in the firm and externally. However, 'believer' firms and top teams operating in alliances such as GFANZ (2021) and PPCA (2021) still provide momentum, ideas, and actions to counter totally hypocritical and dysfunctional financial firm behaviour relative to Net Zero aims.

4.4.1. Changing the Green Focus of Financial Decisions

The green organisational contexts, working conditions, and adapted decision conditions in individuals and their teams are intended to improve each decision team's focus on risk and opportunities relative to Net Zero and financial aims. They change the way financial resources are used in operational activities and financial decisions.

For example, in their financial 'machine', Schroders have used their strategic changes and SBTi (2022) information in Sustainex to align their financial decisions (transactions, portfolios, and whole firm) with Paris COP21 (2015) target to stay below 2.0°C, and preferably 1.5°C. SBTi (2022) confirmed that Schroders' scopes 1 and 2 targets are in line with a 1.5°C trajectory, and relevant assets under management are targeted to be fully aligned with a 1.5°C pathway by 2040 (Schroders, 2022). The targets include Schroders' own operations, as well as their 'financed emissions' from companies they invest in.

The 'Sustainex' changes, SBTi information, and new staff capabilities have allowed Schroders to engage with investors and *investee companies* in much more informed way. Schroders can engage with investee companies how they perform against their science-based targets. This is key to aligning whole sectors and portfolios held by Schroders and their investors. The external SBTi process formally evaluates financial firm compliance with their standard and will rate Net Zero performance of Schroders and other member firms in a voluntary process. This allowed Schroders to produce standardised metrics and targets for scopes 1 to 3 emissions wholly based on GHGs and do not include offsets. This improves pillar four of their reporting in their TCFD (2017) and more broadly improves comparison and accountability within and across specialist financial firms. The SBTi process and Schroders reporting and information sources have allowed Schroders to engage with *institutional and retail investors* in much more informed way on their Net Zero aims, economic rationality, and community concerns.

4.4.2. Changing Financial Decisions

The above green changes influence case international financial firm decisions to save, lend, invest, or insure to reduce harm of GHG emissions. These include production and delivery of green oriented equity investments, bonds, loans, and insurance.

Financial firms thus use firm wide green contexts and conditions to create information, control behaviour, and reduce information asymmetry and transaction costs at the point of financial decision making. They do this to enhance liquidity management, diversification, and risk management. This creates conditions for financial intermediation and hence the transformation of financial capital and its risks (Holland, 2019; Lewis & Davies, 1987) consistent with Net Zero aims. This is the means to deliver complementary Net Zero and financial aims. The financial function remains central but is adapted by the non-financial context to reflect climate change concerns.

For example, the *non-financial contextual resources* (in 'house' and 'community') influence behaviours in the financial 'machine' such as risk appetites, acceptable exposures, and risk taking (Mio et al., 2022). This concern behaviour and policies by top teams. It concerns behaviour by individuals and teams throughout the firm and in markets. The *non-financial contexts* influence the Net Zero and financial orientations of such behaviour when using financial resources. They influence the 'soundness' or resilience of financial resources in the firm and markets to financial risk. Soundness involves green oriented choices about resilience of financial resource risk 'buffers' such as increasing 'adequacy' of capital and cash, asset and liability quality (reduce losses), diversification on balance sheet and in markets, and increasing liquidity levels (Abdallah et al., 2020).

This highlights that adapted non-financial resources, ideas of acceptable financial behaviour, financial soundness in financial resources and their complementary Net Zero and financial orientation and aims are necessary conditions for 'success'. 'Success' involves delivery of financial transactions and services, management of specialist financial portfolios, and financial intermediation process in the financial firm. 'Success' concerns the financial firm fulfilling specialist financial functions, say banking, insurance, and fund management, by delivering financial services and transactions to customers. 'Success' involves achieving Net Zero and sustainability aims, and creating financial value over the short, medium, and long terms.

5. DISCUSSION

In metaphor terms (Morgan, 1997) the green BTFF is made up of 'head', 'house', 'community', and financial 'machine' elements driven by complementary Net Zero and finance purpose. 'Net Zero' involves ongoing reduction of GHG emissions and ensuring that remaining emissions are balanced by implementing methods to absorb carbon dioxide from the atmosphere. Finance purpose concerns the delivery of financial value, risk management, and required green products, as the firm's core function in a sustainable economy and planet. These empirical insights are interpreted in an interdisciplinary frame (De Bakker et al., 2019) for each metaphor and overall interactions in green BTFF. The findings of the previous section show the potential power of interdisciplinary analysis in understanding complex fields of practice (Knights & Willmott, 1997). The cases also show that green BTFF can play a role in developing new forms of theory construction (Section 5.1) and empirical tests in finance (Section 5.2). It also contributes to the ongoing debate about the nature and purpose of international financial capitalism (Section 5.3).

5.1. New Forms of Theory Construction

For example, the *theoretical analysis* and discussion of the wider firm context in the chapter makes use of theory of the firm (Barney, 1991; Hart, 1995; Teece et al., 1997), an embryonic 'social finance' literature (Hirshleifer, 2015), 'behavioural finance' biases (Statman, 1999), and theory of finance (financial intermediation) (Scholtens & van Wensveen, 2003). The chapter develops a theoretical narrative to connect, 'theory of the firm', 'social finance', to 'behavioural finance' and to conventional finance theory. This approach forms a new basis for thinking about theory construction. It suggests that the green BTFF can support development of an academic research programme about a green and international version (Peng, 2001) of the RBV.

The above theories can be connected in a theoretical narrative. In the theory of the firm (Barney, 1991; Peng, 2001), the international financial firms seek to develop unique competitive advantage within their sector in areas of 'social finance' (Hirshleifer, 2015) and 'behavioural finance' (Statman, 1999). They do this within a global sector wide creation of shared co-operation advantages in 'social finance' and 'behavioural finance'. They seek to do this within a new consensus that this is desirable within their sector, the world of finance and wider society. In finance theory terms (Scholtens & van Wensveen, 2003), this is expected to make the 'financial machine' more efficient in individual firms and across sectors in terms of reduced financial intermediation risks, enhanced delivery of green oriented financial services, and generation of profit. The combination of unique competitive advantages (social, behavioural, and finance) is expected to be the basis for financial and Net Zero outperformance relative to peers.

This approach does not seek to 'integrate' traditional theories – of finance, sociology, strategy, and management – in 'meta theory'. The more modest aim is to improve theory content by using the interdisciplinary analysis and theoretical narrative to provide examples of conceptual 'connections' possible between theories (Holland, 2019). The approach provides means to address some of the problems identified with using theory, especially traditional finance theory, in explaining the economic and social role of large international financial firms such as NatWest and Schroders in global climate change. This provides new opportunities to think about theory construction during complex and rapid change.

5.2. Developing New Empirical Tests in Finance

The green BTFF also provides a means to develop new empirical tests in finance. A key empirical question is, will the systemic green change, as proposed in the green BTFF, work for all types of international financial firm? An associated empirical question is, will the systemic green change, as proposed in the green BTFF, involve 'organised hypocrisy' and 'Greenwash' facades (Brunsson, 1989).

In the *first question*, the case data and experience of change (Holland, 2010) indicate that *three* factors will drive the chances of success: (1) *the history of firms,* (2) their dominant social/economic context, (3) plus the *degree of achievement of Net Zero change aims* in *the four metaphor areas* in firms. Firms in the same sector such as banking may have different histories, contexts, culture, awareness of, and commitment to green change and sustainability. They may have different pathways and rates of evolution relative to these issues.

Triodos bank, for instance, started with a socially responsible and sustainable purpose to their banking function. NatWest had a financial 'dominant logic' (Engelmann et al., 2020) from the 1980s onwards. It experienced major problems in the GFC and contributed to major harms. Both banks seek to systematically incorporate Net Zero green change ideas throughout the firm in the *four metaphor areas in the* BTFF. Both banks manage financial risk to survive and make money from financial services. At times financial pressures alone in the world of finance may dominate their thinking and logic.

Triodos with its embeddedness in social settings, is likely to maintain complementary green change and finance aims as its combined 'dominant logic', despite increasing finance pressures. It is likely to manage the ongoing tensions and opportunities between a logic for green change and a financial logic. It is less likely to crowd out prosocial behaviour (Bénabou & Tirole, 2006). NatWest, with a history of finance culture and operating mainly in financial markets domains, may be vulnerable to retreat to a financial logic as financial risks increase and financial returns decline. Under intense time pressures, it may seek to exploit green change as a 'cloak' or facade to pursue financial aims. Both banks may be steered away from Net Zero aims and changes by highly significant events such COVID-19 and the Russia–Ukraine war.

The green BTFF can be used to test for success in systemic green change and how it differs between banks such as Triodos and NatWest. The positive *or* negative role of quantifiable factors: in history, dominant context, and in *each* metaphor area; on green 'success', is a way to test for 'Greenwash' in firms such as Triodos and NatWest. The positive impact of factors in all areas would indicate systemic green change in the whole firm.

In terms of the *second question*, this is a means to test for the authentic or dysfunctional nature of behaviour concerning organised hypocrisy and facades (Brunsson, 1989) by these firms. It can distinguish between the rational, progressive, and reputation façades (Cho et al., 2015) created by these firms relative to Net Zero aims, and whether they have sought to deliver these through strategic actions and communication (Rose, 2020).

As a result, green aspects of quantitative finance research can be rethought using the green BTFF. This chapter hypothesises that the *degree of achievement of Net Zero aims* is a function of *connected* and *combined* changes in *key variables* in all of the empirical themes or metaphors. This firm wide hypothesis (Poterba, 2021) of change in green finance differs from conventional quantitative finance studies by its focus on the whole system rather than parts. The green BTFF could therefore be used as a wider explanatory context to hypothesise which independent variables or factors are expected to predict 'success' as the dependent variable, in a *green, and finance oriented* financial firm.

'Success' as the dependent variable, in achieving complementary financial and Net Zero aims and outcomes, could be defined in many ways. For example, Fig. 9.1 illustrates how 'success' in a variety of metric forms could be hypothesised as being associated with Net Zero changes in independent variables in each

Dependent variables - 'Success'

- Growth in profits from green finance
- Improvement in firm green finance ranking, and in stock price
- Growth in green oriented assets
- Reduction in activity on environmentally irresponsible finance deals or carbon finance deals
- Increase in Net Zero oriented finance transactions – or all of these.

Independent variables

Context
History of firm – Green purpose from the start of the firm's life - or as a recent addition to a long-established firm
Dominant social and economic context - finance community or wider community

'Head' = Top team green leadership, green purpose, metrics, Climate change experts on board,
History of top firm - Green purpose adopted by board and executive team for in past decade, five, two years

'House' = Existence of - Green aware - decision teams and routines (top, middle management, front line),
& of - Green products, £s for new technology for carbon & GHG measurement, & cost/benefit measurement,
& Green oriented incentives, green control systems - complementary to financial value aims

'Community' = New green training £s, increase in meetings to focus on green issues, customer engagement activity,
positive surveys of employees and customers

'Machine' = Changes to green capital bases (debt short and long, and equity), and to a green capital structure
Stability of financial position of the firm relative to climate change risks
Range of green oriented financial products, with offsetting and diversified financial risks,
and aggregate positive climate change impacts

Fig. 9.1. Hypotheses About Change.

metaphor areas. Thus 'success' in the green financial 'machine' can be developed as a dependent variable with this being a function of the above 'context, head, house, and community' factors. The wider theory narrative in Section 5.1 provides a context to form shared meanings and understanding about such empirical results and their impact on Net Zero change.

5.3. Larger debate

The green BTFF and the theoretical analysis in the chapter are also part of *larger policy and academic debates about* the nature and purpose of the world of global finance and financial firms (Collier, 2018; Kay & King, 2020) such as evolving the means to realign value in financial markets with values of wider society (Carney, 2020). The green BTFF also reflects an emerging 'social finance' research programme (Hirshleifer, 2015) of field and qualitative research in finance which focusses on understanding how financial firms function in and exploit expert knowledge in social and economic contexts (cf. Chen et al., 2014, 2018; Holland, 2016, 2017, chapter 8, part 3, 2018, chapter 5, 2019).

6. CONCLUSIONS

This chapter discussed how large international financial firms can become green oriented and Net Zero focussed. This ambition involves changes in complex systems and was explored through empirical narratives and theoretical narratives (Golden-Biddle & Locke, 2007) to build-up a conceptual framework in the form of a green BTFF (Holland, 2019, 2021).

The green BTFF directly addresses issues of uncertainty and complexity. It is part of a knowledge risk management strategy (La Torre, 2020; Zack, 1999) aimed at understanding complex firms and responding to uncertainty (Bratianu, 2020). It closes *knowledge gaps* (Holland, 2010) in both practice and academe. This is part of the evolving discourse on how to realign financial value in international financial markets with values of wider global society (Carney, 2020).

The holistic narrative and metaphor approach in green BTFF is aimed at the popular question 'what is going on here?' (Kay & King, 2020, p. 10). This is a basis to develop integrated thinking and actions in practice (VRF (Value Reporting Foundation), 2021) and academe (Gendron & Smith-Lacroix, 2013.

The four-part metaphors and narrative aid comprehension of the whole financial firm system, and each metaphor area. This reduction of knowledge risks (La Torre, 2020) goes beyond financial risks (Holland, 2010) to include risks with non-financial aspects of firms in support of Net Zero aims.

This approach has potential 'to make a difference' in researching, learning, thinking, and believing about desirable actions and responses to demands for green oriented change in financial firms and wider systems (Shiller, 2019). It forms part of the means to sceptically and critically explore whether international financial firms will deliver authentic Net Zero change and reporting (Brunsson, 1993; Cho et al., 2015; Schoeneborn et al., 2020).

In the *field of practice,* integrated thinking offers many potential ways 'to make a difference' in *actions* in international financial firms, global finance system, and global and national societies. In individual *firms*, the green BTFF can be used to argue that firms must clarify the key components, connections, and dynamics in these Net Zero changes via internal communications and external reporting (VRF (Value Reporting Foundation), 2021). It is used to argue that significant coordinated changes are required in 'deep' or hidden aspects of financial firms and in their specialist roles, to make them 'green oriented' and responsive to climate change problems. The green BTFF is a means to develop reporting to explore if organised hypocrisy behaviour (Brunsson, 1989) is used by specific international financial firms such as NatWest and Triodos and is undermining authenticity and becoming dysfunctional. It is a means in global financial *markets and society* to explore if organised hypocrisy behaviour is used *by alliances* of many international financial firms such as GFANZ (2021).

In the *field of academe*, this enhanced understanding can support development of an academic research programme about 'green finance' using a range of non-finance academic disciplines in management, sociology, and systems. The use of interdisciplinary sources, and development of connected theoretical narrative can encourage a rethink of empirical research and theory development in the field of finance (Gendron & Smith-Lacroix, 2013; Holland, 2019). The green BTFF provides also the basis to enhance collaboration amongst academics to critically analyse how financial firms can and should go green. For example, the green BTFF reflects an international version (Peng, 2001) of *RBV* of the firm (Barney, 1991; Hart, 1995; Teece et al., 1997) whereby the large international case financial firms such as NatWest/RBS, Lloyds, Schroders, and Baillie Gifford, develop unique firm specific competitive advantages concerning Net Zero change within a global and sector wide alliance creation of shared co-operation advantages.

REFERENCES

Abdallah, S. B., Saïdane, D., & Slama, M. (2020). CSR and banking soundness: A causal perspective. *Business Ethics: A European Review*, *29*, 706–721.

Aden, N., Chang, S., & Murray, J. (2022, March 29). *Setting science-based targets for financial institutions (SBTi*, 13:30-13:55). In *Net Zero Finance Conference*. https://app.swapcard.com/event/net-zero-finance/planning/UGxhbm5pbmdfODYyMDA3

Barile, S., Orecchini, F., Saviano, M., & Farioli, F. (2018). People, technology, and governance for sustainability: The contribution of systems and cyber-systemic thinking. *Sustainability Science*, *13*, 1197–1208.

Barney, J. (1991). Firm resources and sustained competitive advantage. *Journal of Management*, *17*(1), 99–120.

Bénabou, R., & Tirole, J. (2006). Incentives and prosocial behavior. *American Economic Review*, *96*(5), 1652–1678.

Beusch, P., Frisk, E., Rosén, M., & Dilla, W. (2016, January). *Management control for sustainability: The development of a fully integrated strategy* [Conference presentation]. 2016 Management accounting section meeting: Research and case conference, Dallas, TX, USA.

Bourdieu, P. (1990). *The logic of practice*. Polity Press.

Bratianu, C. (2020). A knowledge management approach to complex crises. *Management Dynamics in the Knowledge Economy*, *8*(4), 345–356. https://doi.org/10.2478/mdke-2020-0022. ISSN 2392-8042

Brunsson, N. (1989). *The organization of hypocrisy talk, decisions, and actions in organizations*. Chichester John Wiley & Sons Inc.

Brunsson, N. (1993, August). Ideas and actions: Justification and hypocrisy as alternatives to control. *Accounting, Organizations and Society*, *18*(6), 489–506

Buckle, M., & Beccali, E. (2011). *Principles of banking and finance* (FN1024, 2790024). https://www.coursehero.com/file/83165270/1-PBF-FN1024-VLE-full-version-2011pdf/

Burnes, B. (2004, September). Kurt Lewin and the planned approach to change: A re-appraisal. *Journal of Management Studies*, *41*(6), 977–1002. https://onlinelibrary.wiley.com/doi/full/10.1111/j.1467-6486.2004.00463.x

Carney, M. (2015, September 29). *Breaking the tragedy of the horizon – Climate change and financial stability*. Speech given by Mark Carney, Governor of the Bank of England, Lloyd's of London. https://www.bankofengland.co.uk/speech/2015/breaking-the-tragedy-of-the-horizon-climate-change-and-financial-stability

Carney, M. (2020, December). *How we get what we value*. Reith Lecture, BBC Radio 4. https://www.bbc.co.uk/programmes/m000qkms

Chen, L., Danbolt, J., & Holland, J. (2014). Rethinking bank business models: The role of intangibles. *Accounting, Auditing, and Accountability Journal*, *27*(3), 563–589.

Chen, L., Danbolt, J., & Holland, J. (2018). Information about bank intangibles, analyst information intermediation, and the role of knowledge and social forces in the 'market for information'. *Accounting Forum*, *42*(3), 261–276. https://doi.org/10.1016/j.accfor.2018.08.002

Child, J. (1972). Organizational structure, environment and performance: The role of strategic choice. *Sociology*, *6*(1), 1–22. https://doi.org/10.1177/003803857200600101

CGFI. (2019, November 19). *Climate risk and green finance regulatory forum*. America Square Conference Centre.

Cho, C., Laine, M., Roberts, R., & Rodrigue, M. (2015, January). Organized hypocrisy, organizational façades, and sustainability reporting, *Accounting, Organizations and Society*, *40*, 78–94.

City Week. (2020, September, 21). *Greening the financial system and making green financial products mainstream* [Conference presentation]. 10th annual international financial services forum 2020 2020 | Guildhall, London, UK. https://www.fsdc.org.hk/en/events/london-city-week-2020

Collier, P. (2018). *The future of capitalism; facing the new anxieties*. HarperCollins.

Cyert, R., & March J. (1963). *A behavioral theory of the firm*. Prentice-Hall.

Dasgupta, P. (2021). *Final report – The economics of biodiversity: The Dasgupta review*. https://www.gov.uk/government/publications/final-report-the-economics-of-biodiversity-the-dasgupta-review

Dashwood, H. S. (2020). The rise of corporate social responsibility as a global norm informing the practices of economic actors. In H. Hansen-Magnusson & A. Vetterlein (Eds.), *The rise of responsibility in world politics* (pp. 167–187). Cambridge University Press.

De Bakker, F., Crane, A., Henriques, I., Husted, B. W. (Eds.). (2019). Publishing interdisciplinary research in *Business & Society*. *Business & Society*, *58*(3), 443–452.

Diaz-Rainey, I., Robertson, B., & Wilson, C. (2017). Stranded research? Leading finance journals are silent on climate change. *Climatic Change*, *143*, 243–260. https://doi.org/10.1007/s10584-017-1985-1

EBA. (2021). *EBA report on management and supervision of ESG risks for credit institutions and investment firms (EBA/REP/2021/18)*. https://www.eba.europa.eu/eba-publishes-its-report-management-and-supervision-esg-risks-credit-institutions-and-investment

Engelmann, A., Kump, B., & Schweiger, C. (2020). Clarifying the dominant logic construct by disentangling and reassembling its dimensions. *International Journal of Management Reviews*, *22*, 323–355.

Esteban-Sanchez, P., Cuesta-Gonzalez, M. D. L., & Paredes-Gazquez, J. D. (2017, September). Corporate social performance and its relation with corporate financial performance: International evidence in the banking industry. *Journal of Cleaner Production*, *162*, 1102–1110.

Ethical Finance Conference. (2018, October 22–23). *GEFI: Regaining trust and demonstrating impact*, Edinburgh. https://www.globalethicalfinance.org/wp-content/uploads/2019/02/Ethical-Finance-2018-Summary.pdf

Ethical Finance Conference. (2019, October 8–9). *GEFI: Finance for positive change*, Edinburgh. https://www.globalethicalfinance.org/wp-content/uploads/2019/11/EF19-Summary.pdf

Ethical Finance Conference. (2020, October 5–8). *GEFI: Talk to action*, Edinburgh. https://www.globalethicalfinance.org/our-work/ethical-finance-summit/ethical-finance-2020/

Ethical Finance Conference. (2021, June 8–10). *GEFI: Financing a sustainable future: Climate and beyond*, Edinburgh. https://www.globalethicalfinance.org/our-work/ethical-finance-summit/ethical-finance-2021/

Ethical Finance Conference. (2022, September 6). *GEFI: ESG in a volatile world: Profit, purpose or politics?* Edinburgh. https://www.globalethicalfinance.org/our-work/ethical-finance-summit/global-2022/#agenda

Ethical Finance Conference. (2023, September 20). *GEFI: ESG emergency – From promises to solutions*, Edinburgh. https://www.globalethicalfinance.org/our-work/ethical-finance-summit/ethical-finance-global-2023-highlights/

EU. (2020). *Taxonomy for sustainable activities | European Commission*. https://ec.europa.eu/info/business-economy-euro/banking-and-finance/sustainable-finance/eu-taxonomy-sustainable-activities_en

Falkner, R. (2021). *Environmentalism and global international society*. Cambridge University Press.

Finance for Impact Summit. (2022, July 18). City of London Corporation, Impact Investing Institute, and KPMG. City of London. https://www.theglobalcity.uk/finance-for-impact

Gendron, Y., & Smith-Lacroix, J.-H. (2013). The global financial crisis: Essay on the possibility of substantive change in the discipline of finance. *Critical Perspectives on Accounting*, *30*, 83–101.

GFANZ. (2021). *New financial alliance for Net Zero emissions launches*. https://unfccc.int/news/new-financial-alliance-for-net-zero-emissions-launches

Glasgow COP26. (2021, November 12). *Glasgow agreement on climate change*, UNFCCC. https://unfccc.int/conference/glasgow-climate-change-conference-october-november-2021

Golden-Biddle, K., & Locke, K. (2007). *Composing qualitative research*. Sage.

Gond, J. P., Grubnic, S., Herzig, C., & Moon, J. (2012). Configuring management control systems: Theorizing the integration of strategy and sustainability. *Management Accounting Research*, *23*(3), 205–223

Green Finance Institute. (2021, November 2–12). *Wake up to GHS@COP26* [Programme] by UN Special Envoy, Climate Action and Finance, Mark Carney and Chief Financial Officer of Alphabet, Ruth Porat. YouTube. https://www.youtube.com/watch?v=ac3WhYAiOWs

Green Finance Summit. (2017, May 31–June 1). *The future of green finance*. Guildhall.

Green Finance Summit. (2018, July 17). *Mainstreaming green finance*. Guildhall.

Green Finance Summit. (2019, July 2). *Green finance. Investing in actions*. Guildhall

Green Horizon Summit. (2020, November 9–11). *Green horizon perspectives* [Online live digital magazine shows]. City of London and Green Finance Institute. https://www.theglobalcity.uk/sustainable-finance/green-horizon-events

Green Horizon Summit. (2021, November 2–5 & November 11). *COP26 Glasgow: Rising to the climate challenge* [Online conference]. City of London and Green Finance Institute. https://www.greenfinanceinstitute.com/event/green-horizon-summit-cop26/

Harrison P., in discussion with Leppard, K. (2020, May 21). *Schroders, 'Schroders' Peter Harrison discusses COVID-19 and beyond* [Webinar]. https://www.cazenovecapital.com/uk/financial-adviser/insights/strategy-and-economics/a-conversation-with-schroders-ceo-peter-harrison-webinar/

Harrison P., in discussion with Romer-Lee, R. (2019, March 19). *Talking with Peter Harrison.* Square Mile. https://www.squaremileresearch.com/Insights/entryid/535/Default

Hart, S. L. (1995). A natural-resource-based view of the firm. *Academy of Management Review*, *20*(4), 966–1014.

Heyden, M., Fourne, S., Koene, B., Werkman, R., & Ansari, S. (2017). Rethinking 'top-down' and 'bottom-up' roles of top and middle managers in organizational change: Implications for employee support. *Journal of Management Studies*, *54*(7), 961–985. https://doi.org/10.1111/joms.12258

Hirshleifer, D. (2015). Behavioral finance. *Annual Review of Financial Economics*, *7*, 133–159.

Hirsch, W. (2016, June 29) *What is change management?*. https://wendyhirsch.com/blog/change-management-concepts

Holland, J. B. (2010). Banks, knowledge and crisis: A case of knowledge and learning failure. *Journal of Financial Regulation and Compliance*, *18*(2), 87–105.

Holland, J. B. (2014). *Complexity; A very short introduction.* Oxford University Press.

Holland, J. B. (2016). A behavioural theory of the fund management firm. *European Journal of Finance*, *22*(11), 1004–1039.

Holland, J. B. (2017). Rethinking models of banks and financial institutions. In J. Guthrie, F. Ricceri, J. Dumay, & C. Nielsen (Eds.), *The Routledge companion to intellectual capital* (pp. 113–129). Routledge.

Holland, J. B. (2018). Changing financial firms relative to ESG issues. In S. Arvidsson (Ed.), *Challenges in managing sustainable business: Reporting, taxation, ethics and governance* (pp. 111–133). Palgrave Macmillan. ISBN 9783319932651

Holland, J. B. (2019). A new approach to research and theory development for financial firms. *Journal of Financial Regulation and Compliance*, *27*(2), 215–242.

Holland, J. B. (2021). *How can financial firms go green?* [Working paper]. The Adam Smith Business School, University of Glasgow.

Hong, H., & Scheinkman, J. A. (2020). Climate finance. *Review of Financial Studies, Society for Financial Studies*, *33*(3), 1011–1023. https://academic.oup.com/rfs/issue/33/3

Howard, A. (2019, April 11). *SustainEx: Quantifying the hidden costs of companies' social impacts.* https://www.asianinvestor.net/article/schroders-sustainex-a-new-measure-to-quantify-social-and-environmental-impacts-o/454248.

IIRC. (2013). The international framework. Integrated Reporting Council. https://www.integratedreporting.org/wp-content/uploads/2013/12/13-12-08-THE-INTERNATIONAL-IR-FRAMEWORK-2-1.pdf

IPCC. (2021, August 9). *Sixth assessment report (AR6)*. https://www.ipcc.ch/assessment-report/ar6/

Kay, J., & King, M. (2020, March). *Radical uncertainty: Decision-making for an unknowable future.* The Bridge Street Press.

Knights D., & Willmott, H. (1997, March). The hype and hope of interdisciplinary management studies. *British Journal of Management*, *8*(1), 9–22.

Larsen, M. (2017). *Applying the integrated reporting concepts of outcomes and social and relationship capital in the banking industry*. Banking Network. https://www.integratedreporting.org/wp-content/uploads/2016/03/SR-Capital-and-Outcomes-IR-Banking-Network.pdf

Larsen, M. (2021). Net Zero Finance Conference.

Lave, J., & Wenger, E. (1991). *Situated learning: Legitimate peripheral participations.* Cambridge University Press.

La Torre, M. (2020). Knowledge management, risk management, knowledge risk management: What is missing (or messed) in financial and banking sectors. In *Risk in banking*. Palgrave Pivot. https://doi.org/10.1007/978-3-030-54498-0_3

Lejarraga, J., & Pindard-Lejarraga, M. (2020, September). Bounded rationality: Cognitive limitations or adaptation to the environment? The implications of ecological rationality for management learning. *Academy of Management Learning & Education*, *19*(3), 289–306.

Levy, B (2014). *Working with the grain: Integrating governance and growth in development strategies.* Oxford University Press.

Lewin, K (1947). Frontiers of group dynamics: Concept, method and reality in social science, social equilibria, and social change. *Human Relations, 1*, 5–41. https://doi.org/10.1177/001872674700100103

Lewis, M., & Davies, K. T. (1987). *Domestic and international banking*. MIT Press.

Lueg, R., & Radlach, R. (2016, April). Managing sustainable development with management control systems: A literature review. *European Management Journal, 34*(2), 158–171.

Mathews, A. (2021). Net Zero Finance Conference.

Mio, C., Agostini, M., & Panfilo, S. (2022). Bank risk appetite communication and risk taking: The key role of integrated reports. *Risk Analysis, 42*(3), 634–652. https://doi.org/10.1111/risa.13776

Mirfenderesk, H., & Corkill, D. (2009). Sustainable management of risks associated with climate change. *International Journal of Climate Change Strategies and Management, 1*, 146–159.

Mitleton-Kelly, E. (2003). *Ten principles of complexity and enabling infrastructures*. In E. Mitleton-Kelly (Ed.), *Complex systems and evolutionary perspectives on organisations* (pp. 3–20). Elsevier.

Moon, J., Gond, J. P., Grubnic, S., & Herzig, C. (2011, September). *Management control for sustainability strategy*. In Research Executive Summary Series (Vol. 7, Issue 12). CIMA.

Morgan, G. (1997). *Images of organisation*. Sage.

Morgan, G. (2013). *Riding the waves of change – Developing managerial competencies for a turbulent world*. https://yorkspace.library.yorku.ca/xmlui/handle/10315/27895

Mumford, E. A. (2000). Socio-technical approach to systems design. *Requirements Engineering, 5*, 125–133.

Net Zero Conference. (2023, October 31–November 1). Business Green. Business Design Centre. London. https://www.netzerofestival.com/

Net Zero Finance Conference. (2021, March 16). Business Green, London. https://app.swapcard.com/event/businessgreen-pathway-finance/plannings/RXZlbnRWaWV3XzEzNTgwNg%3D%3D?

Net Zero Finance Conference. (2021, October 21). *Banking on the future - how can banks help accelerate the net zero transition?* Business Green, London. https://www.businessgreen.com/interview/4038770/net-zero-festival-banking-future-banks-help-accelerate-net-zero-transition

Net Zero Finance Conference. (2022, March 29). *We're going to be all in': Bank of America executive calls for stronger net zero push in response to geopolitical instability*. Business Green, London. https://www.businessgreen.com/news/4047455/bank-america-executive-calls-stronger-net-zero-push-response-geopolitical-instability

Net Zero Nature. (2021, May 27). Business Green, London. https://app.swapcard.com/event/business-green-net-zero-nature/planningslive/RXZlbnRWaWV3XzIwODczMw%3D%3D

Ortiz-de-Mandjona, N., & Bansal, T. (2015, June). The long-term benefits of organizational resilience through sustainable business practices. *Strategic Management Journal, 37*(8), 1615–1631.

Paris COP21. (2015, December 12). *Paris agreement on climate change*. UNFCCC. https://unfccc.int/process-and-meetings/the-paris-agreement

Pedler, M., Burgoyne, J., & Boydell, T. (1997). *The learning company* (2nd ed.). McGraw Hill.

Peng, M. W. (2001, November–December). The resource-based view and international business. *Journal of Management, 27*(6), 803–829.

Poterba, J. (2021, August 5). *Discussion of "How can financial firms go green?" by John Holland* [Conference presentation]. World Finance Conference – Norway.

Porter, M. (1985). *Competitive advantage: Creating and sustaining superior performance*. Free Press.

PPCA. (2021). *New PPCA members tip the scales towards 'consigning coal to history' at COP26*. https://www.poweringpastcoal.org/news/press-release/new-ppca-members-tip-the-scales-towards-consigning-coal-to-history-at-cop26

PRB. (2019). *Principles of responsible banking (PRB)*. UNEP Financial Initiative. https://www.unepfi.org/banking/bankingprinciples/

Rose, A. (2020 September, 21). *CEO of NatWest/RBS, 'Greening the financial system and making green financial products mainstream'* [Annual forum]. 10th annual international financial services forum. https://www.youtube.com/watch?v=k-7fUV8q15A&feature=youtu.be

Schoeneborn, D., Morsing, M., & Crane, A. (2020). Formative perspectives on the relation between CSR communication and CSR practices: Pathways for walking, talking, and t(w)alking. *Business & Society, 59*(1), 5–33.

Scholtens, B., & van Wensveen, D. (2003). *The theory of financial intermediation: An essay on what it does (not) explain*. SUERF –The European Money and Finance Forum. ISBN 3-902109-15-7. https://www.suerf.org/studies/143/the-theory-of-financial-intermediation-an-essay-on-what-it-does-not-explain

Schroders. (2022). *Schroders' greenhouse gas emission reduction goals formally validated by the science based targets initiative*. Media Relations. https://www.schroders.com/en/media-relations/newsroom/all_news_releases/schroders-greenhouse-gas-emission-reduction-goals-formally-approved-by-the-science-based-targets-initiative/

Schein, E. H. (2004). *Organizational culture and leadership* (3rd ed.). Jossey-Bass.

SBTi. (2022, February). *Financial sector science-based targets – Guidance version 1.0*. https://sciencebasedtargets.org/resources/files/Financial-Sector-Science-Based-Targets-Guidance.pdf

Shiller, R. (2019). *Narrative economics: How stories Go viral and drive major economic events*. Princeton University Press.

Silverman, D. (1970). *The theory of organisations: A sociological framework*. Heinemann.

Statman, M. (1999). Behaviorial finance: Past battles and future engagements. *Financial Analysts Journal*, *55*(6), 18–27.

Strauss, A., & Corbin, J. (1998). *Basic of qualitative research: Techniques and procedures for developing grounded theory* (2nd ed.). Sage.

TCFD. (2017, June). *Technical supplement: The use of scenario analysis in disclosure of climate-related risks and opportunities*. Task Force on Climate related Financial Disclosures. https://www.fsb-tcfd.org/publications/final-technical-supplement/

Teece, D. J. (2007). Explicating dynamic capabilities: The nature and microfoundations of (sustainable) enterprise performance. *Strategic Management Journal*, *28*(13), 1319–1350.

Teece, D. J., Pisano, G., & Schuen, A. (1997). Dynamic capabilities and strategic management. *Strategic Management Journal*, *18*(7), 509–533.

Tversky, A., & Kahneman, D. (1974). Judgment under uncertainty: Heuristics and biases. *Science*, *185*(4157), 1124–1131.

UKSIF. (2018, March 22–23). *UK Sustainable Investment and Finance Association conference 2018* [Conference presentation]. The Royal Society of Edinburgh, Edinburgh. https://www.eurosif.org/event/uksif-edinburgh-conference-2018/.

UNEP FI. (2022). *About us*. United Nations Environment Programme Finance Initiative. https://www.unepfi.org/about/

VRF (Value Reporting Foundation). (2021). *Integrated thinking principles: Supporting holistic decision-making*. http://www.integratedreporting.org/wp-content/uploads/2021/12/VRF_ITP-Main-120721.pdf

Whelan-Berry, K. S., Gordon, J. R., & Hinings, C. R. (B.). (2003). Strengthening organizational change processes: Recommendations and implications from a multilevel analysis. *The Journal of Applied Behavioral Science*, *39*(2), 186–207.

Yin, R. K. (2018). *Case study research and applications: Design and methods* (6th ed.). Sage.

Zack, M. H. (1999). Developing a knowledge strategy. *California Management Review*, *41*(3), 125–145.

PART III

WALKING THE TALK AT THE INTERFACE WITH GOVERNMENT POLICIES

CHAPTER 10

THE INVOLVEMENT OF FIRMS IN CHINESE FOREIGN AID PROJECTS: A DATA OVERVIEW AND RESEARCH AGENDA

Guus Hendriks

Amsterdam Business School, Strategy & International Business Section, University of Amsterdam, Amsterdam, The Netherlands

ABSTRACT

China's foreign aid efforts in Africa remain contentious. Chinese foreign aid tends to be different from "traditional" development assistance in that it frequently involves firms as the implementing agents of projects. Firms bring unique resources to public–private partnerships (PPPs) formed with government agencies, but their possible self-interested nature also gives rise to concerns over their development impact. Yet, on a larger scale, little is known about the characteristics of Chinese PPPs in foreign aid. Using project-level data available for 1,308 Chinese aid projects in 50 countries across Africa, the author characterizes the projects undertaken by firms and government agencies in a PPP and contrasts them to those executed by Chinese government agencies without firm involvement. This exploratory data analysis suggests that important differences apply, as Chinese PPPs tend to target different sustainable development goals (SDGs), work on the basis of distinct aid conditions, and implement projects that tend to be larger than those that are solely run by government agencies. Such observations raise important questions of

Walking the Talk? MNEs Transitioning Towards a Sustainable World
Progress in International Business Research, Volume 18, 169–196

ISSN: 1745-8862/doi:10.1108/S1745-886220240000018010

an ethical, theoretical, and international nature, and warrant further research. The author develops a research agenda that aims at issues particularly important for business ethics scholars, organization theorists, and international business scholarship.

Keywords: Africa; business ethics; Chinese multinational enterprises; foreign aid; sustainable development goals; public–private partnerships

INTRODUCTION

Lingering controversy seems to surround China's foreign aid efforts in Africa. On the one hand, there are some who argue that China is bringing much-needed capital, infrastructure development, and business activity to the continent (Dreher & Fuchs, 2015; The Economist, 2017). Others remain more skeptical, and not only attribute China's interest in Africa to its desire for secure access to natural resources, but also to its broader strategic and geopolitical aims which match its rise as a global superpower (Bräutigam, 2011; Sun, 2014; Tan-Mullins et al., 2010).

Why are China's aid efforts in Africa so contentious? One characteristic that sets apart Chinese foreign aid from "traditional" development assistance is that it frequently involves firms as *implementing agents* of aid projects (Dreher et al., 2018; Bluhm et al., 2018). Under such circumstances, the Chinese government will typically agree with their African counterparts on the general conditions of an aid project but will also form a partnership with a firm that oversees and manages the implementation of that project. Although it is sometimes argued that firms bring unique resources to the table, and may be more experienced in bringing infrastructure projects to a successful end (Ballesteros et al., 2017; McEwan et al., 2017), critics voice concern that they primarily operate with their own interests at heart rather than those of African communities affected by the project (Bräutigam, 2009; Dreher et al., 2018).

Still others attribute the controversy over African aid projects to the practices and intentions of the Chinese government in general, rather than those specific to the firms that implement and oversee the projects (Harchaoui et al., 2021; Naím, 2007). That is, Chinese firms are thought to follow guidelines and stipulations expressed by their government. It is these guidelines that conflict with the Western view of foreign aid which traditionally regarded it more as one-directional flows of development assistance. The "Chinese model" of aid conflicts stands in stark contrast to such a view, with its emphasis on creating benefits not only for the recipients of such assistance, but also for the donor state (Morgan & Zheng, 2019; Strauss, 2009).

More research, however, is needed to better understand what roles Chinese firms and government agencies play in African foreign aid projects to get at the heart of the controversy that seems to characterize the phenomenon. An interesting starting point involves looking at the differences between government-run projects and those managed by Chinese firms as implementing agents. Those who view the participation of profit-seeking firms as the driving force behind unwanted social outcomes would likely expect relatively large differences between both of these types of projects. Should there be few differences, however, this may suggest more that Chinese foreign aid projects tend to share similar characteristics, regardless of who oversees the project as an implementing agent, and that the observed practices and outcomes might not necessarily be shaped by the participation of firms. As such, detailed comparisons at the project level will likely shed new light on Chinese foreign aid and the unique role that firms play in it.

Complementing case studies and within-country studies (Bräutigam, 2011; Gu, 2009; Tan-Mullins et al., 2010), my study performs a large-scale analysis of 1,308 Chinese aid projects in 50 countries across Africa. Such an analysis allows for a broader comparison of projects that are principally managed by Chinese government agencies and those that are implemented by firms. My exploratory data analysis suggests that important differences tend to apply. Chinese aid projects with firms as implementing agents target different sustainable development sectors, work on the basis of distinct aid conditions, and tend to be larger than those that are run solely by government agencies. Furthermore, my study also suggests that firm-led aid projects themselves tend to be different in nature, dependent on whether the implementing agent is state owned or privately owned. This serves as a further indication that a firm-level focus is needed to get at the heart of the controversy.

Previous literature on Chinese foreign aid, developed mostly in the fields of international relations and development studies, has more rarely applied such a firm-level perspective (for a notable exception, see Gu, 2009). As such, my study is among the first to use large-scale data to perform an analysis of private sector involvement in Chinese foreign aid. Although exploratory in nature, my study highlights important differences between government-led and firm-run projects. These differences raise important questions of an ethical, theoretical, and international nature, and warrant further research on a phenomenon whereby (international) management scholars are uniquely positioned to make a contribution (Doh, 2015; Howard-Grenville et al., 2019). To guide that research effort, I develop a research agenda that takes aim at issues particularly important for business ethics scholars, organization theorists, and international business scholarship.

CHINESE AID IN AFRICA

Although China's most recent endeavors have received the bulk of scholars' and journalists' attention, its history of aid in Africa already goes back six decades (for detailed historical overviews, see Bräutigam, 2009; Morgan & Zheng, 2019; Tan-Mullins et al., 2010). China concluded the first formal aid relationships with a small number of countries in Africa that typically ranked better on per capita gross domestic product (GDP) and the life expectancy of its citizens (Morgan & Zheng, 2019). While foreign aid gradually increased in volume throughout the 1970s and 1980s, China's involvement with Africa grew in real significance after President Jiang Zemin's African tour in 1996 (Strauss, 2009). The principles behind the extension of foreign aid also evolved over time, as the contemporary mix between traditional official development assistance (ODA) and commercial engagement can be seen as a relatively recent phenomenon that is observed only since the 1990s (Morgan & Zheng, 2019). Before that period, China's approach, as a relatively poor country in economic terms, was piecemeal and experimental, and focused on a more limited set of African countries. From about 20 in the 1970s, the number of African countries with which China had aid relations grew to 50 out of 53 countries by 1994. Perhaps somewhat surprisingly, health and agriculture remained the most important sectors for China throughout much of its initial decades of involvement in Africa (Lin et al., 2016). Yet, by the 1990s, a diversification could be observed, as an increasing number of projects focused on business services and trade. This change went together with China's broader "Go Global" initiative and developed into a distinct "Chinese model" of aid, which is known for mixing ODA-like aid with commercial capital (Morgan & Zheng, 2019).

From China's perspective, this model was built on five main principles: reliable friendship, respect for sovereign equality, non-intervention in domestic affairs, mutually beneficial relations, and a focus on sustainable development (Strauss, 2009). That focus seems to correspond with an initial dominance of projects in social sectors such as health care and education, even though an increasing number of later projects were developed in production sectors and infrastructure. Moreover, China's aid model was also built around its willingness to work with any state, in particular with countries from which Western organizations shied away, such as Sudan, Somalia, or Ghana (Bräutigam, 2009; Tan-Mullins et al., 2010), for example, because of armed conflicts (Lis, 2018). From a Western perspective, Chinese aid contrasts with that of conventional Organisation for Economic Co-operation and Development donors (Morgan & Zheng, 2019). That is, Chinese officials are seen to give a substantial amount of diplomatic attention and support to large aid projects in particular, with a supposed emphasis on those aimed at resource extraction, infrastructure, and other productive sectors (Morgan & Zheng, 2019). Such projects involve a low-cost building strategy that is centered on making use of low-skilled laborers, many of whom come from China, rather than from Africa itself, which often contributes to local unemployment. To support such a strategy, officials draw on a mix of ODA-like finance and commercial capital, by giving a large role to Chinese firms as implementing agents of a project. Although recent scholarly work has started to provide a quantitative overview of Chinese aid projects more in general (e.g., Bluhm et al., 2018; Morgan, 2018), little is known about the role of such firms in particular, leaving the consequences of involving them poorly understood.

Firm Involvement in Foreign Aid

Foreign aid traditionally represented a relationship between donor and recipient governments, whereby the former provides development assistance through the voluntary transfer of goods, money, and services (Hattori, 2001). Often, governments team up with non-governmental organization and draw on their skills and experience to fulfill important goals in relation to the humanitarian or economic development of recipient countries (Banks et al., 2015; Lister, 2000). Increasingly, however, governments now turn to for-profit companies and give them key roles in the implementation of foreign aid projects (Aaronson, 2005; Abramov, 2009). In doing so, they form PPPs in foreign aid, or collaborative efforts between government agencies and for-profit firms (Lawson, 2011). Firms bring unique resources to these partnerships formed with government agencies, allowing PPPs to supply resources that the state is less able to provide, but firms' possible self-interested nature also gives rise to concerns over PPPs' development impact. Agency problems, for example, could arise because of incongruent goals and information asymmetry (e.g., Martimort & Pouyet, 2008). As such, different behavior and outcomes are typically expected when comparing PPPs in foreign aid to projects managed by governmental agencies (e.g., Boardman & Vining, 2012; Kolk et al., 2008).

Such expectations are primarily shaped by findings from studies of "traditional" development aid. While until recently the vast majority of foreign aid came from Western economies, the new climate looks different, with actors from distinctly different institutional settings coming to the scene, most notably from China. This new situation may also come with revised expectations regarding the behavior and outcomes of PPPs in comparison with government-run foreign aid. Especially in the context of Chinese foreign aid, the firms with which the government forms partnerships may either be private enterprises or state-owned enterprises (SOEs). The Chinese government forms partnerships with both these types of companies to further their foreign aid objectives (Bräutigam, 2011; Zhang & Smith, 2017). Yet, for Chinese foreign aid specifically, little is known about the characteristics of projects undertaken by PPPs, and whether they differ from those of projects managed by government agencies only. A better understanding of these characteristics may be considered important, however, given the controversy that is associated with for-profit firms' participation in Chinese foreign aid projects, and the doubts that surround the human rights orientation of Chinese SOEs in particular (Whelan & Muthuri, 2017). The decisions that public–private foreign aid partnerships make likely have consequences for a great range of stakeholders residing at home, and especially, in the recipient country (McEwan et al., 2017). Although foreign aid is traditionally associated with positive development consequences, the decisions made in Chinese aid projects are often criticized, with some arguing that they cause local unemployment, environmental damage, and resource depletion (Morgan, 2018; Tan-Mullins et al., 2010). To better understand the characteristics of projects undertaken by PPPs, I contrast them to projects executed by Chinese government agencies without firm involvement. The aim is to understand whether there are differences between these types of projects, and if such differences may relate to the ownership and governance mode of the involved firms.

METHODOLOGY

To understand the characteristics of foreign aid projects undertaken by firms as implementing agents, and those of projects run only by governments, I use AidData's Geocoded Global Chinese Official Finance dataset (Bluhm et al., 2018).

This dataset provides detailed information on Chinese aid projects that were implemented between 2000 and 2014. As it is primarily used by scholars in fields such as international relations, data on private firm participation have only been scarcely analyzed. However, given that it provides information on the types of actors involved in aid projects, the implementation agents, as well as overall project characteristics, it is particularly suitable for an exploratory analysis of the involvement of firms in Chinese foreign aid. From this large-scale dataset on Chinese foreign aid it can be distilled that project partners need to make strategic and tactical decisions on key issues that define the aid project in which they are involved. These decisions imply that such projects differ in five main respects, namely by (1) the targeted development sector, (2) the aim of the project, (3) the aid conditions set, (4) the size of the project, and (5) the type of host country chosen (Bluhm et al., 2018; Dreher et al., 2018). For 1,522 projects, this dataset provides the most detailed information, which includes a project description, the target sector, the recipient country, flow type (i.e., loan or grant), flow class (i.e., official development aid or other official flows), loan conditions, the implementation agency, and the amount of aid.

First, I linked the targeted development sector to the most closely associated SDG to better understand the sustainable development impact of Chinese foreign aid partnerships. Adopted by the United Nations' General Assembly in December 2015, the SDGs succeeded the Millennium Development Goals as the principal set of targets aimed at promoting prosperity and peace for all member nations. The first 16 SDGs set concrete targets for a broad spectrum of widely accepted development issues, ranging from poverty reduction (SDG 1), to climate action (SDG 13), and gender equality (SDG 5). Somewhat distinct from the other goals, SDG 17 is aimed at promoting cooperation between the private sector and governmental organizations to achieve all the SDGs, for example, through PPPs (Witte & Dilyard, 2017). For 214 out of the 1,522 projects, the sector targeted by the Chinese government in their foreign aid efforts was not specified. I linked the remaining 1,308 projects to the dominant SDG targeted, based on the sector listed in the dataset, and the project descriptions. For example, a project in the "Health" sector was linked to SDG 3 "Good Health and Well-being." In a similar vein, projects in the sector of "Water Supply and Sanitation" were linked to SDG 6 "Clean Water and Sanitation." This classification was corroborated by three graduate students who acted as independent coders, whereby 98% of projects were allocated to the same SDG by three or more coders, and 100% of projects by at least two coders.

Second, the foreign aid projects differ by aim, whereby international development scholars seek to classify them as purely developmental or commercial in nature, or as a project that combines these objectives to a various degree (Bräutigam, 2011; Sears, 2019). AidData divides the projects into six categories by their aim: commercial, developmental, mixed, mixed (no development), mixed (some development), and representational. For one foreign aid project, the aim was not identified (Bluhm et al., 2018).

Third, Chinese foreign aid projects also differ in the conditions that are tied to the provided aid. Again, AidData's dataset is a good source to better understand

Chinese foreign aid partnerships in this respect, as it categorizes projects by the conditions linked to the transfer of finances, goods, or services from the donor to the project recipient. Most Chinese foreign aid projects are either loan-based or grant-based, but further categories are identified, including debt forgiveness and free-standing technical assistance.

Fourth, most of Chinese officials' diplomatic attention and support tends to go to large aid projects (Morgan & Zheng, 2019), but AidData's dataset indicates that there are numerous smaller-sized projects as well. I created five categories based on the overall amount of aid that was pledged by Chinese authorities and thereby distinguish between projects with a value below 1 million USD, and those with amounts between 1 to 10 million USD, 10 to 100 million USD, 100 to 500 million USD, and large projects that involve amounts over 500 million USD.

Fifth, based on AidData's dataset, I provide an overview of projects by a recipient country's "Rule of Law" score and its "Human Development Index." I selected these two key characteristics of recipient countries as both indices can be used as indicators of development progress that African countries have made, whereby the focus is not solely put on economic growth, as some traditional measures have done. By contrast, these measures reflect how well corruption is combatted in a country and what the standard of living in society is. Specifically, World Bank's Rule of Law score captures "the quality of contract enforcement, property rights, the police, and the courts, as well as the likelihood of crime and violence" (Kaufmann et al., 2010). As such, the Rule of Law score is often used in international business studies as a representative measure of the general degree of development of a country's formal institutional environment (Hendriks et al., 2018; Liu et al., 2011). United Nations' Human Development Index is a composite measure of average achievement in key dimensions of human development, and captures life expectancy, health and well-being, education progress, and society's standard of living. In addition, I also distinguish between projects managed by government agencies, SOEs, and private firms by a recipient country's GDP share of total natural resources rents, a figure that is taken from World Bank's World Development Indicators database. This figure is calculated as the difference between commodity prices and the average cost of producing them, before it is expressed as a percentage of a country's GDP. As such, total natural resources rents are the sum of coal rents, forest rents, mineral rents, oil rents, and natural gas rents. All three measures are used to explore what type of countries are targeted by Chinese foreign aid initiatives, to study whether they indeed favor some countries over others for commercial reasons, as some scholars have argued, based on case evidence or more limited amounts of data (Foster et al., 2009; Jiang, 2009; Tan-Mullins et al., 2010).

To establish whether there are differences between the projects undertaken by firms and government agencies in a PPP, and those executed by government agencies without firm involvement, I perform ANOVA tests. ANOVA is a simple but effective technique to test whether two population means are equal. With these tests, it can be verified whether government-run and firm-run projects significantly differ with respect to each of the five categories of indicators outlined in this section.

DATA OVERVIEW AND ANALYSIS

Table 10.1 shows an overview of the sixteen SDGs and the number of foreign aid projects that specifically target that SDG. Both panels (A) and (B) of Table 10.1 give insight into the types of SDGs targeted by projects that are solely government-run, and those that represent a collaboration, either with an SOE or a private firm as the implementing agent of the project. Panel (B) of Table 10.1 shows the number of projects as a percentage of all projects run by either government agencies, SOEs or private firms. For example, 9.3% of all government-run projects target the SDG on poverty eradication, namely SDG 1, whereas there are no projects executed by a firm, either state owned or private, which have the same objective. Table 10.1 reveals more differences in the stated objectives of the projects that are run by either government agencies or by firms. Interestingly, only 21% of all projects executed by these government agencies target SDG 9 on infrastructure development, but for projects that involve a firm as a partner this figure is roughly 40%. In a similar vein, projects executed by firms seem to disproportionally target SDG 7, a goal which aims to ensure affordable and clean energy. This particularly holds for SOEs, as 19.9% of all projects that are executed by such firms target this objective. Government agencies, on the other hand, appear more likely to target the SDGs on health care and well-being (SDG 3) and the set of goals aimed at promoting peace and strong institutions (SDG 16), where private firms, and SOEs in particular, seem to be comparatively underrepresented.

Further differences between government-run projects and those executed by firms become apparent when other characteristics are studied. Table 10.2 presents a similar overview as the one provided above, but now shows the number of projects by stated intent, as these projects can be classified as purely developmental or commercial in nature, or as a combination of these objectives (Bluhm et al., 2018). As panel (B) of Table 10.2 indicates, there appears to be little variation in the projects executed by government agencies, as 90.2% of them are wholly developmental in nature. Projects with a firm as implementing agent appear to be different, in that a far greater share can be classified as mixed with only some aim of promoting development, while 3.6% (SOEs) to 5.7% (private firms) of firm-run projects are even purely commercial in nature.

Table 10.3 shows an overview of the number of projects managed by government agencies, private firms, and SOEs, by the conditions linked to the transfer of finances, goods or services from the donor to the project recipient. Again, panel (B) of Table 10.3 shows the number of projects as a percentage of all projects run by either government agencies, SOEs or private firms. Interestingly, 62.2% of all government-run projects are on the basis of a grant. Grants are different from loans in that they do not "carry the burden of future repayment" (Clements et al., 2004, p. 46). Loans, on the other hand, may induce "policymakers to use funds wisely and to mobilize taxes or, at least, to maintain current levels of revenue collection" (Clements et al., 2004, p. 46). A striking 64.7% of all projects run by SOEs are on the basis of loans, whereas only 24.7% of government-run projects are on such basis. Similarly, private firms seem to be involved more often in projects financed by export credits, as well as those on the basis of loans.

Table 10.1. Overview of the Number of Projects Managed by Government Agencies, Private Firms, and SOEs, by Targeted SDG.

Panel (A)

SDG	Government	Firm		ANOVA[a]
		SOE	Private	
1 – No Poverty	83	0	0	*43.1 (0.000)*
2 – Zero Hunger	40	1	1	*14.9 (0.001)*
3 – Good Health	152	26	7	*20.2 (0.000)*
4 – Quality Education	84	27	6	*0.86 (0.354)*
5 – Gender Equality	7	1	0	*1.41 (0.234)*
6 – Clean Water & Sanitation	23	23	2	*9.26 (0.002)*
7 – Affordable & Clean Energy	31	66	11	*88.9 (0.000)*
8 – Decent Work & Ec. Growth	64	9	9	*4.08 (0.044)*
9 – Industry & Infrastructure	187	126	36	*46.8 (0.000)*
10 – Reduced Inequalities	1	0	0	*0.45 (0.502)*
11 – Sustainable Communities	68	24	5	*0.22 (0.649)*
12 – Responsible Production	32	7	3	*1.35 (0.247)*
13 – Climate Action	1	0	0	*0.45 (0.502)*
14 – Life Below Water	2	3	1	*3.33 (0.068)*
15 – Life on Land	3	2	1	*0.90 (0.343)*
16 – Peace, Justice, & Institutions	111	16	6	*16.5 (0.000)*
Total	889	331	88	

Panel (B)

SDG	Government (%)	Firm	
		SOE (%)	Private (%)
1 – No Poverty	9.3	0.0	0.0
2 – Zero Hunger	4.5	0.3	1.1
3 – Good Health	17.1	7.9	8.0
4 – Quality Education	9.4	8.2	6.8
5 – Gender Equality	0.8	0.3	0.0
6 – Clean Water & Sanitation	2.6	6.9	2.3
7 – Affordable & Clean Energy	3.5	19.9	12.5
8 – Decent Work & Economic Growth	7.2	2.7	10.2
9 – Industry & Infrastructure	21.0	38.1	40.9
10 – Reduced Inequalities	0.1	0.0	0.0
11 – Sustainable Communities	7.6	7.3	5.7
12 – Responsible Production	3.6	2.1	3.4
13 – Climate Action	0.1	0.0	0.0
14 – Life Below Water	0.2	0.9	1.1
15 – Life on Land	0.3	0.6	1.1
16 – Peace, Justice, & Institutions	12.5	4.8	6.8
Total	100	100	100

[a]Column reports ANOVA *F*-statistic from comparison between government and all firms (SOE and private).

p-Values in parentheses.

Table 10.2. Overview of the Number of Projects Managed by Government Agencies, Private Firms, and SOEs, by Project Aim.

Project Aim	Government	Firm		ANOVA[a]
		SOE	Private	
Commercial	20	12	5	*3.40 (0.066)*
Development	802	242	57	*80.5 (0.000)*
Mixed	1	2	1	*3.48 (0.062)*
Mixed (No development)	0	1	1	*4.31 (0.038)*
Mixed (Some development)	50	73	24	*93.9 (0.000)*
Representational	15	1	0	*4.96 (0.026)*
Unset	1	0	0	*0.45 (0.502)*
Total	889	331	88	

Panel (A)

Project Aim	Government (%)	Firm	
		SOE (%)	Private (%)
Commercial	2.2	3.6	5.7
Development	90.2	73.1	64.8
Mixed	0.1	0.6	1.1
Mixed (No development)	0.0	0.3	1.1
Mixed (Some development)	5.6	22.1	27.3
Representational	1.7	0.3	0.0
Unset	0.1	0.0	0.0
Total	100	100	100

Panel (B)

[a]Column reports ANOVA *F*-statistic from comparison between government and all firms (SOE and private).
p-*V*alues in parentheses.

Table 10.3. Overview of the Number of Projects Managed by Government Agencies, Private Firms, and SOEs, by Project Aid Conditions.

Project Aid Conditions	Government	Firm		ANOVA[a]
		SOE	Private	
Debt forgiveness	49	1	0	*21.8 (0.000)*
Debt rescheduling	4	1	1	*0.01 (0.940)*
Export credits	4	14	5	*28.0 (0.000)*
Foreign direct investment	6	1	1	*0.17 (0.681)*
Free-standing technical assistance	30	6	3	*1.46 (0.226)*
Grant	553	81	20	*188.9 (0.000)*
Joint venture with recipient	6	3	4	*2.90 (0.089)*
Loan (excluding debt rescheduling)	220	214	51	*217.7 (0.000)*
Scholarships/training in the donor country	6	0	0	*2.80 (0.094)*
Undisclosed	11	10	3	*5.48 (0.019)*
Total	889	331	88	

Panel (A)

Table 10.3 (*Continued*)

Project Aid Conditions	Government (%)	Firm	
		SOE (%)	Private (%)
Debt forgiveness	5.5	0.3	0.0
Debt rescheduling	0.4	0.3	1.1
Export credits	0.4	4.2	5.7
Foreign direct investment	0.7	0.3	1.1
Free-standing technical assistance	3.4	1.8	3.4
Grant	62.2	24.5	22.7
Joint venture with recipient	0.7	0.9	4.5
Loan (excluding debt rescheduling)	24.7	64.7	58.0
Scholarships/training in the donor country	0.7	0.0	0.0
Undisclosed	1.2	3.0	3.4
Total	100	100	100
	Panel (B)		

[a]Column reports ANOVA F-statistic from comparison between government and all firms (SOE and private).
p-Values in parentheses.

Table 10.4 provides information on the size of the foreign aid projects and the percentage share of such projects managed by government agencies, SOEs, and by private firms. Panel (B) indicates that government-run projects appear to be smaller on average. When a firm is chosen as implementing agent, a relatively greater share of projects run by them is large in size. That is, 77.3% of all projects managed by private firms involve an amount greater than 10 million USD. For government-run projects this is only 38.4%. Firms seem to be disproportionally involved especially in projects valued between 100 and 500 million USD, as 27.2% of all SOE-run projects are in this category, but for government agencies this category represents only 7.9% of all projects.

Table 10.4. Overview of the Number of Projects Managed by Government Agencies, Private Firms, and SOEs, by Project Size.

Project Size	Government	Firm		ANOVA[a]
		SOE	Private	
Below 1 million USD	260	10	4	*126.4 (0.000)*
1–10 million USD	288	54	16	*36.2 (0.000)*
10–100 million USD	242	152	46	*53.2 (0.000)*
100–500 million USD	70	90	18	*82.4 (0.000)*
Over 500 million USD	29	25	4	*9.05 (0.003)*
Total	889	331	88	
		Panel (A)		

(*Continued*)

Table 10.4 (*Continued*)

Project Size	Government (%)	Firm	
		SOE (%)	Private (%)
Below 1 million USD	29.2	3.0	4.5
1–10 million USD	32.4	16.3	18.2
10–100 million USD	27.2	45.9	52.3
100–500 million USD	7.9	27.2	20.5
Over 500 million USD	3.3	7.6	4.5
Total	100	100	100

Panel (B)

[a]Column reports ANOVA F-statistic from comparison between government and all firms (SOE and private).
p-Values in parentheses.

Lastly, firms' participation in foreign aid projects may also mean that activities are developed in specific types of countries, but not in others. Table 10.5 explores this premise by presenting the number of projects managed by government agencies, SOEs, and private firms by a recipient country's "Rule of Law" score and its "Human Development Index." The Rule of Law indicator ranges from –2.5 to 2.5, with most developed countries reporting a score of 1.3 or higher. United Nations' Human Development Index consists of four categories of countries, namely those characterized by low human development, low-to-medium human development, medium-to-high human development, and high human development. Table 10.5 reveals fewer pronounced differences between government-run projects and projects managed by firms than the findings above indicate. Yet, panel (B) seems to indicate that a greater share of company-run projects takes place in countries characterized by a relatively poor rule of law. That is, 35.6% of SOE-run projects are in countries with a rule of law score between –1.5 and –1 such as Liberia or Sudan. Only 25.3% of government-run projects are in this category of countries. When countries are categorized by their Human Development Index score, however, few differences can be observed between company-run projects and those that are managed by government agencies, as indicated by panels (C) and (D) of Table 10.5.

Some scholars have expressed concern about China's role in Africa, arguing that the increasing amount of attention going to this continent may be related to its aim of securing access to natural resources (Jiang, 2009). This while the potential of MNE-led development in the extractive sector is weak, as typically only few linkages are established with local firms and the communities they represent (Narula, 2018). Table 10.6 presents the number of projects managed by government agencies, SOEs, and private firms by a recipient country's GDP share of total natural resources rents. Both panels (A) and (B) of Table 10.6 do not suggest great differences between government-run projects and those managed by private companies in a partnership. However, it appears that SOEs as implementing agents are more frequently involved in aid projects that target resource-rich countries. That is, 26.9% of their aid projects take place in countries where the natural resources sector makes up at least 25% of GDP. Interestingly, for government-run projects and those managed by private firms the equivalent figure is much lower, 13.6% and 12.5% respectively.

Table 10.5. Overview of the Number of Projects Managed by Government Agencies, Private Firms, and SOEs, by Key Characteristics of Recipient Countries.

Recipient Country Rule of Law	Government	Firm		ANOVA[a]
		SOE	Private	
Below −2	8	1	0	*1.80 (0.180)*
Between −2 and −1.5	85	33	12	*0.50 (0.478)*
Between −1.5 and −1	225	118	28	*13.5 (0.000)*
Between −1 and −0.5	261	78	18	*5.69 (0.017)*
Between −0.5 and 0	202	63	19	*1.52 (0.218)*
Between 0.0 and 0.5	62	12	8	*2.26 (0.132)*
Over 0.5	33	18	2	*0.87 (0.351)*
Total	876	323	87	

Panel (A)

Recipient Country Rule of Law	Government (%)	Firm	
		SOE (%)	Private (%)
Below −2	0.9	0.3	0.0
Between −2 and −1.5	9.7	10.2	13.8
Between −1.5 and −1	25.7	36.5	32.2
Between −1 and −0.5	29.8	24.1	20.7
Between −0.5 and 0	23.1	19.5	21.8
Between 0.0 and 0.5	7.1	3.7	9.2
Over 0.5	3.8	5.6	2.3
Total	100	100	100

Panel (B)

Recipient Country Human Development Index	Government	Firm		ANOVA[a]
		SOE	Private	
Low human development	135	50	12	*0.03 (0.857)*
Low-to-medium human development	460	177	50	*0.70 (0.404)*
Medium-to-high human development	244	85	22	*0.53 (0.467)*
High human development	29	10	3	*0.02 (0.881)*
Total	868	322	87	

Panel (C)

Recipient Country Human Development Index	Government (%)	Firm	
		SOE (%)	Private (%)
Low human development	15.6	15.5	13.8
Low-to-medium human development	53.0	55.0	57.5
Medium-to-high human development	28.1	26.4	25.3
High human development	3.3	3.1	3.4
Total	100	100	100

Panel (D)

[a]Column reports ANOVA *F*-statistic from comparison between government and all firms (SOE and private). *p*-Values in parentheses.

Table 10.6. Overview of the Number of Projects Managed by Government Agencies, Private Firms, and SOEs, by Recipient Countries' GDP Share of Natural Resource Rents.

Recipient Country – Natural Resource Rents as % of GDP	Government	Firm		ANOVA[a]
		SOE	Private	
0–5	178	62	14	*0.56 (0.455)*
5–10	217	67	24	*1.01 (0.315)*
10–15	163	46	17	*2.02 (0.155)*
15–20	140	46	17	*0.08 (0.780)*
20–25	57	13	4	*2.87 (0.009)*
25–30	20	6	2	*0.14 (0.706)*
30–35	29	10	1	*0.37 (0.545)*
35–40	16	9	0	*0.20 (0.654)*
40–45	9	10	1	*4.98 (0.026)*
45–50	23	44	6	*49.5 (0.000)*
Over 50	24	10	1	*0.00 (0.951)*
Total	876	323	87	

Panel (A)

Recipient Country – Natural Resource Rents as % of GDP	Government (%)	Firm	
		SOE (%)	Private (%)
0–5	20.0	18.7	15.9
5–10	24.4	20.2	27.3
10–15	18.3	13.9	19.3
15–20	15.7	13.9	19.3
20–25	6.4	3.9	4.5
25–30	2.2	1.8	2.3
30–35	3.3	3.0	1.1
35–40	1.8	2.7	0.0
40–45	1.0	3.0	1.1
45–50	2.6	13.3	6.8
Over 50	2.7	3.0	1.1
Total	100	100	100

Panel (B)

[a]Column reports ANOVA *F*-statistic from comparison between government and all firms (SOE and private).
p-Values in parentheses.

Besides reporting the (relative) numbers of projects managed by either government agencies or by firms, Tables 10.1–10.6 also show the results of ANOVA tests that compare the means of these two groups. In Table 10.1, for example, the F-statistic of 14.9 indicates that there is a significant difference between the proportion of projects targeting SDG 2 (Zero Hunger) managed by government agencies (4.5%) and the collective amount executed by private firms *and* SOEs (0.5%) ($p = 0.001$). In line with the observations discussed above, these ANOVA tests suggest that there are particularly stark differences when it comes to projects targeting SDG 3 (Good Health), SDG 7 (Affordable and Clean Energy), SDG 9 (Industry and Infrastructure) and SDG 16 (Peace, Justice, and Institutions). In a similar vein, the ANOVA test results reported in Tables 10.2–10.4 confirm earlier observations by indicating significant differences between government-managed and firm-run projects in terms of their aim (developmental or mixed with only some aim of development), their conditions (grant- versus loan-or credit-based), and their size. In contrast, Tables 10.5 and 10.6 provide fewer reasons to suspect differences between government- and firm-run projects, although for some categories the ANOVA test results are significantly different, that is, when the recipient country's rule-of-law score ranges from –1.5 to –0.5 and when resource rents constitute 40 to 50% of that country's GDP.

RESEARCH AGENDA

The preliminary analysis of data pertaining to Chinese foreign aid projects and their implementing agents provides interesting insights, but raises many questions, too. PPPs in development policy have attracted an increasing amount of attention from scholars in various fields (Brogaard & Petersen, 2018; Kolk et al., 2008). Prior studies have thereby shown that agency problems could arise because of incongruent goals and information asymmetry, primarily through analyses of development-oriented PPPs from Western countries (Martimort & Pouyet, 2008; Parker et al., 2018). Agency problems are likely substantially reduced when an agent depends on its principal for resources, as deviating behavior may carry much greater consequences for that agent (Klijn, 2022; Kort et al., 2016). Under such circumstances, it is expected that PPPs will differ much less in behavior and outcomes from "traditional" government-controlled foreign aid, as these PPPs are expected to align more with government policy. Chinese aid to Africa provides such a context, but the findings of this study hint at relatively large differences since Chinese PPPs target different SDGs, work with distinct aid conditions, and implement larger projects than those solely run by government agencies. That is, the findings displayed in Tables 10.1–10.6 raise the question whether PPPs in foreign aid have a disproportionate tendency to implement projects that target certain SDGs at the expense of others, and similarly, whether these partnerships are characterized by a different intent and work on the basis of different aid conditions. Hence, when firms take on an important role in foreign aid projects there may be significant consequences for the various stakeholders involved in these projects.

Such implications can be approached from different angles. First, the presented data raise ethical concerns and questions. Increasingly, some governments seem willing to allocate important roles to companies in foreign aid projects, as can be observed in the case of China. Is that trend accompanied by a shift in focus toward particular SDGs, too? Furthermore, can progress on all SDGs be made through foreign aid projects when firms are in charge? Second, the findings raise questions from an organization theory perspective, particularly in relation to the complexity that implementing agents face when managing projects with multiple stakeholders whose interests may not be aligned (Bode et al., 2019). Third, issues specifically international in nature emerge as important. Specifically, to what extent is it important that companies are *foreign* when they are involved in foreign aid projects? I will consider these questions and issues as I set a research agenda in relation to each of the five themes that were explored above. Table 10.7 shows the preliminary framework that guides this research agenda.

Sectors of Sustainable Development

The findings displayed in Table 10.1 predominantly raise the question if Chinese foreign aid partnerships with firms as agents have a disproportionate tendency to implement projects in the areas of infrastructure, sanitation, and energy, and less in areas that target poverty eradication and health care. Further research is needed to better understand this tendency, in China, but also in other contexts (Sato et al., 2011; Stallings & Kim, 2017). This is particularly relevant given that health and education tended to be the sectors in which the largest proportion of aid projects were developed throughout much of China's history of involvement in Africa (Lin et al., 2016; Morgan & Zheng, 2019). In a similar vein, more research is needed to comprehend the ethical implications, especially given the urgency and complexity of Agenda 2030 and the limited progress on SDGs that has been made in some countries (Allen et al., 2018; George et al., 2016). Is progress toward *all* the SDGs possible through PPPs such as those observed in Chinese foreign aid projects? As attention increasingly turns to companies and their contribution to the SDGs, do expectations about their development impact need to be adjusted downwards, particularly those coming from non-munificent institutional environments (Bhatt et al., 2019)? Relatedly, does an underrepresentation of certain development sectors suggest that limits may apply to corporate social responsibility? Given the complexity of Agenda 2030, how do PPPs prioritize development sectors and projects?

From an organization theory perspective, it may be said that PPPs in foreign aid represent a relatively understudied context where more work can be undertaken to test boundary conditions of various theories (cf. Scherer & Palazzo, 2011). When firms are involved in foreign aid projects, and specific SDGs are targeted, scholars may study the reasons why this behavior is observed. Do PPPs target sectors where host-country legitimacy matters to a different degree? From a resource-based view standpoint, could the SDG focus be explained by the specific resources that firms bring to the table, or alternatively, their interest in gaining access to novel ones?

Table 10.7. Research Agenda and Framework of Research Issues.

	Ethical Issues	Organization Theory Issues	International Issues
Do Chinese foreign aid partnerships have a disproportionate tendency to implement projects in infrastructure, sanitation, and energy, at the expense of poverty eradication and health care?	Fulfilment of all SDGs Complexity of Agenda 2030 Limits to CSR	Provision and recombination of resources Host-country legitimacy	Transferability and recombination of resource bundles: "hard" versus "soft" resources
Do Chinese foreign aid partnerships have a disproportionate tendency to implement projects with at least some form of commercial intent?	Balance between developmental and commercial aims	Self-interest versus legitimacy Power dynamics and resource dependence in partnerships	Applies to "Belt and Road Initiative"?
Do Chinese foreign aid partnerships have a disproportionate tendency to work on the basis of loans rather than grants?	Stability of income for developing countries with weak institutions	Principal–agent problem and incentive alignment	Foreign relations and "no strings attached" aid policy
Do Chinese foreign aid partnerships have a disproportionate tendency to implement projects that are larger in size?	SDG potential but dependence on projects with multiple stakeholders and interests	Complexity of managing project: wicked problem	Combination of "liability of foreignness" with difficulties of managing large projects
Do Chinese foreign aid partnerships have a disproportionate tendency to implement projects in resource-rich countries and in those with a poor rule of law?	Potential problem of corruption "Quid pro quo" in foreign aid	Institutional voids and non-market resource bundles	Perverse risk incentive and foreign aid project risks Multi-country involvement and temporal dynamics

Firms may contribute to foreign aid projects with unique resources, but the necessity of leveraging them across country borders raises questions about issues distinctly international in nature. Much research has been undertaken in the field of international business on the transferability and recombination of resource bundles (e.g., Grøgaard et al., 2022; Verbeke & Kano, 2016). Yet, it remains to be studied if activity in certain sustainable development sectors may facilitate these processes, in that "hard" resources for infrastructure projects are more easily deployable than "soft" resources such as those needed for socio-economic projects.

Project Aim

The data displayed in Table 10.2 also raise questions of an ethical, theoretical, and international nature. Mainly, it suggests that Chinese PPPs in foreign aid may have a disproportionate tendency to implement projects with at least some form of commercial intent. As such, it is worth studying how such PPPs balance developmental and commercial aims, and what that balance implies for countries' progress on SDGs.

For an in-depth understanding of the possibly divergent interests of firms and foreign aid stakeholders, and for insights about the alignment of such interests, a combination of theories may be needed. Drawing on institutional theory, it may be expected that projects involving firms are more focused on development, compared to those that are solely run by government agencies, particularly for legitimacy reasons (Beddewela & Fairbrass, 2016; Hah & Freeman, 2014). Rational choice theory would dictate the opposite, as firm-run projects are more likely to be commercial in nature given the implementing agent's self-interested behavior (Pettit, 1990). When considering what types of firms are involved, it could be expected that private firms are more focused on development than state-owned firms, when drawing on the same legitimacy argument from institutional theory, and vice versa when rational choice theory is consulted (Cuervo-Cazurra et al., 2014). As such, foreign aid projects appear to be an excellent context for scholars to test under what conditions reasons of self-interest and legitimacy-seeking behavior prevail (Cobb et al., 2016; Scherer et al., 2006), or how firms incorporate competing institutional logics (Pache & Santos, 2013). Furthermore, resource dependence theory and power dynamics perspectives seem particularly relevant in the context of foreign aid partnerships (Ashraf et al., 2017). Research may address to what extent there is resource *interdependence* and how possible power imbalances shape the aim of foreign aid projects. That is, a dependency on the resources that private firms bring may imply that such firms hold greater power, which could influence the nature of a foreign aid project and the conditions associated with it.

Discussions about the commercial and developmental intent of Chinese foreign aid projects also connect well to the current probe into China's Belt and Road Initiative, primarily undertaken by scholars in the international business literature (e.g., Buckley, 2020; Li et al., 2019, 2022). It remains to be studied if this initiative can be seen as an extension of China's foreign aid policy that characterized the period up to 2013, for which data were analyzed in this study.

Project Conditions

Table 10.3 raises questions about the nature of project conditions, which may be set in a different way for public–private foreign aid partnerships than for projects run solely by government agencies. Specifically, it is worth investigating if such foreign aid partnerships have a disproportionate tendency to work on the basis of loans rather than grants. Such a tendency may have several ethical implications. Scholars studying foreign aid have long been interested in the effects on recipient countries of extending either loan-based or grant-based aid (Gupta et al., 2004). Grants tend to work as one-off sources of income, while loans stimulate public officials to maintain revenue collection through taxes, in order to repay them at a future point in time (Brech & Potrafke, 2014). As such, loans may positively affect the stability of income in developing countries. However, weak institutions significantly hamper domestic revenue mobilization, so that countries characterized by such institutions may actually be worse off when receiving loans rather than grants (Gupta et al., 2004).

From an organization theory point-of-view, scholars may study whether foreign aid projects represent a "traditional" principal–agent problem (Eisenhardt, 1989). That is, donor governments can be seen as principals that appoint firms as implementing agents of foreign aid projects. Which instruments are used to align divergent interests, especially when multiple agents, including the recipient authority, need to be monitored? For example, do donors as principals stipulate that agent compensation is partially conditional on loan repayment?

Whether foreign aid is loan-based or grant-based may affect the foreign relations between donor and recipient countries and change their dependency relationship. Scholars have shown a marked interest in China's "no strings attached" aid policy, questioning among other aspects its motivations (Lin et al., 2016; Tseng & Krog, 2017). Table 10.3 suggests that such a policy is mainly reflected in the projects managed solely by government agencies, as PPPs in foreign aid work more often on the basis of non-concessionary loans. More research on China's foreign relations with recipient countries is welcomed, specifically studies which aim to understand how the conditions of foreign aid projects impact on the dependency of recipients, as well as those that seek to link that dependency to patterns of foreign investment and bilateral trade (e.g., Duanmu, 2014).

Project Size

The data presented in Table 10.4 raise the question if Chinese public–private foreign aid partnerships have a disproportionate tendency to implement projects that are larger in size. Large projects may have a similarly great potential to positively impact developing countries' progress toward the SDGs. However, when a large project is managed through a PPP it means that multiple stakeholders are involved, which may affect how benefits are distributed and appropriated, especially when such partnerships serve as vehicles for state aid to preferred enterprises (Bovis, 2010). Hence, ethical concerns may apply when an increasing number of projects that are larger in size are managed by (private) companies. Scholars are thus advised to study how benefits are divided between stakeholders in public–private foreign aid projects, especially in those larger in size.

In a similar vein, more research is needed to better understand the complexity associated with the management of large projects that bring together multiple stakeholders (Bode et al., 2019; Van Tulder et al., 2016; Van Tulder & Keen, 2018). Such research may take a systems perspective to gain a deeper understanding of that complexity and its consequences for the success of foreign aid projects. For example, scholars may address how stakeholders interact in an aid project system and whether the size of the project affects how many complex interactions need to be managed.

International business scholars have long sought to clarify the difficulties that firms face when operating internationally, through research on the so-called "liability of foreignness" (e.g., Cao & Alon, 2021; Klossek et al., 2012; Newenham-Kahindi & Stevens, 2018). To overcome these difficulties, firms often seek to expand gradually, and carefully increase their commitment in a foreign country, particularly when distances need to be bridged and when multiple country expansions need to be managed (Hendriks, 2020a). When involved in large foreign aid projects, firms may not have that option, as the scale of it implies that they make a relatively large commitment early on in the internationalization process. Moreover, differences between China and African countries, in cultural and institutional terms, tend to be large (Liu et al., 2020; Müller et al., 2017; Rui et al., 2016). Scholars may draw on international business theories to establish if the difficulties of overcoming cross-country differences interact with the complexity of managing large aid projects in a foreign country, thereby possibly affecting the success of such a project.

Recipient Country Characteristics

Even though Table 10.5 does not indicate that differences between company-run foreign aid projects and those managed by government agencies are as pronounced as observed previously, Table 10.7 suggests that more scholarly attention toward possible country biases is warranted, particularly for the associated ethical implications. That is, it is important to study if PPPs in foreign aid have a disproportionate tendency to implement projects in less-developed countries and those characterized by a poor rule of law. Although less-developed countries potentially stand to benefit the most from foreign aid projects, problems may emerge if a poor rule of law is associated with corruption and misappropriation of project funds (Abramov, 2009; Nwabuzor, 2005). Additionally, foreign aid is generally thought to be more effective when countries with strong institutions are targeted (Dollar & Levin, 2006). More research is needed to understand if and why PPPs target countries with a poor rule of law, and what the consequences are for societies' progress on SDGs. Moreover, it is unclear why government agencies, private firms, and SOEs mainly carry out foreign aid projects in countries characterized by low-to-medium human development. It may be the case that commercial opportunities are best pursued in these countries that may be energy-producing economies (Dreher et al., 2018; Lee, 2019), more so than in nations with (very) low levels of human development. Only about 15%

of Chinese foreign aid projects are carried out in this latter set of countries, while close to a quarter of all African nations can typically be characterized as such. SDG progress is most urgent in countries with low levels of human development, which makes it important to know if and why Chinese foreign aid partnerships shy away from these places. Another ethical issue is the possible tendency of Chinese PPPs to disproportionally target resource-rich countries with their foreign aid projects. Should this foreign aid be seen as "quid pro quo" assistance, where natural resources are traded for goods, money, and services? Considering the data presented in Table 10.6, scholars may also explore if SOEs are used as vehicles to secure access to natural resources. And importantly, what does "quid pro quo" aid mean for the sustainable development potential of developing countries?

Foreign aid projects may also serve as an interesting context for organization theory scholars to study how companies fill institutional voids, particularly if they operate in countries characterized by a poor rule of law through their PPPs. Are they able to leverage non-market resource bundles that they have developed at home, and could these bundles explain their tendency to implement projects in some countries but not in others (Cuervo-Cazurra & Genc, 2008; Khanna & Palepu, 2006)?

Conditional on further corroborating evidence, the possible observation that Chinese PPPs disproportionately target countries with a poorer rule of law would be in line with earlier studies in international business. These studies found, somewhat counterintuitively, that Chinese companies seem to be attracted to, or at least have a greater tolerance toward, host-country risk (Buckley et al., 2007). More recent work has distinguished between host-country controllable institutional risk and non-controllable risk and found that Chinese companies only show greater tolerance toward the latter when they have slack resources (Buckley et al., 2018). Being involved in foreign aid projects introduces firms to novel types of controllable and non-controllable risks, such as risks associated with political allegiance and reputation losses, and it remains to be studied how tolerant companies exactly are toward these risks and how they seek to manage them.

Taking a country-level perspective, international business scholars could also study the interlinkages between Chinese foreign aid and the aid of other countries, and the role that firms play in it. More research is needed to understand if China develops projects in high-risk countries on its own, or possibly in collaboration with other countries and their firms. In a similar vein, the temporal interlinkages between projects need to be better understood. It may be the case that early projects in recipient countries are principally led by the Chinese government, possibly in collaboration with "traditional" foreign donators before private firms or SOEs are increasingly involved on a greater scale. By building on the expanding body of literature aimed at understanding temporal changes in foreign operation-modes (Putzhammer et al., 2020), and by applying this knowledge to the context of firm involvement in foreign aid, international business scholarship could contribute to this promising area of research.

DISCUSSION

Findings and Contributions

The participation of firms in foreign aid projects, and thereby their possible development impact through such projects, remains a relatively understudied phenomenon. Firms may bring valuable resources to these projects, potentially allowing them to generate positive externalities or reduce negative externalities (Montiel et al., 2021). Yet, their participation in projects in relatively unfamiliar foreign environments, together with stakeholders with possibly diverging aims, may also give rise to various ethical concerns over their behavior. Despite that *foreign* aid entails cross-border activity by definition, surprisingly little attention has been devoted in the international business literature to the topic, even though there is evidence that firms increasingly take on important roles in the implementation of aid projects (Lawson, 2011; Samii et al., 2002). Comparatively more attention went to the topic of foreign aid in the business ethics literature (Abramov, 2009; Robertson, 2009), but more studies are still needed to understand the ethical implications associated with bringing together multiple stakeholders in aid projects that each may have very diverging interests. This study aims to contribute to three distinct literatures. First, it inspires business ethics scholars to more closely examine foreign aid projects, and especially PPPs in foreign aid, as a phenomenon that raises ethical concerns and questions (cf. Bardy et al., 2012; Van Tulder et al., 2016), particularly in the African context (Adeleye et al., 2020). As a distinct phenomenon, foreign aid has received relatively little attention in the broader field of business studies, apart from notable exceptions that focus on the examination of development partnerships (e.g., Abramov, 2009; Reed & Reed, 2009). This is despite the existence of work in adjacent fields such as development studies and international relations, where scholars voiced concerns about MNEs' impact on development through aid, particularly in the context of Africa (Bräutigam, 2011; Gu, 2009; Tan-Mullins et al., 2010). This study served as a large-scale but exploratory study of Chinese PPPs in foreign aid, to understand if their goals, aims, and conditions of aid projects differ in comparison to projects undertaken solely by government agencies. The initial findings raise important ethical questions and concerns. As such, business ethics scholars are encouraged to study the main themes that emerge from the analyses, which are captured by the research agenda presented in Table 10.7. For example, these questions connect to existing debates on corporate motivations behind engaging in social initiatives (e.g., Simcic Brønn & Vidaver-Cohen, 2009) and performance shortfalls of global sustainable development initiatives (Bowen et al., 2017; Prasakh Sethi & Schepers, 2014).

Second, this study encourages organization theorists to approach foreign aid projects as contexts where multiple stakeholders and their possibly diverging interests and objectives come together in PPPs for development (George et al., 2016). Among other findings, my exploratory data overview suggests that Chinese PPPs tend to target different SDGs, work on the basis of distinct aid conditions, and implement projects that tend to be larger than those that are run by governments only. More research is needed to establish if the participation of for-profit firms is at the basis of these differences and what motivations precisely govern their behavior. By studying these questions, organization theorists may discover novel ways of reinvigorating long-lasting but unsettled debates on the nature of firm behavior, for example, by addressing the balance between self-interested and legitimacy-seeking behavior (Pache & Santos, 2013; Scherer & Palazzo, 2011), by approaching foreign aid projects as a principal–multiple agent problem (Greve et al., 2010), and by studying resource interdependence and power imbalances in foreign aid partnerships (Ashraf et al., 2017). An interesting observation coming from this study is the difference between SOEs and private firms, particularly in terms of the SDGs they target and the types of countries they are active in. However, the situation is complex as the distinction between government agencies and SOEs often is not clear-cut (Bruton et al., 2015). For example, government officials may sit on boards of firms even if the percentage of state ownership is small, thereby still exerting an influence over the strategic choices that such firms make. In China, in particular, this may also affect private firms, and more in-depth research in the domain of corporate governance is needed to understand the behavioral differences between SOEs and private firms in foreign aid partnerships.

Third, given that foreign aid implies cross-border activity by various stakeholders, including firms, international business scholars may benefit from exploring this phenomenon in greater detail. Such a focus may help to better understand the contribution of MNEs to the SDGs, a topic that received widespread attention of late (Kolk et al., 2017; Montiel et al., 2021). On this topic, scholarly efforts mainly focused on MNEs from advanced economies, while "IB scholarship has not addressed the potential contribution of emerging economy MNEs to host-country economic development, or their sociological effects on host-country societies" (Buckley et al., 2017, p. 1048), barring several exceptions (e.g., Hendriks, 2017). Addressing the participation of such MNEs in foreign aid projects may help to fill that void. This study has developed a research agenda to provide guidance to that effort and encourages scholars to further study the sustainable development interests and priorities of emerging market multinationals. On the basis of some initial evidence provided by this and other studies, scholars are encouraged to also study emerging market multinationals' possibly negative impact, thereby taking a balanced view of their contribution to the SDGs (Hendriks, 2020b).

Limitations and Future Research

The primary purpose of this study was to characterize the projects undertaken by Chinese firms and government agencies in a PPP and contrast them to those executed by government agencies without firm involvement. Having conducted an exploratory study, my approach was inductive, drawing observations from a comprehensive dataset on all Chinese foreign aid projects between 2000 and 2014. These observations inspired a research agenda and highlighted areas where theory development is most welcomed. In addition to this inductive method, scholars are encouraged to also take a hypothetico-deductive approach based on a combination of relevant theories such as those referred to in Table 10.7.

Having conducted a large-scale study on the differences between government-run foreign aid projects and those managed by firms, I used relatively aggregated data as the basis for my analyses. Future studies should aim to collect more detailed data on specific projects, thereby possibly following a qualitative approach to better understand the diverging objectives of multiple stakeholders in foreign aid partnerships and how this divergence affects sustainable development progress. Moreover, scholars are encouraged to extend the analysis to non-profit organizations and their role in Chinese foreign aid projects (Zheng et al., 2019).

Furthermore, AidData's dataset provided information on the dominant SDG sector target by firms and government agencies in their foreign aid projects, but more research is needed to understand how they target a combination of SDGs, and how they prioritize contributions to one goal over the other. By studying these questions, scholars may provide more insight into the "wickedness" of Agenda 2030 and the interrelatedness of the various SDGs (Bowen et al., 2017; Liou & Rao-Nicholson, 2021).

Lastly, this study has focused on the specific context of Chinese foreign aid in Africa, to better understand if Chinese PPPs tend to target different SDGs, work on the basis of distinct aid conditions, and implement projects that tend to be larger than those that are run solely by government agencies. As such, it chose a relatively controversial phenomenon that was subject to a heated debate and surrounded by a range of ethical questions and dilemmas (Sun, 2014). Although it can be expected that the involvement of (private) firms in aid projects has general implications beyond the nature of the donating country, more research is needed to establish if my initial findings are generalizable to other contexts, including those of other Asian countries (Sato et al., 2011; Stallings & Kim, 2017), and the "traditional" donor countries in the Western world (Brogaard & Petersen, 2018).

REFERENCES

Aaronson, S. A. (2005). "Minding our business": What the United States government has done and can do to ensure that U.S. multinationals act responsibly in foreign markets. *Journal of Business Ethics*, *59*, 175–198.

Abramov, I. (2009). Building peace in fragile states – Building trust is essential for effective public–private partnerships. *Journal of Business Ethics*, *89*, 481–494.

Adeleye, I., Luiz, J., Muthuri, J., & Amaeshi, K. (2020). Business ethics in Africa: The role of institutional context, social relevance, and development challenges. *Journal of Business Ethics*, *161*, 717–729.

Allen, C., Metternicht, G., & Wiedmann, T. (2018). Initial progress in implementing the sustainable development goals (SDGs): A review of evidence from countries. *Sustainability Science*, *13*, 1453–1467.

Ashraf, N., Ahmadsimab, A., & Pinkse, J. (2017). From animosity to affinity: The interplay of competing logics and interdependence in cross-sector partnerships. *Journal of Management Studies*, *54*(6), 793–822.

Ballesteros, L., Useem, M., & Wry, T. (2017). Masters of disasters? An empirical analysis of how societies benefit from corporate disaster aid. *Academy of Management Journal*, *60*(5), 1682–1708.

Banks, N., Hulme, D., & Edwards, M. (2015). NGOs, states, and donors revisited: Still too close for comfort? *World Development*, *66*, 707–718.

Bardy, R., Drew, S., & Kennedy, T. F. (2012). Foreign investment and ethics: How to contribute to social responsibility by doing business in less-developed countries. *Journal of Business Ethics, 106*, 267–282.

Beddewela, E., & Fairbrass, J. (2016). Seeking legitimacy through CSR: Institutional pressures and corporate responses of multinationals in Sri Lanka. *Journal of Business Ethics, 136*, 503–522.

Bhatt, B., Qureshi, I., & Riaz, S. (2019). Social entrepreneurship in non-munificent institutional environments and implications for institutional work: Insights from China. *Journal of Business Ethics, 154*, 605–630.

Bluhm, R., Dreher, A., Fuchs, A., Parks, B., Strange, A., & Tierney, M. (2018). *Connective financing: Chinese infrastructure projects and the diffusion of economic activity in developing countries* [AidData Working Paper #64]. AidData at William & Mary, Williamsburg, VA.

Boardman, A. E., & Vining, A. R. (2012). The political economy of public–private partnerships and analysis of their social value. *Annals of Public and Cooperative Economics, 83*(2), 117–141.

Bode, C., Rogan, M., & Singh, J. (2019). Sustainable cross-sector collaboration: Building a global platform for social impact. *Academy of Management Discoveries, 5*(4), 396–414.

Bovis, C. (2010). State aid and public private partnerships – Containing the threat to free markets and competition. *European Public Private Partnership Law Review, 5*(2), 167–190.

Bowen, K. J., Cradock-Henry, N. A., Koch, F., Patterson, J., Häyhä, T., Vogt, J., & Barbi, F. (2017). Implementing the "sustainable development goals": Towards addressing three key governance challenges – Collective action, trade-offs, and accountability. *Current Opinion in Environmental Sustainability, 26–27*, 90–96.

Bräutigam, D. (2009). *The Dragon's gift: The real story of China in Africa*. Oxford University Press.

Bräutigam, D. (2011). Aid 'with Chinese characteristics': Chinese foreign aid and development finance meet the OECD-DAC aid regime. *Journal of International Development, 23*(5), 752–764.

Brech, V., & Potrafke, N. (2014). Donor ideology and types of foreign aid. *Journal of Comparative Economics, 42*(1), 61–75.

Brogaard, L., & Petersen, O. H. (2018). Public–private partnerships (PPPs) in development policy: Exploring the concept and practice. *Development Policy Review, 36*(S2), 729–747.

Bruton, G. D., Peng, M. W., Ahlstrom, D., Stan, C., & Xu, K. (2015). State-owned enterprises around the world as hybrid organizations. *Academy of Management Perspectives, 29*(1), 92–114.

Buckley, P. J. (2020). China's belt and road initiative: Changing the rules of globalization. *Journal of International Business Studies, 51*, 279–281.

Buckley, P. J. Chen, L., Clegg, L. J., & Voss, H. (2018). Risk propensity in the foreign direct investment location decision of emerging multinationals. *Journal of International Business Studies, 49*, 153–171.

Buckley, P. J., Clegg, L. J., Cross, A. R., Liu, X., Voss, H., & Zheng, P. (2007). The determinants of Chinese outward foreign direct investment. *Journal of International Business Studies, 38*, 499–518.

Buckley, P. J., Doh, J. P., & Benischke, M. H. (2017). Towards a renaissance in international business research? Big questions, grand challenges, and the future of IB scholarship. *Journal of International Business Studies, 48*(9), 1045–1064.

Cao, M., & Alon, I. (2021). Overcoming the liability of foreignness – A new perspective on Chinese MNCs. *Journal of Business Research, 128*, 611–626.

Clements, B., Gupta, S., Pivovarsky, A., & Tiongson, E. R. (2004). Foreign aid: Grants versus loans. *Finance & Development*, September 2004, 46–49.

Cobb, J. A., Wry, T., & Zhao, E. Y. (2016). Funding financial inclusion: Institutional logics and the contextual contingency of funding for microfinance organizations. *Academy of Management Journal, 59*(6), 2103–2131.

Cuervo-Cazurra, A., & Genc, M. (2008). Obligating, pressuring, and supporting dimensions of the environment and the non-market advantages of developing-country multinational companies. *Journal of Management Studies, 48*(2), 441–455.

Cuervo-Cazurra, A., Inkpen, A., Musacchio, A., & Ramaswamy, K. (2014). Governments as owners: State-owned multinational companies. *Journal of International Business Studies, 45*, 919–942.

Doh, J. (2015). From the Editor: Why we need phenomenon-based research in international business. *Journal of World Business, 50*(4), 609–611.

Dollar, D., & Levin, V. (2006). The increasing selectivity of foreign aid, 1984–2003. *World Development*, *34*(12), 2034–2046.

Duanmu, J. L. (2014). State owned MNCs and host country expropriation risk: The role of home state soft power and economic gunboat diplomacy. *Journal of International Business Studies*, *45*(8), 1044–1060.

Dreher, A., & Fuchs, A. (2015). Rogue aid? An empirical analysis of China's aid allocation. *The Canadian Journal of Economics*, *48*(3), 988–1023.

Dreher, A., Fuchs, A., Parks, B., Strange, A. M., & Tierney, M. J. (2018). Apples and dragon fruits: The determinants of aid and other forms of state financing from China to Africa. *International Studies Quarterly*, *62*(1), 182–194.

Eisenhardt, K. M. (1989). Agency theory: An assessment and review. *Academy of Management Review*, *14*(1), 57–74.

Foster, V., Butterfield, W., Chen, C., & Pushak, N. (2009). *Building bridges: China's growing role as infrastructure financier for Sub-Saharan Africa*. The World Bank.

George, G., Howard-Grenville, J., Joshi, A., & Tihanyi, L. (2016). Understanding and tackling societal grand challenges through management research. *Academy of Management Journal*, *59*(6), 1880–1895.

Gupta, S., Clements, B., Pivovarsky, A., & Tiongson, E. R. (2004). Foreign aid and revenue response: Does the composition of aid matter? In S. Gupta, B. Clements, & G. Inchauste (Eds.), *Helping countries develop: The role of fiscal policy* (pp. 385–406). International Monetary Fund.

Greve, H. R., Palmer, D., & Pozner, J. E. (2010). Organizations gone wild: The causes, processes, and consequences of organizational misconduct. *Academy of Management Annals*, *4*(1), 53–107.

Grøgaard, B., Colman, H. L., & Stensaker, I. G. (2022). Legitimizing, leveraging, and launching: Developing dynamic capabilities in the MNE. *Journal of International Business Studies*, *53*, 636–656.

Gu, J. (2009). China's private enterprises in Africa and the implications for African development. *European Journal of Development Research*, *21*, 570–587.

Hah, K., & Freeman, S. (2014). Multinational enterprise subsidiaries and their CSR: A conceptual framework of the management of CSR in smaller emerging economies. *Journal of Business Ethics*, *122*, 125–136.

Harchaoui, T. M., Maseland, R. K. J., & Watkinson, J. A. (2021). Carving out an empire? How China strategically uses aid to facilitate Chinese business expansion in Africa. *Journal of African Economies*, *30*(2), 183–205.

Hattori, T. (2001). Reconceptualizing foreign aid. *Review of International Political Economy*, *8*(4), 633–660.

Hendriks, G. (2017). The sustainable development effects of investment by emerging market multinationals: Shaping beneficial outcomes for home and host country. *Transnational Corporations*, *24*(3), 73–88.

Hendriks, G. (2020a). How the spatial dispersion and size of country networks shape the geographic distance that firms add during international expansion. *International Business Review*, *29*(6), 101738.

Hendriks, G. (2020b). How outward investment from emerging markets affects economic development at home: Using the eclectic paradigm to synthesize two IB literatures. *Multinational Business Review*, *28*(4), 463–482.

Hendriks, G., Slangen, A. H. L., & Heugens, P. P. M. A. R. (2018). How a firm's domestic footprint and domestic environmental uncertainties jointly shape added cultural distances: The roles of resource dependence and headquarters attention. *Journal of Management Studies*, *55*(6), 883–909.

Howard-Grenville, J., Davis, G. F., Dyllick, T., Miller, C. C., Thau, S., & Tsui, A. S. (2019). Sustainable development for a better world: Contributions of leadership, management, and organizations. *Academy of Management Discoveries*, *5*(4), 355–366.

Jiang, W. (2009). Fuelling the Dragon: China's rise and its energy and resources extraction in Africa. *The China Quarterly*, *199*, 585–609.

Kaufmann, D., Kraay, A., & Mastruzzi, M. (2010). *The worldwide governance indicators: Methodology and analytical issues* [World Bank Policy Research Working Paper No. 5430]. http://papers.ssrn.com/sol3/papers.cfm?abstract_id=1682130

Khanna, T., & Palepu, K. (2006). Emerging giants: Building world class companies in developing countries. *Harvard Business Review*, *84*, 60–69.

Klijn, E. H. (2022). Theories of public–private partnerships. In G. A. Hodge & C. Greve (Eds.), *A research agenda for public–private partnerships and the governance of infrastructure* (pp. 33–52). Edward Elgar.

Klossek, A., Linke, B. M., & Nippa, M. (2012). Chinese enterprises in Germany: Establishment modes and strategies to mitigate the liability of foreignness. *Journal of World Business*, *47*(1), 35–44.

Kolk, A., Kourula, A., & Pisani, N. (2017). Multinational enterprises and the sustainable development goals: What do we know and how to proceed? *Transnational Corporations*, *24*(3), 9–32.

Kolk, A., van Tulder, R., & Kostwinder, E. (2008). Business and partnerships for development. *European Management Journal*, *26*, 262–273.

Kort, M., Verweij, S., & Klijn, E. H. (2016). In search for effective public–private partnerships: An assessment of the impact of organizational form and managerial strategies in urban regeneration partnerships using fsQCA. *Environment and Planning C: Politics and Space*, *34*(5), 777–794.

Lawson, M. L. (2011). *Foreign assistance: Public–private partnerships (PPPs)*. Congressional Research Service.

Lee, C. (2019). China's energy diplomacy: Does Chinese foreign policy favor oil-producing countries? *Foreign Policy Analysis*, *15*(4), 570–588.

Li, J., Assche A. van, Fu, X., Li, L., & Qian, G. (2022). The belt and road initiative and international business policy: A kaleidoscopic perspective. *Journal of International Business Policy*, *5*, 135–151.

Li, J., Liu, B., & Qian, G. (2019). The belt and road initiative, cultural friction and ethnicity: Their effects on the export performance of SMEs in China. *Journal of World Business*, *54*(4), 350–359.

Lin, S., Gao, L., Reyes, M., Cheng, F., Kaufman, J., & El-Sadr, W. M. (2016). China's health assistance to Africa: Opportunism or altruism? *Globalization and Health*, *12*, 83.

Liou, R. S., & Rao-Nicholson, R. (2021). Multinational enterprises and Sustainable development goals: A foreign subsidiary perspective on tackling wicked problems. *Journal of International Business Policy*, *4*, 136–151.

Lis, P. (2018). The impact of armed conflict and terrorism on foreign aid: A sector-level analysis. *World Development*, *110*, 283–294.

Lister, S. (2000). Power in partnership? An analysis of an NGO's relationships with its partners. *Journal of International Development*, *12*(2), 227–239.

Liu, R., Feils, D. J., & Scholnick, B. (2011). Why are different services outsourced to different countries? *Journal of International Business Studies*, *42*, 558–571.

Liu, A., Lu, C., & Wang, Z. (2020). The roles of cultural and institutional distance in international trade: Evidence from China's trade with the belt and road countries. *China Economic Review*, *61*, 101234.

Martimort, D., & Pouyet, J. (2008). To build or not to build: Normative and positive theories of public–private partnerships. *International Journal of Industrial Organization*, *26*(2), 393–411.

McEwan, C., Mawdsley, E., Banks, G., & Scheyvens, R. (2017). Enrolling the private sector in community development: Magic bullet or sleight of hand? *Development and Change*, *48*(1), 28–53.

Montiel, I., Cuervo-Cazurra, A., Park, J., Antolín-López, R., & Husted, B. W. (2021). Implementing the United Nations' sustainable development goals in international business. *Journal of International Business Studies*, *52*, 999–1030.

Morgan, P. (2018). Ideology and relationality: Chinese aid in Africa revisited. *Asian Perspective*, *42*(2), 207–238.

Morgan, P., & Zheng, Y. (2019). Old bottle new wine? The evolution of China's aid in Africa 1956–2014. *Third World Quarterly*, *40*(7), 1283–1303.

Müller, M., Hendriks, G., & Slangen, A. H. L. (2017). How the direction of institutional distance influences foreign entry mode choices: An information economics perspective. In A. Verbeke, J. Puck, & R. van Tulder (Eds.), *Progress in International Business Research* (Vol. 12, pp. 271–296). Emerald Group Publishing Ltd.

Naím, M. (2007). Rogue aid. *Foreign Policy*, *159*, 95–96.

Narula, R. (2018). Multinational firms and the extractive sectors in the 21st century: Can they drive development? *Journal of World Business*, *53*(1), 85–91.

Newenham-Kahindi, A., & Stevens, C. E. (2018). An institutional logics approach to liability of foreignness: The case of mining MNEs in Sub-Saharan Africa. *Journal of International Business Studies*, *49*, 881–901.

Nwabuzor, A. (2005). Corruption and development: New initiatives in economic openness and strengthened rule of law. *Journal of Business Ethics*, *59*, 121–138.

Pache, A. C., & Santos, F. (2013). Inside the hybrid organization: Selective coupling as a response to competing institutional logics. *Academy of Management Journal*, *56*(4), 972–1001.

Parker, D. W., Dressel, U., Chevers, D., & Zeppetella, L. (2018). Agency theory perspective on public–private-partnerships: International development project. *International Journal of Productivity and Performance Management*, *67*(2), 239–259.

Pettit, P. (1990). Virtus Normativa: Rational choice perspectives. *Ethics*, *100*(4), 725–755.

Prasakh Sethi, S., & Schepers, D. H. (2014). United Nations Global Compact: The promise–performance gap. *Journal of Business Ethics*, *122*, 193–208.

Putzhammer, M., Puck, J., & Lindner, T. (2020). Changes in foreign operation modes: A review and research agenda. *International Business Review*, *29*(1), 101619.

Reed, A. M., & Reed, D. (2009). Partnerships for development: Four models of business involvement. *Journal of Business Ethics*, *90*, 3–37.

Robertson, D. C. (2009). Corporate social responsibility and different stages of economic development: Singapore, Turkey, and Ethiopia. *Journal of Business Ethics*, *88*, 617–633.

Rui, H., Zhang, M., & Shipman, A. (2016). Relevant knowledge and recipient ownership: Chinese MNCS' knowledge transfer in Africa. *Journal of World Business*, *51*(5), 713–728.

Samii, R., Van Wassenhove, L. N., & Bhattacharya, S. (2002). An innovative public–private partnership: New approach to development. *World Development*, *30*(6), 991–1008.

Sato, J., Shiga, H., Kobayashi, T., & Kondoh, H. (2011). "Emerging donors" from a recipient perspective: An institutional analysis of foreign aid in Cambodia. *World Development*, *39*(12), 2091–2104.

Sears, C. (2019). What counts as foreign aid: Dilemmas and ways forward in measuring China's overseas development flows. *The Professional Geographer*, *71*(1), 135–144.

Scherer, A. G., & Palazzo, G. (2011). The new political role of business in a globalized world: A review of a new perspective on CSR and its implications for the firm, governance, and democracy. *Journal of Management Studies*, *48*(4), 899–931.

Scherer, A. G., Palazzo, G., & Baumann, D. (2006). Global rules and private actors: Toward a new role of the transnational corporation in global governance. *Business Ethics Quarterly*, *16*(4), 505–532.

Simcic Brønn, P., & Vidaver-Cohen, D. (2009). Corporate motives for social initiative: Legitimacy, sustainability, or the bottom line? *Journal of Business Ethics*, *87*, 91–109.

Stallings, B., & Kim, E. M. (2017). *Promoting development: The political economy of East Asian foreign aid*. Palgrave Macmillan.

Strauss, J. C. (2009). The past in the present: Historical and rhetorical lineages in China's relations with Africa. *The China Quarterly*, *199*, 777–795.

Sun, Y. (2014). China's aid to Africa: Monster or Messiah? *Brookings East Asia Commentary*. https://www.brookings.edu/opinions/chinas-aid-to-africa-monster-or-messiah/

Tan-Mullins, M., Mohan, G., & Power, M. (2010). Redefining 'aid' in the China–Africa context. *Development and Change*, *41*(5), 857–881.

The Economist. (2017, October 14). *Despite its reputation, Chinese aid is quite effective*. https://www.economist.com/china/2017/10/12/despite-its-reputation-chinese-aid-is-quite-effective

Tseng, H. K., & Krog, R. (2017, January 6–8). *No strings attached: Chinese foreign aid and regime stability in resource-rich recipient countries* [Paper presentation]. 2017 annual meeting of the American Economic Association, Chicago, IL.

Van Tulder, R., & Keen, N. (2018). Capturing collaborative challenges: Designing complexity-sensitive theories of change for cross-sector partnership. *Journal of Business Ethics*, *150*(2), 315–332.

Van Tulder, R., Seitanidi, M. M., Crane, A., & Brammer, S. (2016). Enhancing the impact of cross-sector partnerships: Four impact loops for channeling partnership studies. *Journal of Business Ethics*, *135*, 1–17.

Verbeke, A., & Kano, L. (2016). An internalization theory perspective on the global and regional strategies of multinational enterprises. *Journal of World Business*, *51*(1), 83–92.

Whelan, G., & Muthuri, J. (2017). Chinese state-owned enterprises and human rights: The importance of national and intra-organizational pressures. *Business & Society*, *56*(5), 738–781.

Witte, C., & Dilyard, J. (2017). Guest editors' introduction to the special issue: The contribution of multinational enterprises to the sustainable development goals. *Transnational Corporations*, *24*(3), 1–8.

Zhang, D., & Smith, G. (2017). China's foreign aid system: Structure, agencies, and identities. *Third World Quarterly*, *38*(10), 2330–2346.

Zheng, W., Ni, N., & Crilly, D. (2019). Non-profit organizations as a nexus between government and business: Evidence from Chinese charities. *Strategic Management Journal*, *40*(4), 658–684.

CHAPTER 11

DE-INTERNATIONALIZATION OR RE-INTERNATIONALIZATION? THE ROLE OF ENVIRONMENTAL SUSTAINABILITY ORIENTATION AT FIRM AND COUNTRY LEVEL

Martina Barbaglia[a], Roberto Bianchini[a,b], Vincenzo Butticè[a] and Stefano Elia[a]

[a]*Department of Management, Economics and Industrial Engineering, Politecnico di Milano, Milan, Italy*
[b]*Centre for Applied Research on International Markets, Banking, Finance and Regulation, Baffi CAREFIN, Università Commerciale Luigi Bocconi, Milan, Italy*

ABSTRACT

This study investigates how firms' awareness of sustainability affects the revision of their internationalization strategy. Adopting a resource-based view (RBV) approach, the authors argue that sustainable-oriented firms have a higher propensity to de-internationalize (i.e., to go back to their home country) when confronted with the need to relocate foreign manufacturing subsidiaries, as the shortening of value chains would allow the reduction of transportation emissions and enhanced corporate image as green-oriented entities. Furthermore, the authors explore the role exerted by a stringent regulatory setting in the home country on the likelihood of de-internationalization. The empirical test conducted on a sample of relocations performed across European nations in 2002–2014 reveals that multinational enterprises (MNEs) – regardless of their

Walking the Talk? MNEs Transitioning Towards a Sustainable World
Progress in International Business Research, Volume 18, 197–221

ISSN: 1745-8862/doi:10.1108/S1745-886220240000018015

sustainability orientation – have a higher probability to de-internationalize when their home countries have strict institutional contexts in place.

Keywords: De-internationalization; relocation; sustainability; environmental policy stringency; competitive advantage; back-shoring

1. INTRODUCTION

For decades, the international business (IB) literature has experienced a concentration of scholars' attention on the Uppsala type of internationalization strategy, which delineates internationalization as a linear and gradual process, comprising incremental international involvement (Johanson & Vahlne, 1977, 1990). Nevertheless, companies often adopt non-linear internationalization patterns, which are marked by uneven decisions and fluctuations in the international commitment (McIvor & Bals, 2021; Ozkan, 2020; Vissak et al., 2020). These asymmetrical cycles and divestments deserve much more attention, exacerbated by some macro-trends that are completely altering the international configuration of globally spread value chains and urging a contraction of firms' international footprint (Petricevic & Teece, 2019). First and foremost, the heightening emphasis on sustainability, the fight against global warming, and the race for the accomplishment of climate neutrality are increasingly driving a rationalization of value chains and a pull-back from international production (UNCTAD, 2020). Rising is also the pressure toward de-globalization, as attention has been called upon the societal downsides brought about by the unprecedented globalization wave that occurred in the past decades, namely inequalities and unbalances in trade and profit creation (Cuervo-Cazurra et al., 2020; Milberg, 2008). These developments experienced by the global landscape are going hand in hand with a revision of the traditional business paradigm toward a more inclusive, environmentally friendly way of operating. Scholars have already pointed out how the integration of the natural environment into corporations' mission and vision could be a source of environmental, social, and economic value creation and competitive advantage (Hart, 1995; Hart & Milstein, 2003). However, the interplay between the progressive consolidation of sustainability as a strategic move and the modification of firms' international footprint is far less researched and understood. Indeed, the IB literature studying MNEs' relocation choices – either returning to their domestic markets or entering new cross-border ones – has focused on the drivers encouraging firms to renege their previous offshoring decisions, divesting value chain activities that were formerly located in distant environments (e.g., Benito & Welch, 1997; Berry, 2013; Chung et al., 2013; Schmid & Morschett, 2020). Nevertheless, among the motives addressed by the literature, a gap can be pinpointed in relation to sustainability issues (Fratocchi & Di Stefano, 2019; Orzes & Sarkis, 2019).

To fill this gap, in this study, we inquire whether MNEs committed to engaging sustainability in their value creation strategy might favor de-internationalization (i.e., going back to their home country) over re-internationalization (i.e., moving to a new host country) when confronted with the need to relocate foreign subsidiaries. In addition, we also investigate whether the formal institutional difference between home and host country might exert an influence on this alteration of internationalization trajectory. Drawing from the theoretical lenses of the RBV, we argue that MNEs pioneering sustainable practices might be keener to revise their internationalization portfolio by voluntarily repatriating foreign operations, aiming to boost their image as sustainable-oriented agents. As a matter of fact, firms' perception of the relevance played by the sustainability trend might trigger a rethinking of their global manufacturing involvement (Ashby, 2016). Through a reduction of the physical length of their supply chain, companies could iterate their pledge for emission reduction, slow down globalization, and align with consumers' solicitations about pollution abatement. Furthermore, de-internationalization might allow companies to reduce exposure to liabilities and risks deriving from host country factors (e.g., Vissak & Francioni, 2013), including lenient environmental and human rights regulations and consequent scandals and foreign counterparts' misconducts. Thus, sustainability may provide an incentive to repatriate previously offshored activities in order to benefit from lower risk exposure, improved reputation, and therefore increased competitive advantage, meanwhile achieving a reduction of the ecological footprint and confirmation of firms' moral responsibility and credibility.

In addition, we also claim that de-internationalization could be the result of firms' acknowledgment of the divergence among social and environmental regulatory frameworks in the home and host country (Kafouros & Aliyev, 2016; Witt, 2019). The ongoing disparity resulting from the discrepancy between social and environmental laws on a global scale continues to create an uneven playing field. This adds to the challenge of coordinating foreign value chain participants, potentially leading to conflicts among stakeholders that could hinder the sustainability of foreign operations for long-term competitive advantage (Villena & Gioia, 2018). Particularly, we draw from environmental regulatory stringency, defined as a higher, explicit or implicit, cost of polluting or environmentally harmful behavior (Botta & Koźluk, 2014), to argue about the way in which international expansion might be hindered or facilitated by divergent regulatory regimes. Thus, there are two research questions this chapter aims at investigating: (1) Do MNEs that value sustainability as a potential source of competitive advantage have a higher propensity to go back to their home country (i.e., de-internationalization) over moving to a new one (i.e., re-internationalization) after deciding to abandon a country in which they have previously invested? And (2) Do more stringent environmental policies in the home country exacerbate the tendency of companies to opt for de-internationalization over re-internationalization?

To shed light on these two research questions, we perform an empirical analysis on a sample of 130 relocations that occurred within European boundaries in the timespan 2002–2014, comprising 42 de-internationalizations and 88 re-internationalizations. After differentiating corporations based on their commitment to sustainability reporting, we carry out an econometric analysis to inquire whether firm-level sustainability orientation affects the propensity of MNEs to pull-back from foreign production when dealing with the relocation of cross-country subsidiaries. In doing so, we take into account the level of environmental policy stringency (EPS) in the home country to assess the way in which how more demanding environmental regulations might alter such propensity. The results stemming from our analysis put forward the fact that the sustainability dimension of firms seems not to be sufficient for justifying a restructuring of their whole international value chain through repatriation, but they confirm a significant role played by home countries' formal institutional environment on companies' internationalization strategy. Indeed, a higher EPS in the headquarters' country emerges as a factor able to increment the MNEs' tendency to engage in de-internationalization. To the best of our knowledge, this is the first study that addresses the impact of firm and country-level sustainability-related variables on the propensity to undertake a de-internationalization rather than persevere in a progressive internationalization scheme. Our study further contributes to the IB literature by advancing novel insights on potential enabling factors of disorderly internationalization cycles.

The remainder of the chapter is structured as follows. In Section 2, we frame the theoretical background underpinning our study and we develop our research questions; Section 3 describes the data and the empirical methodology; finally, Section 4 contains the discussion and the conclusions arising from our findings.

2. THEORETICAL BACKGROUND

2.1. Firm's Sustainability and Competitive Advantage: An RBV

The idea that firms have a moral responsibility toward the natural environment and the social context they are embedded into is not a new construct and traces back to the conceptualization of the stakeholder theory, according to which firms should embrace stakeholders' claims as well as shareholders' ones (Freeman, 1984, 2010). Despite general consensus around the irreversibility of damages engendered by economic activity on societies and the planet, the reconciliation between profits and the generation of public good might not be straightforward, as managers often fail to grasp the consistency of the competitive advantage resulting from sustainable-oriented strategies (Hart & Milstein, 2003). Adopting the theoretical lenses of the RBV, firms can obtain a sustained competitive advantage by exploiting their tangible and intangible bundle of resources in unique ways, far from being easy for competitors to imitate (Barney, 1991). Owning and deploying such valuable, rare, and non-substitutable resources allows companies to develop distinctive capabilities that can be leveraged to unlock new sources of competitive advantage (Hart, 1995). Thus, resource-based theorists explain sustainable competitive advantage as resulting from the ability of companies to leverage their core competencies and adapt them to the changing external circumstances (Prahalad & Hamel, 1990), including exogenous factors linked to the protection of the natural environment and the growing stakeholders' demand for its preservation. Consequently, extensive literature argues that recognizing the close interdependence between business and society and integrating sustainability within firms' vision presents significant opportunities for developing environmental management capabilities. These capabilities not only deliver value to shareholders but also address the ground-breaking sustainability challenge (Hart, 1995; Hart & Milstein, 2003; Porter & Kramer, 2006). Environmental management capabilities have been defined as those "organizational abilities or skills that enable firms to improve their performance on environmental issues" (Lee & Klassen, 2008, p. 574) by means of proactive environmentally sustainable strategies that might improve organizational performances and shareholders' value (Hart & Milstein, 2003; Orsato, 2006). Numerous are potential ways for firms to profit from sustainable-themed investments, either in the context of improved revenues or cost-cutting (Ambec & Lanoie, 2008). Focusing on internal organizational processes, the easiest way to lower production costs consists of the elimination of waste, by-products, and pollution stemming from manufacturing processes, all signs of inefficient utilization of resources (Porter & Van Der Linde, 1995). By working to prevent pollution from occurring, for example, substituting hazardous raw materials, recycling, minimizing emissions, or using cleaner energy sources, firms will be able to improve resource productivity, remove unneeded activities, and, ultimately, cut costs and increase shareholders' value in the short-term (Hart, 1995; Hart & Milstein, 2003). These are all actions that fall into the "eco-efficiency" domain (e.g., Orsato, 2006). Thinking about the common good might also lead to the development of new, greener technologies for product realization that simultaneously lower firms' environmental footprint and improve process performance and quality of the output (Shrivastava, 1995). Such a radical restructuring of manufacturing processes entails substantial environmental investments that might set the conditions for the attainment of long-term competitiveness (Porter & Kramer, 2006). However, acting at the level of production processes is not the unique lever firms can utilize to transform environmental and social goals into economic opportunities. For instance, a proactive engagement in sustainability since the preliminary phases of the product life cycle is expected to amplify the magnitude of savings and reputational gains. Empathizing with people and engaging communities in the earliest phases of the design process of new services and products are the core of human-centered design thinking, proven to be a source of innovation and value creation (Dell'Era et al., 2020). Moreover, sustainable design and sustainable product development allow firms to exploit first-mover advantages and to grow proprietary information and capabilities that competitors are unable to replicate and copy in a short window of time, thereby providing additional sources of competitive advantage (Kleindorfer et al., 2005; Shrivastava, 1995).

Environmentally responsible actions and integrating consumers' concerns about the natural environment and employees' rights protection also increases corporate image performance and reputation (Sharma et al., 2010). For companies that value customers' perception and strive for building up their credibility as green-oriented agents, a proactive environmental strategy, deployed through beyond compliance adoption of environmental standards or use of eco-friendly materials, might positively influence public opinion and lead to image-related benefits (Dangelico & Pontrandolfo, 2015; Zeng et al., 2005). To achieve credibility and enhance corporate image, developing collaborations and interactions with non-business actors such as non-governmental organizations have proven to be effective as well (Dangelico & Pontrandolfo, 2015).

Adopting a more holistic view that spans outside the boundaries of the focal firm and includes its whole supply chain offers the chance to capture alternatives for sustainable value creation that account for the multiple linkages among different stages of the value chain (Schrettle et al., 2014). Increased concerns about the negative environmental externalities generated by a worldwide fragmentation of production activities are creating the conditions that allow companies to move toward value-adding activities by concurrently reducing their environmental harm (Golini et al., 2018). Extending the adoption of standards and environmental certifications to first and second-tier suppliers, knowledge transfer to less-experienced partners, and new product design and specifications that enable the use of greener raw materials are just some examples of approaches tackled by lead firms to create more value (e.g., lower manufacturing costs and niche differentiation) and drive greening processes (De Marchi, Di Maria, & Micelli, 2013). Regardless of the competitive focus of corporations, adopting green-oriented strategies might not just strengthen intra-firm value creation, but also serves as a mean for upgrading firms' position in globally spread value chains, thereby obtaining an economic advantage from the redesign of inter-firm relationships by virtue of greater bargaining power over customers and deeper relationships with suppliers/clients, coupled with a reduction of the ecological footprint (De Marchi, Di Maria, & Ponte, 2013).

Therefore, companies with an eye toward sustainability, that incorporate environmental and societal needs within their strategy and are prone to contribute to solving sustainable issues will be able to enact interventions valuable for both the public and the corporation itself.

2.2. Linking Firms' Sustainability to De-internationalization

Embarking on the sustainability challenge and capitalizing on the sustainability momentum is a strategic decision that must progressively be iterated to be promoted as a source of sustained competitive advantage over time (Hart, 1995). In the past, certifications of environmental management systems such as the ISO 14000 series were sufficient to improve the quality of organizational processes and stand out from less committed competitors (Orsato, 2006). However, the escalating concern that the whole world is displaying toward environmental degradation, and workers exploitation and inequalities calls for a clearer embeddedness of sustainability and a more profound and broader articulation of firms' business model around sustainability to outclass competition. In this respect, the choice about where to locate manufacturing production, relationships with foreign counterparts, and more generally, internationalization and supply chain-related interventions cannot be disregarded from sustainable corporations' attempts to fulfill their dual scope (Kolk & van Tulder, 2010). Indeed, depending on the strategic salience organizations attach to sustainable value creation, sustainability might play the role of driver, motivator, barrier, or enabling factor in firms' decision to revise their internationalization path and worldwide presence (Fratocchi & Di Stefano, 2019).

Supply chains that span outside domestic boundaries are notable potential sources of legal, reputational, and operational risks connected to environmental and safety accidents (Kleindorfer et al., 2005), which MNEs located throughout the world are particularly exposed to (Klassen & Vereecke, 2012). The global dispersion of subsidiaries and the consequent fragmentation of production activities often give rise to challenges in the management of environmental and social strategies that firms might not always be able to solve, which translates into a reduced competitive advantage in the form of higher costs, damages to firms' image, and revenue loss (Grimm et al., 2016). Disruptive subsidiaries' behaviors are often the undesirable consequence of substantial formal and informal institutional distance between the home and the host country (Dhanaraj & Beamish, 2009). Entities operating in culturally distant markets endure complications in sociocultural integration and the creation of a sense of trust, shared identity, and compatible values (Stahl & Voigt, 2008), thus undermining the transfer of corporate social responsibility ideologies overseas. As the cultural distance between domestic and foreign countries increases, the efforts multinationals must undertake to deal with information asymmetries, uncertainty, and risks arising from operations conducted abroad must rocket as well (e.g., Hamilton & Kashlak, 1999; Padmanabhan & Cho, 1996). The greater an organization's foreign presence, the greater the struggle for effective coordination and communication with transnational subsidiaries (e.g., Kostova et al., 2016) and the lower the control over foreign counterparts' performances, including their social and environmental ones (Foerstl et al., 2016). International exposure is indeed a source of pressure for MNEs and their adoption of sustainable-oriented practices, which is challenged by the inherent heterogeneity of the multiple institutional environments in which they operate (Marano & Kostova, 2016). In this perspective, managers striving to realize both profit and sustainable improvement could find in de-internationalization a conduit to diminish risk exposure and vulnerability, nurture competitive advantage creation, and gain credibility in the eyes of external agents (Ashby, 2016).

As internationalization may put at risk the path to the achievement of a global reduction of greenhouse gas emissions, a pull-back from international production might additionally promote firms' environmental integrity. Sirilertsuwan et al. (2018) provide evidence of the fact that the shortening of globally spread value chains through the repatriation of foreign subsidiaries allows the reduction of carbon emissions, as a consequence of lower transportation requirements and usage of cleaner technologies and energy sources. Thanks to the exploitation of greener, advanced technologies, moving some stages of the production process back to the home country advocates for both a reduction of environmental harm and an increase in responsiveness and resource productivity (Martínez-Mora & Merino, 2020). Consistently with this predicament, Ashby (2016) points out how having a local supply chain could provide firms with opportunities to access skills and competencies essential for enacting sustainable strategies, minimizing environmental impact, and increasing managerial control. More recently, Nippa et al. (2021) have discovered that MNEs' plants located in the same country as their headquarters exhibit superior carbon performance compared to the ones situated in foreign countries, likely as a result of closer positioning to R&D, which allows a prompt application of carbon emission reduction strategies.

To simultaneously achieve economic, social, and environmental goals, an MNE aiming to impose itself as a conscientious corporation that cares about environmental and social welfare should be more geared toward a repatriation of globally dispersed manufacturing activities than a further re-internationalization to unknown countries. Following the RBV line of reasoning, de-internationalization becomes a solution for firms to promote their credibility as sustainability-oriented agents, access valuable resources, and develop additional social and environmental management capabilities, thereby unlocking new sources of sustainable competitive advantage. Thus, in conclusion, our first research question is the following:

> *RQ1*. In front of a decision to relocate foreign manufacturing operations, will MNEs lean toward de-internationalization over re-internationalization when they are sustainability-oriented?

2.3. Linking EPS to De-internationalization

Although companies' awareness about environmental-related issues and green-oriented investments might boost and create firms' advantages (Kolk & Pinkse, 2008) and affect location decisions (Tate, 2014), the interplay between economic and non-economic value creation cannot be disentangled from the international regulatory landscape firms have to cope with (Rugman & Verbeke, 1998). In response to global sustainability challenges, such as global warming, countries consistently issue environmental laws and regulations that vary according to nation-level political and social preferences, thus magnifying the unpredictability of responsible business activity outcomes (Wijen & Van Tulder, 2011). Concerning the internationalization processes enacted by firms, scholarly debate about the role of environmental regulations traces back to the pollution-haven hypothesis, according to which companies in developed countries tend to exploit lax environmental regimes by transferring their highly polluting activities to such regions (McGuire, 1982). For decades, literature has contrasted this "race to the bottom" perspective to the Porter hypothesis, which posits that appropriately designed environmental policies facilitate investments in pollution-saving technologies and increase firms' competitiveness by triggering the development of green capabilities, quality, and productivity improvement (Porter & Van Der Linde, 1995). Nonetheless, country-specific regulatory pressure is an external institutional factor MNEs must confront (Kafouros et al., 2021; Rugman & Verbeke, 1998) and an external source able to reshape organizational decision-making and internationalization propensity (Gaur et al., 2014; Huang & Sternquist, 2007; Tate et al., 2009). Given the increased scrutiny of business sustainability by various stakeholders, the significant misalignment among environmental policies within national and international regulatory regimes is not only a source of turbulence and uncertainty but also an intriguing opportunity for value-seeking corporate strategies. However, the implications that disparate regulatory frameworks have on the internationalization strategy put on by MNEs are not clear-cut and their impact is contingent on the extent to which they contribute to competitive advantage (Berger et al., 2007; Wijen & Van Tulder, 2011). Leveraging on the different degrees of stringency of environmental regulations in home and host countries, we argue that MNEs will be more incentivized to repatriate cross-border production activities when addressing a relocation decision if their home base displays stricter environmental laws. This would reflect the growing public demand for low-carbon initiatives due to increased attention to social and environmental concerns. As companies are increasingly placing attention on value-capturing opportunities (Ellram, 2013; Gereffi & Lee, 2012) and considering that government regulatory pressure figures as the foremost determinant for firms to integrate environmental issues into their business plan (Henriques & Sadorsky, 1996), the enforcement of tight environmental policies might persuade the relocation of multinational companies in countries they regard as favorable ones. Specifically, in locations with stronger enforcement of environmental laws, firms are expected to gain greater benefits from legitimacy-seeking activities that enhance their credibility and perception among stakeholders. This is because strong regulatory settings often align with societal interests in environmental safeguarding (Young & Makhija, 2014). A de-internationalization toward more relaxed institutional contexts would be unsuitable for a fruitful communication of an organization and therefore, it would not be effective for sustainable value creation. Former studies investigating the influence of the level of environmental regulatory stringency in the home and host country on the internationalization approach adopted by companies also suggest that moving production to nations with lenient environmental policies with the purpose of exploiting lower labor costs brings about negative repercussions for the environment worldwide, raising the level of global pollution and carbon emissions (Branger & Quirion, 2014). Moreover, companies operating in countries with tighter environmental regulations tend to emit less CO_2 compared to firms subjected to lenient regulatory frameworks (Ben-David et al., 2021; Cheng et al., 2017). Despite being resource-intensive and often associated with spending, efforts toward the improvement of carbon performance can be rewarded by increased competitive advantage (Ghose et al., 2022). Following environmental ameliorations, companies may experience several beneficial outcomes, including lowered perceived risk and cost of capital, that eventually lead to higher firm value, thus representing a win–win situation for both the firm and the society (Jung et al., 2018; Zhang & Zhao, 2022).

Besides, environmental-oriented policies do not only foster the reduction of carbon emissions but green and sustainable innovations as well (Nesta et al., 2014). Particularly, higher regulatory pressure from governments through stringent environmental laws encourages firms to conceive new environmental methods and technologies. This would create the conditions that motivate firms to actively enhance their environmental responsibility, stand out through unique sustainable interventions (Berrone et al., 2013), and reap differentiation advantages. For instance, Garrone et al. (2018) show that formal institutions have the power to spur the adoption of energy-efficiency innovations, which ultimately causes a reduction of operational costs. Additionally, the enforcement of draconian, green-oriented regulations is not only a symptom of the public acknowledgment of the carbon pledge, but it is also one of the cardinal ingredients for green market development (Pinkse & Kolk, 2012) from which firms can profit.

Moreover, a de-internationalization toward stronger regulative settings could be an attempt to grow a sustainable-oriented knowledge base and to build the necessary know-how to comply with environmental regulations that become sterner and tighter (Schrettle et al., 2014). Lower will be the compliance costs compared to those competitors unable to anticipate more demanding requirements and to develop green skills and capabilities needed to perform value-added activities in such contexts (Rugman & Verbeke, 1998). It is also worth bearing in mind that environmental-oriented policy initiatives often yield subsidies that make sustainable-related investments more viable (Okereke & Russel, 2010) and incentivize managerial commitment to pursuing benign strategic choices (Hartmann et al., 2020), without bearing their burden.

Henceforth, we conceptualize a favorable role played by home country environmental regulatory stringency on firms' decision to de-internationalize, as ampler would be the competitive advantage they would be able to attain. Therefore, our second research question is the following:

> *RQ2*. In front of the decision to relocate foreign manufacturing operations, will MNEs lean toward de-internationalization over re-internationalization when the relative EPS is higher in the home than in the host country?

3. EMPIRICAL METHODOLOGY

3.1. Data Sources

We test the relationship between sustainability and the choice of de-/re-internationalize employing a dataset that collects relocations that occurred in Europe in the timespan 2002–2014. Data were retrieved from the European Restructuring Monitor (ERM).[1] The information recorded by the ERM is composed of cross-border relocations in one of the EU 27 nations plus Norway that entail a reduction of at least 100 jobs or, in the case of sites employing more than 250 people, 10% of the workforce. The final sample is constituted of 130 events implemented by 69 different MNEs (i.e., firms could have performed more than one relocation in the timeline considered). It should be noted that such displacements of foreign activities include both a complete and partial divestment of operations that were previously offshored by the MNE in a first host country. Consequently, every single observation recorded in the dataset refers to the announcement of a resettlement of activities that were formerly internationalized in a new country. Such a new country either overlaps with the MNE's home one or conversely, it is a novel European nation different from the MNE's headquarters base. Knowing the nation chosen as the target for the relocation decision allows us to classify de-/re-internationalization phenomena and to explore how firm-level sustainability and country-specific environmental regulation interfere with organizations' decisions to de-/re-internationalize. Particularly, in this empirical analysis, we model de-internationalization as the decision of MNEs to move partially or entirely activities conducted abroad, along with a given share of their employees, in their home country. On the other hand, when the relocation takes place in a new European country, meaning that the MNE decides to carry on its own internationalization process, we talk about re-internationalization.

As regards the sector, the sample employed as the basis of our analysis includes manufacturing firms operating in multiple industries at the NACE Rev. 2 2-digit level.

Overall, the companies comprised in the dataset are manufacturing MNEs, that had offshored operations in a first European host country, and which had eventually chosen to divest partially or entirely their foreign activities, moving back to their domestic market or persisting in their internationalization strategy in a new European nation.

3.2. Variables

Dependent variable. In our model, the dependent variable is *De-internationalization*, a dummy variable that takes the value of 1 when the MNE has performed a relocation to its own home country, displacing part or its whole foreign production activity and putting in place a non-sequential scheme. Conversely, it takes the value of 0 when the company has decided to relocate to a new European country, so when the new host country differs from the headquarters one.

Explanatory variables. Being sustainability at the core of this study, the first explicative variable identified in the model captures the level of sustainable commitment of the MNE announcing its decision to relocate, so as to perform a de-/re-internationalization. Defining enterprise sustainability is not an easy task (Searcy, 2016), and numerous proxies have been identified in the literature to measure environmental sustainability performance at the firm level. To account for companies' commitment toward sustainable matters, this study looks at the eventual publication of a non-financial report by the focal MNE, the year prior to the relocation announcement. Precisely, a firm was deemed attentive toward sustainability at the time of the relocation if it had disclosed a non-financial report the year before the announcement date. The independent variable employed is *Sustainability report*, a dummy variable that equals 1 when the relocating MNE has published a non-financial statement the year before the announcement date, and 0 otherwise. Despite the mixed evidence about the relationship between the quality of environmental disclosure and environmental performance, the lack of sustainability reporting can be interpreted as lower managerial commitment and attention toward sustainability matters. When managers do not perceive the business advantages of environmental reporting or they are not subjected to legal requirements and stakeholder pressure, they seem to doubt the potential benefits arising from it, weighting way more the resulting cost burden (Martin & Hadley, 2008; Stubbs et al., 2013). It should also be noted that Directive 2014/95/EU, a landmark intervention for mandatory non-financial disclosure in Europe, was ratified in 2014. This regulation, also known as Non-Financial Reporting Directive, requires large undertakings to disclose information about the policies, outcomes, and risks related to environmental, social, and corporate governance matters and it came into effect in 2018, referring to the financial year of 2017 (EU, 2014). Furthermore, EU Member States were allowed to locally transpose the EU Directive and impose state-specific requirements about the pool of undertakings that must adhere to the regulation.[2] Since our empirical study has been conducted on relocations that took place before 2014, we can infer that most of the publications on behalf of focal MNEs were voluntary ones, thus strengthening the power of sustainability reporting in expressing higher environmental and sustainable commitment.

To answer our second research question, the second independent variable employed in this study accounts for countries' environmental regulatory frameworks. A study conducted by Brunel and Levinson (2013) reviews the main proxies employed in environmental economics literature to measure the level of countries' regulatory commitment, including private-sector pollution abatement expenditures, direct assessment of the regulations, composite indexes, measures based on ambient pollution, emissions or energy use and pollution-control efforts by the government. For the purpose of this analysis, we refer to the OECD EPS indicator (Kruse et al., 2022). It aggregates 13 policy instruments, including both policies aiming at punishing pollution and subsidies into a final score that goes from 0 (non-existing) to 6 (more stringent). Therefore, our second explicative variable (*Environmental policy stringency*) is expressed as the difference between the OECD EPS index in the home country and the OECD EPS index in the first host. Both values are taken the year prior to the relocation announcement. We consider the delta between the home country and the first host to account for MNEs' choice to relocate toward a tighter regulatory regime. A positive figure of this variable expresses a greater EPS in the country of origin compared to the nation from which the firm decides to leave.

Control variables. Leveraging on the kinds of location advantages identified by Dunning (1998), our model comprised a set of control variables that reflect the alternative motivations underlying the choice to offshore production activities. The objective is to capture the market-seeking, efficiency-seeking, resource-seeking, and strategic asset-seeking location advantages of the home country and the first host. For instance, we leveraged the GDP per capita in US$ at Purchasing Power Parity, constant 2011 to address market-seeking benefits in the home country (*Home country market-seeking advantage*) and first host (*First host market-seeking advantage*). We then included the reverse coded Unit Labor Cost in the home (*Home country cost-saving advantage*) and the first host country (*First host cost-saving advantage*) and the ratio of GDP per person employed in US$ at Purchasing Power Parity, constant 2011 in the home country (*Home country productivity-seeking advantage*) and first host (*First host productivity-seeking advantage*). Finally, we incorporated the number of researchers in R&D per million people. In the home country (*Home country strategic asset-seeking advantage*) and first host (*First host strategic asset-seeking advantage*). Data have been gathered by the World Bank database and values are computed, for every single observation, as the average of the three years prior to the relocation announcement.

Further, we introduced *Cultural distance* to control for the cultural distance between the home and the first host country, measured by means of the index developed by Kogut and Singh (1988).[3] *Size* has been inserted to control for the size of the MNE, computed as the average total assets for the 10 years prior to the relocation announcement, expressed in thousands of US$ and retrieved from Orbis – Bureau Van Dijk. All values have been standardized.

Additionally, to reduce the risk of omitted-variable bias and control for time-varying and industry-specific factors, we employed industry and year dummies.

3.3. Descriptive Statistics

Table 11.1 exhibits correlations among the variables inserted in the model. Summary statistics are reported in Table 11.2. Our sample is made of 130 relocation events, broken down into 42 cases of de-internationalization and 88 cases of re-internationalization. From an industry perspective, Table 11.3 shows that the majority of the relocation announcements have been made by companies operating in the "Manufacture of motor vehicles, trailers, and semitrailers" industry (24 events), followed by "Manufacture of chemicals and chemicals products" (17 events) and "Manufacture of food products" (17 events).

Looking at the geographical direction of the relocation events, Table 11.4 displays the distribution of de-/re-internationalization according to the respective home country, the first host country (the one from which the relocation takes place), and the second host country (the destination of the relocation). For what concern the headquarters' country, the most frequent nation is Germany (39 events), followed by Sweden (22 events). Relocations took place mainly in France (22 events) and Germany (14 events). Finally, the preferred destination choices for relocations of production activities are Germany (23 events) and Poland (21 events).

An interesting statistic emerges when looking at the mean values assumed by *Environmental policy stringency* in correspondence with de-internationalization movements. Specifically, de-internationalization occurs, on average, toward more stringent nations (mean value *Environmental policy stringency* = 0.078). This evidence lays the foundations for investigating our second research question.

3.4. Results

Being our dependent variable (*De-internationalization*) a dichotomous one, we perform a probit regression analysis to examine the relationship between firm-level sustainability, EPS, and the propensity to de-/re-internationalize.

Table 11.5 shows the results and the estimated coefficients. We employed *Sustainability report* and *Environmental policy stringency* as explanatory variables. Our estimates show that the variable *Sustainability report* displays a negative, but not statistically significant coefficient. Sustainable-oriented MNEs are not more likely to turn back home rather than re-enter a new foreign market. To answer our first research question, we can conclude that enterprises' focus on sustainability issues seems not to impact the likelihood of de-internationalization.

Table 11.1. Correlation Matrix.

Variables	(1)	(2)	(3)	(4)	(5)	(6)	(7)	(8)	(9)	(10)	(11)	(12)	(13)
(1) De-internationalization	1.000												
(2) Environmental policy stringency	0.047	1.000											
(3) Sustainability report	0.020	0.228	1.000										
(4) Home country cost-saving advantage	−0.133	−0.224	−0.158	1.000									
(5) First host cost-saving advantage	−0.139	−0.016	−0.109	0.621	1.000								
(6) Home country productivity-seeking advantage	0.037	0.101	0.036	−0.656	−0.730	1.000							
(7) First host productivity-seeking advantage	0.143	−0.076	−0.097	−0.380	−0.568	0.489	1.000						
(8) Home country market-seeking advantage	0.026	0.095	0.028	−0.611	−0.623	0.886	0.477	1.000					
(9) First host market-seeking advantage	0.160	−0.069	−0.149	−0.320	−0.575	0.468	0.907	0.498	1.000				
(10) Home country strategic asset-seeking advantage	−0.151	0.145	0.021	0.000	−0.153	0.252	0.222	0.295	0.287	1.000			
(11) First host strategic asset-seeking advantage	0.017	−0.163	−0.181	0.051	−0.084	0.114	0.374	0.151	0.530	0.293	1.000		
(12) Cultural distance	−0.028	−0.039	0.125	0.110	0.021	−0.133	0.068	−0.039	0.077	0.046	0.051	1.000	
(13) Size	0.210	0.157	0.292	−0.065	0.027	−0.111	−0.169	−0.087	−0.215	−0.177	−0.191	−0.063	1.000

Table 11.2. Summary Statistics.

	N	Mean	SD	Min	Max
De-internationalization	130	0.323	0.469	0	1
Environmental policy stringency	130	0.039	0.577	−1.333	1.694
Sustainability report	130	0.462	0.500	0	1
Home country cost-saving advantage	130	0.022	0.866	−1.740	1.530
First host cost-saving advantage	130	0.044	0.821	−1.983	1.632
Home country productivity-seeking advantage	130	−0.078	0.917	−3.115	2.641
First host productivity-seeking advantage	130	−0.043	0.828	−2.335	2.043
Home country market-seeking advantage	130	−0.117	0.799	−2.860	2.780
First host market-seeking advantage	130	−0.096	0.783	−2.340	2.616
Home country strategic asset-seeking advantage	130	0.233	1.123	−1.979	2.654
First host strategic asset-seeking advantage	130	−0.046	0.941	−1.474	2.80
Cultural distance	130	−0.056	1.036	−1.181	6.602
Size	130	0.084	1.305	−0.539	8.446

Table 11.3. Distribution of Relocation Announcements by Industry.

NACE Rev. 2	Re-internationalization	De-internationalization	Total
10 - Manufacture of food products	12	5	17
13 - Manufacture of textiles	1	2	3
15 - Manufacture of leather and related products	2	1	3
17 - Manufacture of articles of paper and paperboard	6	1	7
20 - Manufacture of chemicals and chemical products	9	8	17
21 - Manufacture of basic pharmaceutical products and pharmaceutical preparations	4	1	5
22 - Manufacture of rubber and plastic products	4	1	5
24 - Manufacture of basic metals	1	1	2
26 - Manufacture of computer, electronic and optical products	5	1	6
27 - Manufacture of electrical equipment	12	1	13
28 - Manufacture of machinery and equipment n.e.c.	8	4	12
29 - Manufacture of motor vehicles, trailers and semi-trailers	14	10	24
30 - Manufacture of other transport equipment	4	2	6
31 - Manufacture of furniture	1	3	4
32 - Other manufacturing	5	1	6
Total	88	42	130

Table 11.4. Distribution of Relocation Announcements by Home, First Host, and Second Host Country.

	Home Country	First Host	Second Host
Austria	1	4	4
Belgium	6	10	3
Bulgaria	0	0	1
Czech Republic	0	6	8
Denmark	2	5	3
Estonia	0	1	1
Finland	17	5	4
France	10	26	6
Germany	39	14	23
Greece	0	0	2
Hungary	0	8	7
Ireland	0	4	1
Italy	6	11	5
Lithuania	0	0	1
Netherlands	4	3	6
Norway	3	1	0
Poland	0	2	21
Portugal	0	0	4
Romania	0	0	9
Slovakia	1	1	5
Spain	2	9	3
Sweden	22	11	5
Switzerland	4	0	1
United Kingdom	13	9	7
Total	130	130	130

Table 11.5. Regression Results.

The dependent variable is *De-internationalization.*	
	(1)
Environmental policy stringency	0.637**
	(0.305)
Sustainability report	−0.307
	(0.390)
Home country cost-saving advantage	0.806*
	(0.423)
First host cost-saving advantage	1.057*
	(0.629)
Home country productivity-seeking advantage	−0.807
	(0.662)
First host productivity-seeking advantage	−0.220
	(0.555)
Home country market-seeking advantage	−0.623
	(0.478)
First host market-seeking advantage	1.097*
	(0.662)
Home country strategic asset-seeking advantage	−0.123
	(0.192)

Table 11.5. (*Continued*)

The dependent variable is *De-internationalization.*	
	(1)
First host strategic asset-seeking advantage	−0.153
	(0.227)
Cultural distance	−0.093
	(0.170)
Size	0.576***
	(0.193)
_cons	
Observations	130
R-squared	0.333
Industry dummy	YES
Year dummy	YES

Standard errors in parentheses.
*$p<0.1$, **$p<0.05$, and ***$p<0.01$.

Moving to *RQ2*, the coefficient of our second explicative variable, *Environmental policy stringency*, is positive and significant ($p < 0.05$), thus revealing that MNEs have a higher propensity to reduce their international exposure and move activities to their headquarters' country when it enforces stricter environmental regulations compared to the first host one. According to this finding, MNEs seem to recognize the competitive advantage stemming from a tighter domestic normative pressure in the environmental domain. As originally envisaged, multinational corporations do not perceive more rigorous environmental laws as a cost, but on the contrary as an opportunity to operate in a legislative setting that welcomes and promotes environmental-oriented business routines. Consequently, when their home country enforces tighter environmental regulations, they find in de-internationalization an opportunity to gain a much greater competitive advantage, thus recognizing the benefits hidden in social value creation and in higher accountability and transparency toward a large array of stakeholders. Indeed, repatriation to more stringent domains could be regarded as an option to magnify one's credibility as an environmentally committed agent and cultivate an eco-friendly image.

Concerning the control variables, our dependent variable is just positively affected by cost-saving advantages in the home ($p < 0.1$) and first host country ($p < 0.1$). Moreover, the likelihood of de-internationalization increases significantly ($p < 0.01$) with the size of the firm (*Firm size*), highlighting the resource-intensive nature of relocating value chain operations back to the home country.

4. DISCUSSION AND CONCLUSIONS

The central argument of this study concerns the impact of firm-level sustainability orientation and environmental regulatory frameworks on multinationals' decisions to revise the international distribution of their production activities. Our argument posits that both a heightened awareness of social and environmental concerns and more stringent regulatory frameworks within firms' home countries can facilitate a reduction in the length of value chains through de-internationalization. Our research findings indicate that firms' effort to draft a non-financial statement is not associated with a shift toward de-/re-internationalization. This observation suggests that reporting sustainability information either does not fully reflect MNEs' orientation toward sustainability or may not significantly drive alone domestic relocation or further expansion to distant locations. We add to this result by exploring the role played by the institutional environment in relocation decisions (Kostova, 1999; Wijen & Van Tulder, 2011; Witt, 2019). We refer to environmental regulations' degree of stringency as a potential institutional factor affecting the choice to de-/re-internationalize by MNEs confronting a revision of their international blueprint. As expected, this country-level factor does alter MNEs' international strategy and it does so by increasing the likelihood of returning to domestic regions rather than moving to a foreign new nation when more demanding environmental regulations are in place back home. A first explanation apropos this influence is related to the eventual development of green resources and capabilities supported by laws and policies that deliberately aim at stimulating a low-carbon change in enterprises' business paradigm (Rugman & Verbeke, 1998). Multinationals are more inclined to de-internationalize when they recognize in their headquarters' country the locus where to build a nature-friendly image and to seize consequent economic reward from their ethical standpoint. This opportunity materializes when on top of a shortening of value chains, de-internationalization occurs toward countries with more environmentally conscious governments compared to the nation where the first relocation decision was made. In these environments, green technologies and innovations are more likely to be refined, therefore resource productivity and superior competitiveness are more easily attainable (Porter & Van Der Linde, 1995). A second viable explanation is strictly correlated to the previous one, as properly designed environmental laws also generate a supportive jurisdiction for the development of local green markets and consumers' demand for eco-friendly products, forging avenues for mushroomed market share. Furthermore, if MNEs' headquarters are situated in nations detached from global issues, a pull-back from international manufacturing would not be welcomed and appraised by thoughtful consumers as a proactive ethical maneuver. In a world dominated by the climate crisis, firms are recognizing in more stringent contexts a conduit to take full advantage of the economic value originating from an improved corporate image, resource efficiency, and implementation of cleaner activities.

4.1. Theoretical Contributions

This study holds meaningful contributions to the IB literature. First, by addressing firms' later internationalization strategy, our analysis suggests that de-/re-internationalization pathways might be driven by the incorporation of the sustainability pillar into organizational practices and processes. We provide a first assessment of the relevance of the sustainability thematic for internationalization rethinking, particularly for back-shoring, which can be interpreted as a specific form of de-internationalization aimed at restricting the international fingerprint of firms' portfolio of activities, pulling for regionalization of operations (Di Stefano & Fratocchi, 2019; Orzes & Sarkis, 2019). Consistently with previous literature addressing the interplay among environmental regulations and value creation (Porter & Van Der Linde, 1995; Rugman & Verbeke, 1998), our findings confirm the critical role of the home country institutional domain in understanding how the configuration of firms' strategic advantages can lead to a shift in offshored operations back to the home country. By discussing the sustainability theme, this chapter additionally answers the call to expand IB literature with the incorporation of grand sustainability challenges, like climate change and environmental exploitation, into internationalization decision frameworks (Montiel et al., 2021).

4.2. Practical Implications

The current business and political panorama revolve around the need to embark on a profound transition toward a greener, low-carbon economy. In contrast with the pollution-haven hypothesis, which stresses the burden of regulatory compliance costs, our results indicate that decision-makers should have a solid interest in implementing ambitious environmental laws that answer to public concerns about global warming as they do not necessarily drive off MNEs revising their international distribution. Governments are also advised to provide stimuli and incentives, creating a policy mix that facilitates the development of environmental capabilities and sustainability-themed investments that enable firms to simultaneously lower their environmental impact and upgrade their competitive positioning. Considering the cardinal role political institutions will play in the fight against climate change in the years ahead, this study provides a point of departure for the devising of new environmental instruments nurturing both a just, green transition and economic development.

The present study also puts forward some relevant practical implications for managers. Since we examine internationalization from a dynamic and non-linear perspective, our results could support managers in disentangling the decision of where to relocate cross-border manufacturing subsidiaries up for divestment, highlighting what factors might be crucial for an appropriate redesign of the international configuration of operations. Managers should appropriately balance the competitive opportunities resulting from environmental regulations and incentives, such as access to more developed green markets (Pinkse & Kolk, 2012), sustainable product/process innovation (Porter & Van Der Linde, 1995), and preservation of brand image.

Moreover, managers should closely monitor the snowballing de-globalization and regionalization waves. These trends are expected to interact with the sustainability movement, potentially leading to a reduction of the geographical reach of operations to achieve supply chain resilience and minimize the environmental impact of long-distance transportation (UNCTAD, 2020). Therefore, as in the future global scrutiny of the sustainability mission is supposed to escalate, accounting for the positive repercussions that high-quality environmental regulations in home countries engendered on the natural landscape might open a new de-internationalization drift that will uncover alternative resources and capabilities for competitive advantage proliferation.

4.3. Limitations and Direction for Future Research

This chapter is subjected to some limitations, which represent promising avenues for future research.

First, our empirical analysis is limited to cross-country relocation announcements that were exclusively made within Europe during the period 2002–2014. To enhance the applicability and generalizability of our findings, it would be beneficial to extend the geographic and temporal scope of our study. On one hand, including a wider range of jurisdictions worldwide, like developing countries would allow the consideration of relocation patterns within environments that exhibit more profound differences in terms of regulatory, political, and cultural factors. On the other hand, the current evolving environmental legislative landscape poses the foundations to address how novel sustainability requirements could affect internationalization strategies.

Another development relates to the explicit inclusion of firms' market orientation in the analysis. Indeed, being more sensible about the evolution of consumers' interests and preferences could influence the propensity to exit from foreign markets (Yayla et al., 2018) and to reconcentrate operations in more regulated and less turbulent domestic surroundings with the aim to strengthen one's brand identity.

Furthermore, our sample comprised MNEs that have already announced their intention to relocate previously offshored activities. One could argue that the substantial expenses associated with relocating to more stringent institutional environments might dissuade firms from transferring their overseas manufacturing operations, as the cost of de-globalization could outweigh the potential competitive advantage gained. Adding a counterfactual sample of non-relocating companies would shed light on the probability of static or dynamic internationalization responses.

Future studies are also encouraged to operationalize firm-level sustainability differently, for example, collecting survey-based information about MNEs' environmental, social, and governance culture as well as data about CO_2 emissions, waste, or water management to discriminate firms based on their environmental and social performance. A multidimensional measure of firm-level sustainability aptitude could allow a deeper understanding of its potentiality as a driver guiding de-/re-internationalization.

NOTES

1. https://www.eurofound.europa.eu/observatories/emcc/european-restructuring-monitor
2. Information about the state of implementation and transposition of the Directive 2014/95/EU across European countries were retrieved from the publication made by the Global Reporting Initiative (GRI) and CSR Europe, in collaboration with Accountancy Europe (https://www.accountancyeurope.eu/wp-content/uploads/1711-NFRpublication-GRI-CSR-Europe.pdf).

3. Kogut and Singh's (1988) composite index encompasses the four cultural dimensions identified by Hofstede (2001), namely power distance, uncertainty avoidance, masculinity or femininity, and individualism or collectivism. The control variable *Cultural distance* was computed as the difference between the value of Kogut and Singh's index in the home country and the one in the first host country.

ACKNOWLEDGMENT

Project funded under the National Recovery and Resilience Plan (NRRP), Mission 4 Component 2 Investment 1.3 - Call for tender No. 341 of 15/03/2022 of Italian Ministry of University and Research funded by the European Union – NextGenerationEU. Award Number: PE00000004, Concession Decree No. 1551 of 11/10/2022 adopted by the Italian Ministry of University and Research, CUP D43C22003120001, MICS (Made in Italy - Circular and Sustainable).

REFERENCES

Ambec, S., & Lanoie, P. (2008). Does it pay to be green? A systematic overview. *Academy of Management Perspectives*, *22*(4), 45–62. https://doi.org/10.5465/amp.2008.35590353

Ashby, A. (2016). From global to local: Reshoring for sustainability. *Operations Management Research*, *9*(3–4), 75–88. https://doi.org/10.1007/s12063-016-0117-9

Barney, J. (1991). Firm resources and sustained competitive advantage. *Journal of Management*, *17*(1), 99–120. https://doi.org/10.1177/014920639101700108

Ben-David, I., Jang, Y., Kleimeier, S., & Viehs, M. (2021). Exporting pollution: Where do multinational firms emit CO_2? *Economic Policy*, *36*(107), 377–437. https://doi.org/10.1093/epolic/eiab009

Benito, G. R., & Welch, L. S. (1997). De-internationalization. *Management International Review*, *37*, 7–25.

Berger, I. E., Cunningham, P. H., & Drumwright, M. E. (2007). Mainstreaming corporate social responsibility: Developing markets for virtue. *California Management Review*, *49*(4), 132–157. https://doi.org/10.2307/41166409

Berrone, P., Fosfuri, A., Gelabert, L., & Gomez-Mejia, L. R. (2013). Necessity as the mother of "green" inventions: Institutional pressures and environmental innovations. *Strategic Management Journal*, *34*(8), 891–909. https://doi.org/10.1002/smj.2041

Berry, H. (2013). When do firms divest foreign operations? *Organization Science*, *24*(1), 246–261. https://doi.org/10.1287/orsc.1110.0724

Botta, E., & Koźluk, T. (2014). *Measuring environmental policy stringency in OECD countries* [Working Paper No. 1177]. OECD Economics Department (p. 47).

Branger, F., & Quirion, P. (2014). Climate policy and the "carbon haven" effect. *Wiley Interdisciplinary Reviews: Climate Change*, *5*(1), 53–71. https://doi.org/10.1002/wcc.245

Brunel, C., & Levinson, A. (2013). Measuring environmental regulatory stringency. In: OECD Trade and Environment Working Papers, No. 2013/05. OECD Publishing, Paris. https://doi.org/10.1787/5k41t69f6f6d-en.

Cheng, Z., Li, L., & Liu, J. (2017). The emissions reduction effect and technical progress effect of environmental regulation policy tools. *Journal of Cleaner Production*, *149*, 191–205. https://doi.org/10.1016/j.jclepro.2017.02.105

Chung, C. C., Lee, S. H., Beamish, P. W., Southam, C., & Nam, D. D. (2013). Pitting real options theory against risk diversification theory: International diversification and joint ownership control in economic crisis. *Journal of World Business*, *48*(1), 122–136. https://doi.org/10.1016/j.jwb.2012.06.013

Cuervo-Cazurra, A., Doz, Y., & Gaur, A. (2020). Skepticism of globalization and global strategy: Increasing regulations and countervailing strategies. *Global Strategy Journal*, *10*(1), 3–31. https://doi.org/10.1002/gsj.1374

Dangelico, R. M., & Pontrandolfo, P. (2015). Being "green and competitive": The impact of environmental actions and collaborations on firm performance. *Business Strategy and the Environment*, *24*(6), 413–430. https://doi.org/10.1002/bse.1828

De Marchi, V., Di Maria, E., & Micelli, S. (2013a). Environmental strategies, upgrading and competitive advantage in global value chains. *Business Strategy and the Environment*, *22*(1), 62–72. https://doi.org/10.1002/bse.1738

De Marchi, V., Di Maria, E., & Ponte, S. (2013b). The Greening of global value chains: Insights from the furniture industry. *Competition & Change*, *17*(4), 299–318. https://doi.org/10.1179/1024529413Z.00000000040

Dell'Era, C., Magistretti, S., Cautela, C., Verganti, R., & Zurlo, F. (2020). Four kinds of design thinking: From ideating to making, engaging, and criticizing. *Creativity and Innovation Management*, *29*(2), 324–344. https://doi.org/10.1111/caim.12353

Dhanaraj, C., & Beamish, P. W. (2009). Institutional environment and subsidiary survival. *Management International Review*, *49*(3), 291–312. https://doi.org/10.1007/s11575-009-0144-y

Di Stefano, C., & Fratocchi, L. (2019). Manufacturing back-shoring and sustainability: A literature review. *Journal of Special Education*, *32*(1), 21–35. https://doi.org/10.31863/jse.2016.08.32.1.21

Dunning, J. H. (1998). Location and the multinational enterprise: A neglected factor? *Journal of International Business Studies*, *29*, 45–66. https://doi.org/10.1057/palgrave.jibs.8490024

Ellram, L. M. (2013). Offshoring, reshoring and the manufacturing location decision. *Journal of Supply Chain Management*, *49*(2), 3–5. https://doi.org/10.1111/jscm.12023

EU. (2014). *Directive 2014/95/EU*. https://eur-lex.europa.eu/legal-content/EN/TXT/?uri=CELEX%3A32014L0095

Foerstl, K., Kirchoff, J. F., & Bals, L. (2016). Reshoring and insourcing: Drivers and future research directions. *International Journal of Physical Distribution and Logistics Management*, *46*(5), 492–515. https://doi.org/10.1108/IJPDLM-02-2015-0045

Fratocchi, L., & Di Stefano, C. (2019). Does sustainability matter for reshoring strategies? A literature review. *Journal of Global Operations and Strategic Sourcing*, *12*(3), 449–476. https://doi.org/10.1108/JGOSS-02-2019-0018

Freeman, R. E. (1984). *Strategic management: A stakeholder approach*. Pitman.

Freeman, R. E. (2010). Managing for stakeholders: Trade-offs or value creation. *Journal of Business Ethics*, *96*, 7–9. https://doi.org/10.1007/s10551-011-0935-5

Garrone, P., Grilli, L., & Mrkajic, B. (2018). The role of institutional pressures in the introduction of energy-efficiency innovations. *Business Strategy and the Environment*, *27*(8), 1245–1257. https://doi.org/10.1002/bse.2072

Gaur, A. S., Kumar, V., & Singh, D. (2014). Institutions, resources, and internationalization of emerging economy firms. *Journal of World Business*, *49*(1), 12–20. https://doi.org/10.1016/j.jwb.2013.04.002

Gereffi, G., & Lee, J. (2012). Why the world suddenly cares about global supply chains. *Journal of Supply Chain Management*, *48*(3), 24–32. https://doi.org/10.1111/j.1745-493X.2012.03271.x

Ghose, B., Makan, L. T., & Kabra, K. C. (2022). Impact of carbon productivity on firm performance: Moderating role of industry type and firm size. *Managerial Finance*, *49*(5), 866–883. https://doi.org/10.1108/MF-07-2022-0319

Golini, R., De Marchi, V., Boffelli, A., & Kalchschmidt, M. (2018). Which governance structures drive economic, environmental, and social upgrading? A quantitative analysis in the assembly industries. *International Journal of Production Economics*, *203*, 13–23. https://doi.org/10.1016/j.ijpe.2018.05.021

Grimm, J. H., Hofstetter, J. S., & Sarkis, J. (2016). Exploring sub-suppliers' compliance with corporate sustainability standards. *Journal of Cleaner Production*, *112*, 1971–1984. https://doi.org/10.1016/j.jclepro.2014.11.036

Hamilton, R. D., & Kashlak, R. (1999). National influences on multinational corporation control system selection. *Management International Review*, *39*, 167–189.

Hart, S. L. (1995). A natural-resource based view of the firm. *Academy of Management Review*, *20*(4), 986–1014. https://doi.org/10.2307/258963

Hart, S. L., & Milstein, M. B. (2003). Creating sustainable value. *Academy of Management Executive*, *17*, 56–67. https://doi.org/10.5465/AME.2003.10025194

Hartmann, J., Inkpen, A. C., & Ramaswamy, K. (2020). Different shades of green: Global oil and gas companies and renewable energy. *Journal of International Business Studies*, *52*(5), 879–903. https://doi.org/10.1057/s41267-020-00326-w

Henriques, I., & Sadorsky, P. (1996) The determinants of an environmentally responsive firm: An empirical approach. *Journal of Environmental Economics and Management*, *30*(3), 381–395. https://doi.org/10.1006/jeem.1996.0026

Hofstede, G. (2001). Culture's Consequences: Comparing Values, Behaviors, Institutions and Organizations Across Nations. Sage Publications.

Huang, Y., & Sternquist, B. (2007). Retailers' foreign market entry decisions: An institutional perspective. *International Business Review*, *16*(5), 613–629. https://doi.org/10.1016/j.ibusrev.2007.06.005

Johanson, J., & Vahlne, J. E. (1977). The internationalization process of the firm: A model of knowledge development and increasing foreign market commitments. *Journal of International Business Studies*, *8*(1), 23–32. https://doi.org/10.1139/y89-072

Johanson, J., & Vahlne, J. E. (1990). The mechanism of internationalisation. *International Marketing Review*, 4(7), 11–24. https://doi.org/10.1108/02651339010137414

Jung, J., Herbohn, K., & Clarkson, P. (2018). Carbon risk, carbon risk awareness and the cost of debt financing. *Journal of Business Ethics*, *150*(4), 1151–1171. https://doi.org/10.1007/s10551-016-3207-6

Kafouros, M., & Aliyev, M. (2016). Institutions and foreign subsidiary growth in transition economies: The role of intangible assets and capabilities. *Journal of Management Studies*, *53*(4), 580–607. https://doi.org/10.1111/joms.12169

Kafouros, M., Cavusgil, S. T., Devinney, T. M., Ganotakis, P., & Fainshmidt, S. (2021). Cycles of de-internationalization and re-internationalization: Towards an integrative framework. *Journal of World Business*, *57*(1), 101257. https://doi.org/10.1016/j.jwb.2021.101257

Klassen, R. D., & Vereecke, A. (2012). Social issues in supply chains: Capabilities link responsibility, risk (opportunity), and performance. *International Journal of Production Economics*, *140*(1), 103–115. https://doi.org/10.1016/j.ijpe.2012.01.021

Kleindorfer, P. R., Singhal, K., & Van Wassenhove, L. N. (2005). Sustainable operations management. *Production and Operations Management*, *14*(4), 482–492. https://doi.org/10.1111/j.1937-5956.2005.tb00235.x

Kogut, B., & Singh, H. (1988). The effect of national culture on the choice of entry mode. *Journal of International Business Studies*, *19*(3), 411–432. https://doi.org/10.1057/palgrave.jibs.8490394

Kolk, A., & Pinkse, J. (2008). A perspective on multinational enterprises and climate change: Learning from "an inconvenient truth"? *Journal of International Business Studies*, *39*(8), 1359–1378. https://doi.org/10.1057/jibs.2008.61

Kolk, A., & van Tulder, R. (2010). International business, corporate social responsibility and sustainable development. *International Business Review*, *19*(2), 119–125. https://doi.org/10.1016/j.ibusrev.2009.12.003

Kostova, T. (1999). Transnational transfer of strategic organizational practices: A contextual perspective. *The Academy of Management Review*, *24*(2), 308–324. https://doi.org/10.2307/259084

Kostova, T., Marano, V., & Tallman, S. (2016). Headquarters-subsidiary relationships in MNCs: Fifty years of evolving research. *Journal of World Business*, *51*(1), 176–184. https://doi.org/10.1016/j.jwb.2015.09.003

Kruse, T., Dechezleprêtre, A., Saffar, R., & Robert, L. (2022). *Measuring environmental policy stringency in OECD countries: An update of the OECD composite EPS indicator* [Working Paper, No. 1703]. OECD Economics Department, OECD Publishing, Paris. https://doi.org/10.1787/90ab82e8-en

Lee, S. Y., & Klassen, R. D. (2008). Drivers and enablers that foster environmental management capabilities in small- and medium-sized suppliers in supply chains. *Production and Operations Management*, *17*(6), 573–586. https://doi.org/10.3401/poms.1080.0063

Marano, V., & Kostova, T. (2016). Unpacking the institutional complexity in adoption of CSR practices in multinational enterprises. *Journal of Management Studies*, *53*(1), 28–54. https://doi.org/10.1111/joms.12124

Martin, A. D., & Hadley, D. J. (2008). Corporate environmental non-reporting – A UK FTSE 350 perspective. *Business Strategy and the Environment*, *17*(4), 245–259. https://doi.org/10.1002/bse.518

Martínez-Mora, C., & Merino, F. (2020). Consequences of sustainable innovations on the reshoring drivers' framework. *Journal of Manufacturing Technology Management*, *31*(7), 1373–1390. https://doi.org/10.1108/JMTM-12-2019-0426

McGuire, M. C. (1982). Regulation, factor rewards, and international trade. *Journal of Public Economics*, *17*(3), 335–354. https://doi.org/10.1016/0047-2727(82)90069-X

McIvor, R., & Bals, L. (2021, February). A multi-theory framework for understanding the reshoring decision. *International Business Review*, *30*(6), 101827 https://doi.org/10.1016/j.ibusrev.2021.101827

Milberg, W. (2008). Shifting sources and uses of profits: Sustaining US financialization with global value chains. *Economy and Society*, *37*(3), 420–451. https://doi.org/10.1080/03085140802172706

Montiel, I., Cuervo-Cazurra, A., Park, J., Antolín-López, R., & Husted, B. W. (2021, May). Implementing the United Nations' sustainable development goals in international business. *Journal of International Business Studies*, *52*(5), 999–1030. https://doi.org/10.1057/s41267-021-00445-y

Nesta, L., Vona, F., & Nicolli, F. (2014). Environmental policies, competition and innovation in renewable energy. *Journal of Environmental Economics and Management*, *67*(3), 396–411. https://doi.org/10.1016/j.jeem.2014.01.001

Nippa, M., Patnaik, S., & Taussig, M. (2021). MNE responses to carbon pricing regulations: Theory and evidence. *Journal of International Business Studies*, *52*, 904–929. https://doi.org/10.1057/s41267-021-00403-8

Okereke, C., & Russel, D. (2010). Regulatory pressure and competitive dynamics: Carbon management strategies of UK energy-intensive companies. *California Management Review*, *52*(4), 100–124. https://doi.org/10.1525/cmr.2010.52.4.100

Orsato, R. J. (2006). Competitive environmental strategies: When does it pay to be green? *California Management Review*, *48*(2), 127–143. https://doi.org/10.2307%2F41166341

Orzes, G., & Sarkis, J. (2019). Reshoring and environmental sustainability: An unexplored relationship? *Resources, Conservation and Recycling*, *141*, 481–482. https://doi.org/10.1016/j.resconrec.2018.11.004

Ozkan, K. S. L. (2020). International market exit by firms: Misalignment of strategy with the foreign market risk environment. *International Business Review*, *29*(6), 101741. https://doi.org/10.1016/j.ibusrev.2020.101741

Padmanabhan, P., & Cho, K. (1996). Ownership strategy for a foreign affiliate: An empirical investigation of Japanese firms. *Management International Review*, *36*, 45–65.

Petricevic, O., & Teece, D. J. (2019). The structural reshaping of globalization: Implications for strategic sectors, profiting from innovation, and the multinational enterprise. *Journal of International Business Studies*, *50*(9), 1487–1512. https://doi.org/10.1057/s41267-019-00269-x

Pinkse, J., & Kolk, A. (2012). Multinational enterprises and climate change: Exploring institutional failures and embeddedness. *Journal of International Business Studies*, *43*(3), 332–341. https://doi.org/10.1057/jibs.2011.56

Porter, M. E., & Kramer, M. (2006). Strategy and society: The link between competitive advantage and corporate social responsibility. *Harvard Business Review*, *84*(12), 78–92.

Porter, M. E., & Van Der Linde, C. (1995). Toward a new conception of the environment–competitiveness relationship. *Journal of Economic Perspectives, American Economic Association*, *9*(4), 97–118. https://doi.org/10.1257/jep.9.4.97

Prahalad, C. K., & Hamel, G. 1990. The core competence of the corporation. *Harvard Business Review*, *68*(3) 79–91.

Rugman, A. M., & Verbeke, A. (1998). Corporate strategies and environmental regulations: An organizing framework. *Strategic Management Journal*, *19*(4), 363–375. https://doi.org/10.1002/(SICI)1097-0266(199804)19:4%3C363::AID-SMJ974%3E3.0.CO;2-H

Schmid, D., & Morschett, D. (2020). Decades of research on foreign subsidiary divestment: What do we really know about its antecedents? *International Business Review*, *29*(4), 101653. https://doi.org/10.1016/j.ibusrev.2019.101653

Schrettle, S., Hinz, A., Scherrer-Rathje, M., & Friedli, T. (2014). Turning sustainability into action: Explaining firms' sustainability efforts and their impact on firm performance. *International Journal of Production Economics*, *147*(PART A), 73–84. https://doi.org/10.1016/j.ijpe.2013.02.030

Searcy, C. (2016). Measuring enterprise sustainability. *Business Strategy and the Environment*, *25*(2), 120–133. https://doi.org/10.1002/bse.1861

Sharma, A., Iyer, G. R., Mehrotra, A., & Krishnan, R. (2010). Sustainability and business-to-business marketing: A framework and implications. *Industrial Marketing Management*, *39*(2), 330–341. https://doi.org/10.1016/j.indmarman.2008.11.005

Shrivastava, P. (1995). The role of corporations in achieving ecological sustainability. *Academy of Management Review*, *20*(4), 936–960. https://doi.org/10.5465/amr.1995.9512280026

Sirilertsuwan, P., Ekwall, D., & Hjelmgren, D. (2018). Proximity manufacturing for enhancing clothing supply chain sustainability. *International Journal of Logistics Management*, *29*(4), 1346–1378. https://doi.org/10.1108/IJLM-09-2017-0233

Stahl, G. K., & Voigt, A. (2008). Do cultural differences matter in mergers and acquisitions? A tentative model and examination. *Organization Science*, *19*(1), 160–176. https://doi.org/10.1287/orsc.1070.0270

Stubbs, W., Higgins, C., & Milne, M. (2013). Why do companies not produce sustainability reports? *Business Strategy and the Environment*, *22*(7), 456–470. https://doi.org/10.1002/bse.1756

Tate, W. L. (2014). Offshoring and reshoring: U.S. insights and research challenges. *Journal of Purchasing and Supply Management*, *20*(1), 66–68. https://doi.org/10.1016/j.pursup.2014.01.007

Tate, W. L., Ellram, L. M., Bals, L., & Hartmann, E. (2009). Offshore outsourcing of services: An evolutionary perspective. *International Journal of Production Economics*, *120*(2), 512–524. https://doi.org/10.1016/j.ijpe.2009.04.005

UNCTAD. (2020). *World investment report 2020*. https://unctad.org/system/files/official-document/wir2020_en.pdf

Villena, V. H., & Gioia, D. A. (2018). On the riskiness of lower-tier suppliers: Managing sustainability in supply networks. *Journal of Operations Management*, *64*, 65–87. https://doi.org/10.1016/j.jom.2018.09.004

Vissak, T., & Francioni, B. (2013). Serial nonlinear internationalization in practice: A case study. *International Business Review*, *22*(6), 951–962. https://doi.org/10.1016/j.ibusrev.2013.01.010

Vissak, T., Francioni, B., & Freeman, S. (2020). Foreign market entries, exits and re-entries: The role of knowledge, network relationships and decision-making logic. *International Business Review*, *29*(1), 101592. https://doi.org/10.1016/j.ibusrev.2019.101592

Wijen, F., & Van Tulder, R. (2011). Integrating environmental and international strategies in a world of regulatory turbulence. *California Management Review*, *53*(4), 23–46. https://doi.org/10.1525/cmr.2011.53.4.23

Witt, M. A. (2019). De-globalization: Theories, predictions, and opportunities for international business research. *Journal of International Business Studies*, *50*(7), 1053–1077. https://doi.org/10.1057/s41267-019-00219-7

Yayla, S., Yeniyurt, S., Uslay, C., & Cavusgil, E. (2018). The role of market orientation, relational capital, and internationalization speed in foreign market exit and re-entry decisions under turbulent conditions. *International Business Review*, *27*(6), 1105–1115. https://doi.org/10.1016/j.ibusrev.2018.04.002

Young, S. L., & Makhija, M. V. (2014). Firms' corporate social responsibility behavior: An integration of institutional and profit maximization approaches. *Journal of International Business Studies*, *45*(6), 670–698. https://doi.org/10.1057/jibs.2014.29

Zeng, S. X., Tam, C. M., Tam, V. W. Y., & Deng, Z. M. (2005). Towards implementation of ISO 14001 environmental management systems in selected industries in China. *Journal of Cleaner Production*, *13*(7), 645–656. https://doi.org/10.1016/j.jclepro.2003.12.009

Zhang, Z., & Zhao, R. (2022). Carbon emission and credit default swaps. *Finance Research Letters*, *50*, 103286. https://doi.org/10.1016/j.frl.2022.103286

CHAPTER 12

DIFFUSION OF TECHNOLOGIES: DELIVERING ON THE PROMISES OF BATTERY ELECTRIC VEHICLES

Alessandro Lampo[a] and Susana C. Silva[b]

[a] *University of Saint Joseph, China*
[b] *Católica Porto Business School & CEGE Universidade Católica Portuguesa, Portugal*

ABSTRACT

Battery electric vehicles (BEVs) are living up to their claims as consumers choose them more frequently. The increasing demand for sustainable vehicles translates into the global need for specific components, materials, and infrastructures and drives the regulatory frameworks in each country. While BEVs offer environmental benefits and global business opportunities, the technology has not yet gained mainstream acceptance. Thus, this work aims to investigate the characteristics of BEV users and their role in the diffusion of products to larger segments, as this may vary from country to country. For this purpose, a survey based on the Unified Theory of Acceptance and Use of Technology 2 (UTAUT-2) (Venkatesh et al., 2012) framework and structural equation modeling (SmartPLS) was adopted. The results indicated that, except for the constructs of effort expectancy (EE) and social influence (SI), the predictors in the model performed well in this context. Current users are satisfied with their vehicles and are supportive of BEVs in the future. The analysis also revealed that in addition to the availability of financial resources, early adopters are attracted by new technologies in a way that leads them to make decisions outside of the traditional influence of the other members of society. It is suggested to leverage the perceived benefits of status, differentiation, or uniqueness motives, to appeal to those seeking to appear trendy and tech-savvy

Walking the Talk? MNEs Transitioning Towards a Sustainable World
Progress in International Business Research, Volume 18, 223–235

ISSN: 1745-8862/doi:10.1108/S1745-886220240000018016

in society. Companies and policymakers should acknowledge the peculiarities of early customers in their communication strategies to reach a wider audience around the globe and encourage the adoption of BEV technology.

Keywords: Early adopters; diffusion of innovations; battery electric vehicles; air quality; clean energy technology; international marketing; consumer behavior

1. INTRODUCTION

Sustainable development has become a critical issue that is confronting many societies (Ghauri et al., 2021). Over the years, the demand for electric vehicles has translated into the need for specific components and materials from different places, as well as driving the legislation of various countries. For instance, the production of advanced batteries, electric motors, and other essential components today is a highly competitive global industry that is estimated to be worth USD 318.47 billion by 2028 (Mordor Intelligence, 2023). China, Japan, and South Korea are the leading producers of lithium-ion batteries (Tanaka et al., 2018), while American and European companies specialize in developing electric vehicle technology (Emergen Research, n.d.). The increasing demand for electric vehicles, particularly in the USA, China, and Europe, has also led to significant investments in the development of charging infrastructure (Singh et al., 2022). As a result, major oil companies such as Shell and BP have been building charging networks globally (Sagapova et al., 2022), enabling BEV owners to drive longer distances with ease. Further, through regulations and incentives, governments are encouraging the shift to sustainable mobility, with emission targets becoming more stringent worldwide as regulations tighten up (McKinsey, 2021), and sustainable development goals becoming an urgent call for action by all countries (Zakari et al., 2022).

The decision to adopt new technologies such as BEVs is typically surrounded by a veil of uncertainty as individuals may hold doubts about how a new product will perform, even if information regarding specifications and prices is clear. Despite the existence of barriers, some people would adopt innovations regardless of their initial disadvantages (Rogers, 2003). These people possess characteristics that set them apart from the rest of the market and play an important role in the diffusion of innovations to larger segments. As a result, it is important for marketers and policymakers to understand what early BEV users require from their vehicles, which may vary from country to country, in order to promote electric mobility as a dominant global technology. In most studies, empirical data were gathered from non-BEV adopters (Hardman et al., 2016), and studies using actual user data are needed (Lampo, 2022; Lampo et al., forthcoming; Rezvani et al., 2015) because they are more insightful and they embody real-world adoption decisions rather than hypothetical ones. Such data can provide a better understanding of the current challenges in global markets and the attitudes of consumers toward BEVs. It can also help to identify potential areas for improvement in technologies, policies, and marketing strategies.

Table 12.1. BEVs Adoption in Macau.

	2015	2016	2017	2018	2019	2020	2021	2022
Tot. Private Vehicles	102,866	102,147	97,989	98,660	99,260	100,975	103,167	104,043
of which BEVs	48	87	140	245	490	980	1,789	3,035
Rate of adoption (%)	0.05	0.09	0.14	0.25	0.49	0.97	1.73	2.91

Source: Authors' table based on DSEC data.

Furthermore, no research has been conducted on the future purchase intentions of BEV owners. This information is crucial to determining whether the diffusion of BEVs can be sustained in society. To fill the gap in the literature, the aim of this chapter is twofold: (1) to understand the characteristics of early BEV users, and (2) to determine whether consumers intend to return to BEV technology in the future. These objectives are addressed by using the context of Macau the UTAUT-2 (Venkatesh et al., 2012), a state-of-the-art model.

As a small territory with significant financial resources, Macau is a good candidate for the transition to BEVs (Lampo & Silva, 2022). Due to the early government efforts to improve sustainable transportation, electric vehicles are now beginning to gain market share, proving that battery-powered technology can deliver what it promises. Indeed, sales are expected to grow in the near future since neighboring China mandated that BEVs will make up 40% of all vehicle sales by 2030, which is likely to result in more sales and lower costs (Stauffer, 2020). Table 12.1 shows BEVs' adoption rate in Macau over the years.

A growing number of consumers are turning to electric cars in Macau, with nearly 3% of vehicles being fully electric in 2022; this represents an increase of 68.2% over the figures for 2021. Innovation diffusion theory (Rogers, 2003) defines early consumers of technology as "Innovators," a relatively small segment of individuals with specific socioeconomic characteristics that are extremely important in the diffusion of technologies to a larger audience. The adoption of electric mobility is particularly relevant in the case of Macau since pollution reduction is a major concern (Lampo & Silva, 2022) that would benefit residents' health and improve the tourist image of the region. Due to vehicle emissions, Macau's air pollution levels often exceed World Health Organization's recommendations (Meneses, 2021). Perhaps this relates to the fact that concerned consumers are increasingly attracted by the claims of cleaner technologies.

2. LITERATURE REVIEW

Diffusion research explains why some innovations successfully spread in society while others do not (Lampo, 2022). Rogers (2003) described "innovation" as an idea, practice, or object that is perceived as new by an individual. In other words, innovation is anything that is considered new by the user. Studying the characteristics of innovations can help explain their diffusion process; in general, innovations that are perceived as having greater advantages and less complexity are more likely to be accepted faster in society.

The first users of any new technology are particularly important because they help to validate a concept before it becomes mainstream. A fundamental assumption in diffusion research is that individuals adopt innovations over a variable time (Lampo, 2022). It is, therefore, possible to categorize users into groups with specific characteristics and different levels of resistance to innovations. Rogers' (2003) criterion for categorization of adopters is "innovativeness," the degree to which an individual is relatively earlier in adopting new ideas. On this basis, five adopters' categories with increasing resistance to innovations are identified: (1) innovators, (2) early adopters, (3) early majority, (4) later majority, and (5) laggards. Innovators and early adopters represent respectively 2.5% and 13.5% of all adopters (Rogers, 2003). Innovators are adventurous individuals who can cope with the high degree of uncertainty and financial risks that come with new ideas. Early adopters are respected by peers and asked for advice and information about the innovation. In aggregate, the first users of innovations have generally higher socioeconomic status and more years of formal education than later adopters. Similarly, Parasuraman and Colby (2015) proposed a categorization based on the individual's propensity to embrace and use new technologies. They identified an early segment of "Explorers" that represents approximately 18% of total consumers. Explorers are characterized by a high score in optimism and innovativeness. It is a segment of young and highly educated consumers that are likely to work in a technology profession and own the largest number of technology gadgets (Parasuraman & Colby, 2015).

In the BEV domain, Hardman et al. (2016) studied "High-end adopters" and "Low-end adopters" based on vehicles' characteristics; Axsen et al. (2016) separated "Pioneers" from "Mainstream buyers," Baik et al. (2019) suggested the categories of "Early adopters," "Near-term buyers," and "Long-term buyers," while Burs et al. (2020) focused on the adopter groups of "Utilitarian saver," "Performance seekers," and "Green Technologists."

Literature on BEV in North America suggests that the initial adopters are mostly young to middle-aged males, with high social status, highly educated, with two or more cars, and have previously owned a hybrid vehicle (Hardman et al., 2016). Also, they tend to have significantly higher concern for the environment and a technology-oriented lifestyle (Axsen et al., 2016). It is also argued that demographic characteristics appear to be insufficient for predicting and understanding the various adopter groups of electric vehicles (Anable et al., 2011). A study in Germany found that early adopters are generally middle-aged men in technical professions, living in rural or suburban multi-person households (Plötz et al., 2014). A study in Spain, however, concluded that early adopters of BEVs are most likely to be young female customers, with higher income (Higueras-Castillo et al., 2020). A McKinsey analysis indicated that potential BEV adopters demand either a type of high-end performance vehicle or a purpose-built sustainability vehicle (Knupfer et al., 2017). Research in China showed that there are several potential user segments for BEVs, that differ regarding the consumers' preferences for product attributes and personal characteristics (Burs et al., 2020), which means that understanding these characteristics would be crucial for BEV manufacturers.

Since all the previous generalizations rely on data obtained mainly from potential buyers, they would require further validation using empirical evidence from actual BEV early adopters. Thus, the next section develops a theoretical framework that helps unravel the research problem and serves as a basis for the subsequent analysis and discussion of the results.

3. CONCEPTUAL MODEL

Our study is grounded in the UTAUT-2 (Venkatesh et al., 2012). The framework was intended as the synthesis of all previous consumer technology research, including innovation diffusion theory concepts (Venkatesh et al., 2003). Lampo (2022) suggests that the model provides a favorable ratio between the variance explained and the required survey items.

The UTAUT-2 includes seven constructs that are theorized to impact the behavioral intention (BI) to use technologies. As a general rule, the stronger the constructs, the greater the impact on the intention. Table 12.2 reports the definition of each construct.

Table 12.2. Definition of UTAUT-2 Constructs.

Performance expectancy	(PE)	The degree to which using a technology will provide benefits to consumers in performing certain activities
Effort expectancy	(EE)	The degree of ease associated with consumers' use of technology
Social influence	(SI)	The extent to which consumers perceive that important others (e.g., family and friends) believe they should use a particular technology
Facilitating conditions	(FC)	The consumers' perceptions of the resources and support available to perform a behavior
Hedonic motivation	(HM)	The fun or pleasure derived from using a technology
Price value	(PV)	The consumers' cognitive trade-off between the perceived benefits and the monetary cost
Habit	(HB)	The extent to which people tend to perform behaviors automatically because of learning

Source: Authors' table based on Venkatesh et al. (2012).

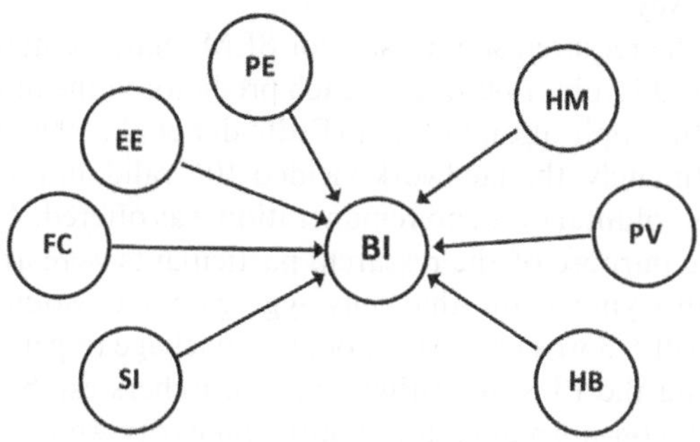

Fig. 12.1. Authors' model based on the UTAUT-2 (Venkatesh et al., 2012).

Table 12.3. Summary of Hypotheses.

H1	A positive and significant relationship exists between PE) and BI
H2	A positive and significant relationship exists between EE and BI
H3	A positive and significant relationship exists between SI and BI
H4	A positive and significant relationship exists between FC and BI
H5	A positive and significant relationship exists between HM and BI
H6	A positive and significant relationship exists between PV and BI
H7	A positive and significant relationship exists between HB and BI

Source: Authors' table based on Venkatesh et al. (2012).

Based on the extensive research and validation of the model in different contexts (Tamilmani et al., 2021; Venkatesh, 2022) the UTAUT-2 predictors were considered suitable for adaptation to the objectives of this study. Literature suggests that each construct impacts future BIs to use technology positively and significantly. Fig. 12.1 illustrates the conceptual model, followed by a summary of the hypotheses in Table 12.3.

4. METHODOLOGY

Our study adopted a descriptive and cross-sectional design to explore the influence of the UTAUT-2 constructs on the BI of BEV owners to use the technology in the future. A self-administrated survey with questions adapted from Venkatesh et al. (2012) was the most practical method of collecting data. The items were grouped according to their latent construct, and assessed using a 7-point Likert scale with anchors ranging from 1 (strongly disagree) to 7 (strongly agree). The target population consisted of Macau residents owning a BEV. Demographic variables were also included to capture respondents' gender, age, education, and income. The survey was initially developed in English and then translated into traditional Chinese using the back-translation technique to facilitate understanding (Son, 2018). To gather insights in a shorter time, the study adopted a convenience sampling technique, assuming that respondents were similar to the target population. Additionally, participants were recruited using the snowball technique (Flick, 2018) to quickly reach participants. An online platform was used to administer the survey.

To determine the required sample size for SEM analysis, standard rules suggest between 10 and 15 observations for each predictor in the model (Hair et al., 2022); similarly, the application GPower (Erdfelder et al., 1996) reported a total of 103 cases. Ultimately, the fieldwork yielded 106 valid and usable responses. Participation was voluntary and no remuneration was offered. Aside from being provided with the purpose of the research, participants were assured that their responses were anonymous and that only aggregate data would be reported in the results. SmartPLS 3 was used as a modeling package to perform data analysis. Before running the PLS algorithm, the researchers established that missing data, outliers, non-normality, and multicollinearity were not the causes of concern.

5. DATA ANALYSIS AND RESULTS

5.1 Details of the Respondents

The analysis reported that the majority of the respondents were female (52%), between 35 and 34 years of age (48%), with a bachelor's degree (44%), and an income between MOP 20,000 and 35,000 (USD 2,500–4,375) monthly (55%) (see Table 12.4).

Table 12.4. Summary of Respondents' Profiles.

Demographic Profile	Description	Frequency	Percentage
Gender	Male	51	48,11
(n = 106)	Female	55	51.88
Age	18–24	6	5.66
(n = 106)	25–34	33	31.13
	35–44	51	48.11
	45 and above	16	15.09
Education	High school	8	7.54
(n = 106)	Bachelor degree	47	44.33
	Higher education	39	36.79
	Other	12	11.32
Monthly income	MOP < 20,000; USD 2,460	5	5.49
(n = 91)	MOP 20,000–35,000; USD 2,460–4,300	50	54.94
	MOP > 35,000; USD 4,300	36	39.56

Source: Researchers' table.

5.2. Evaluation of PLS-SEM Results

Hair et al. (2022) recommended a two-step process to interpret the PLS results: first, the researcher needs to address the measurement model (i.e., outer model), which consists of the relationships between latent constructs and associated indicators. Then the researcher must examine the structural model (i.e., the inner model), which involves testing the hypothesis. To establish whether the construct indicators measured the same concept and were different from other constructs, different tests (construct reliability validity and discriminant validity) were performed on the data.

As reported in Tables 12.5 and 12.6, the results were all within the optimal thresholds, and the measurement model was successfully evaluated.

The analysis of the measurements model suggested that all constructs had acceptable levels of reliability and validity. Before the evaluation of the structural model, VIF statistics were examined to make sure that collinearity did not bias the regression results. In our model, VIF values ranged from 1.264 to 2.917, which was below the recommended value of 3 (Hair et al., 2022).

Our model explained that 78.1% of the target construct variance ($R^2 = 0.781$), which is regarded as a substantial explanatory power (Hair et al., 2022). A closer examination of the structural paths showed that PE ($\beta = 0.356$) had the strongest effect on BI, followed by FC ($\beta = 0.239$), PV ($\beta = 0.180$), HB ($\beta = 0.162$), and HM ($\beta = 0.145$). Bootstrapping routine also validated that these effects were

Table 12.5. Constructs Reliability and Validity.

Construct	Items	Loadings	Cronbach's Alpha	rho_A	C.R.	AVE
Behavior intention	BI1	0.881	0.880	0.881	0.926	0.807
	BI2	0.888				
	BI3	0.925				
Performance expectancy	PE1	0.853	0.882	0.884	0.928	0.811
	PE2	0.915				
	PE3	0.931				
Effort expectancy	EE1	0.946	0.938	0.941	0.956	0.844
	EE2	0.883				
	EE3	0.916				
	EE4	0.916				
Social influence	SI1	0.803	0.901	0.958	0.926	0.806
	SI2	0.957				
	SI3	0.973				
Facilitating conditions	FC1	0.817	0.870	0.875	0.911	0.720
	FC2	0.861				
	FC3	0.890				
	FC4	0.825				
Hedonic motivation	HM1	0.731	0.807	0.821	0.864	0.681
	HM2	0.831				
	HM3	0.905				
Price value	PV1	0.961	0.946	0.965	0.965	0.902
	PV2	0.949				
	PV3	0.938				
Habit	HB1	0.899	0.746	0.787	0.938	0.836
	HB2	0.821				
	HB3	0.844				

Evaluation criteria based on Hair et al. (2022): Loadings >0.70; Cronbach's Alpha: >0.70; rho_A: >0.70; Composite Reliability: >0.70; AVE: >0.50.

Table 12.6. HTMT Ratios.

	BI	EE	FC	HB	HM	PE	PV	SI
BI	–							
EE	0.653	–						
FC	0.736	0.554	–					
HB	0.657	0.534	0.649	–				
HM	0.229	0.236	0.191	0.197	–			
PE	0.809	0.708	0.442	0.407	0.130	–		
PV	0.582	0.136	0.362	0.363	0.256	0.447	–	
SI	0.620	0.624	0.517	0.260	0.134	0.626	0.308	–

Evaluation criterion based on Hair et al. (2022): HTMT<0.90.

significant at the 5% level. Despite the relationships SI→BI ($\beta = 0.118$) and EE→BI ($\beta = 0.107$) being positive, the bootstrap did not substantiate these results.

Additionally, the f^2 effect size was calculated to evaluate whether an omitted construct had a fundamental impact on intention. Values of approximately 0.02, 0.15, and 0.35, represent small, medium, and large effect sizes (Hair et al., 2022). The analysis reported medium effects in the case of dropping PE (0.222) and/

or FC (0.141), while removing either PV (0.089), HM (0.076), HB (0.065), or SI (0.031) would only cause small effects. On the other hand, EE (0.018) could be removed from the model as there was no effect on the dependent variable.

The structural assessment concluded with the standardized root mean square residual (SRMR) measure of approximate model fit. Our analysis reported an SRMR value of 0.082, which was below the recommended threshold of 0.10 (Hair et al., 2022), thus indicating an acceptable fit.

5.3. Analysis of the Constructs

BI is the strongest predictor of the actual use of technology (Venkatesh, 2022). Here, it refers to the likelihood that respondents will use BEV again in the future. In our study, it was found that participants tend to agree ($M = 5.65$, SD = 0.95) with the future use of battery electric technology. The mean value of the constructs FC ($M = 5.92$, SD = 0.80) and HB ($M = 5.84$, SD = 0.82) indicated that respondents in addition to having the resources necessary to utilize the technology, also believed it had become a habit. The pleasure as a result of using BEVs, and measured by the construct HM ($M = 4.48$, SD = 1.10), appeared relevant. The results of EE ($M = 4.19$, SD = 1.50) and PE ($M = 4.01$, SD = 1.23) indicated the use of technology provided benefits to the adopters, and it was generally free from substantial efforts. Also, the assessment of the construct PV ($M = 3.57$, SD = 1.25) indicated that the perceived benefits of BEVs tend to moderately outweigh the monetary costs. Lastly, the analysis of the construct SI ($M = 3.39$, SD = 1.32) indicated that significant others tend not to influence the respondents' decision to use the technology. Having examined the key components of the model, it is now possible to conclude the assessment of the hypotheses.

5.4. Assessment of the Research Hypotheses

To report the results, the standardized path coefficients between latent variables and their significance level were assessed. Out of the seven proposed hypotheses, five were supported. Their path relationships were significant at the 0.05 level, had signs in the expected directions, and possessed path coefficients (β) from 0.145 to 0.356. Table 12.7 reports the findings.

Table 12.7. Assessment of Hypotheses.

Hypothesis	Path	Coefficient	t-Value	*p*-Value	Support
H1	PE→BI	0.356	3.489	0.001	YES
H2	EE→BI	0.107	1.061	0.289	**NO**
H3	SI→BI	0.118	1.550	0.121	**NO**
H4	FC→BI	0.239	3.204	0.001	YES
H5	HM→BI	0.145	2.298	0.022	YES
H6	PV→BI	0.180	2.441	0.015	YES
H7	HB→BI	0.162	2.269	0.023	YES

Hypotheses evaluation criteria: t-Value>1.96. p-Value<0.05.

6. DISCUSSION AND CONCLUSIONS

The growing demand for BEVs has enabled the development of a global market for components and infrastructures and made vehicles more efficient and affordable. In this study, we explored the characteristics of a sample of BEV owners through the lenses of the UTAUT-2 framework. The assessment is relevant because the very first customers to embrace new ideas often shape the direction of the technology, and play a crucial role in the diffusion of products and services to larger market segments.

It appears that BEV owners are satisfied with their choice and intend to continue using the technology under current circumstances. Among the seven proposed hypotheses, five were supported. We found that *EE* and *SI* were not significant predictors of BI. While evidence suggests that using electric vehicles is neither more difficult nor requires effort compared to traditional cars (Jeon et al., 2012; Madigan et al., 2017), the finding related to SI is somehow surprising. In fact, our result differs from the prevailing theory supporting the role of SI as one of the major determinants of behavior (Fishbein & Ajzen, 2009; Venkatesh et al., 2012). Perhaps this could be explained by the consumers' strong interest in new ideas and technologies, which leads them to make purchase decisions outside the typical effects of the members of society. *PE* was the strongest predictor of the model, suggesting that BEVs deliver on their promises and provide benefits that meet customers' expectations. These perceived benefits could be many and diverse, such as the free usage of charging stations, the dedicated areas in public car parks, the exemption of annual road tax, or the awareness of moving around with minimal impact on the environment. Rogers (2003) argued that the possession of financial resources and the ability to deal with technical knowledge can reduce the potential negative impact of innovations. As a result of the respondents' high income and adequate technical knowledge, the construct of *FC* appeared to strongly influence their intentions. Additionally, the assessment of the *PV* was in the same direction, suggesting that the perceived benefits of BEVs tend to outweigh the monetary costs incurred in the acquisition and maintenance of the vehicles. In terms of *HM*, the study found that respondents enjoyed driving their vehicles, for example, due to the ownership of a technologically advanced product, the sophisticated onboard equipment, or the fast acceleration of the vehicles. Similarly, Venkatesh et al. (2012) found HM to be a critical determinant of BI. Lastly, our study indicated that the construct of *HB* was a key factor of intention, with current owners considering that BEVs are well integrated into their lives.

Several theoretical contributions were made. The discussion has helped to define a clearer profile of BEV technology users, including a better understanding of the UTAUT-2 model and its relevance to assessing repurchase intentions. In our study, BEV owners are depicted as being able to make complex purchase decisions independently of what their significant others or society think, a finding that contrasts the prevailing technology research (Fishbein & Ajzen, 2009; Venkatesh et al., 2012) in the Asian context (Venkatesh & Zhang, 2010) where the influence from society is prominent and decisions are collectively made. Our finding implies that the role of society at large is not as influential as previously thought, and instead, technology acceptance is driven by individual interests and motivations that could lead to a new wave of innovation and opportunities in the region. Further, the construct of habit performed well as a predictor of intention. This is relevant because research had often excluded the construct for the reason that respondents had not yet formed a routine behavior about the technology (Tamilmani et al., 2021). Our study brings managerial contributions as well. It is important to recognize that early users serve as role models for other members of a social system and that their preferences may change from country to country when it comes to the introduction and diffusion of innovations. Thus, perceived benefits (e.g., status, differentiation, or uniqueness) could be used in communication strategies to appeal to relatively wealthy groups of customers who wish to appear trendy and technology savvy to their peers. In accordance with Wu and Wu (2014), managers have to maximize the innovation benefits while minimizing the risks of mismatching with local requirements due to a lack of knowledge about local markets. It is also important that marketers and policymakers encourage a positive post-purchase attitude among early users so that BEV technology can be widely adopted on a large scale.

BEVs are indeed living up to their promise of being a viable and sustainable means of transportation that is spreading to different categories of consumers, thus driving global business opportunities. Our research indicates that BEV technology has advanced to a point where electric vehicles can deliver on their promises, thus challenging traditional gasoline-powered cars. One outstanding example of this success in Macau is the Tesla Model 3, which has set new benchmarks for performance and range. Further, the growing electric mobility infrastructure confirms the notion that BEVs are here to stay. In conclusion, BEVs have proven to be living up to their promise of being a sustainable and reliable means of transportation. Their widespread adoption will transform the global automotive industry, making transportation clean, efficient, and cost-effective for everyone. BEVs are expected to create new markets and opportunities in the automotive sector that will create a new wave of economic growth and development around the world. As a matter of fact, manufacturers are already investing heavily in the technology, creating a dynamic market for international suppliers and investors. The future of sustainable transportation is electric, and it is essential that we embrace this transformation to ensure a cleaner and healthier planet for future generations.

7. LIMITATIONS AND FUTURE RESEARCH

Some limitations need to be acknowledged. The design opted for a convenience sampling method under the assumption that respondents were similar to the overall target population, which may not be the case. A larger probabilistic sampling method may be useful as it would allow more statistical tests. Our findings may also differ in places with different socio-cultural characteristics. For this reason, it would be highly interesting to research different contexts and compare among them the obtained results. In addition to addressing the above limitations, future studies should also consider the use of qualitative questions to gain richer insights and deepen the understanding of this key customer segment.

REFERENCES

Anable, J., Skippon, S., Schuitema, G., & Kinnear, N. (2011, June 6). *Who will adopt electric vehicles?: A segmentation approach of UK consumers*. https://abdn.pure.elsevier.com/en/publications/who-will-adopt-electric-vehicles-a-segmentation-approach-of-uk-co

Axsen, J., Goldberg, S., & Bailey, J. (2016). How might potential future plug-in electric vehicle buyers differ from current "Pioneer" owners? *Transportation Research Part D: Transport and Environment*, *47*, 357–370. https://doi.org/10.1016/j.trd.2016.05.015

Baik, Y., Hensley, R., Hertzke, P., & Knupfer, S. (2019). *Making electric vehicles profitable*. McKinsey & Company. Retrieved on January 16, 2023 from https://www.mckinsey.com/industries/automotive-and-assembly/our-insights/making-electric-vehicles-profitable

Burs, L., Roemer, E., Worm, S., & Masini, A. (2020). Are they all equal? Uncovering adopter groups of battery electric vehicles. *Sustainability*, *12*(7), 2815. https://doi.org/10.3390/su12072815

Emergen Research. (n.d.). *Electric vehicles market top companies | EVs industry trends by 2028*. Emergen Research. https://www.emergenresearch.com/blog/top-10-electric-vehicles-companies-focused-on-transforming-the-automotive-industry

Erdfelder, E., Faul, F., & Buchner, A. (1996). GPOWER: A general power analysis program. *Behavior Research Methods, Instruments, & Computers*, *28*(1), 1–11. https://doi.org/10.3758/BF03203630

Fishbein, M., & Ajzen, I. (2009). *Predicting and changing behavior: The reasoned action approach*. Psychology Press. https://doi.org/10.4324/9780203838020

Flick, U. (2018). *An introduction to qualitative research*. SAGE.

Ghauri, P., Strange, R., & Cooke, F. L. (2021). Research on international business: The new realities. *International Business Review*, *30*(2), 101794.

Hair, J., Hult, G. T. M., Ringle, C., & Sarstedt, M. (2022). *A primer on partial least squares structural equation modeling (PLS-SEM)*. https://doi.org/10.1007/978-3-030-80519-7

Hardman, S., Shiu, E., & Steinberger-Wilckens, R. (2016). Comparing high-end and low-end early adopters of battery electric vehicles. *Transportation Research Part A: Policy and Practice*, *88*, 40–57. https://doi.org/10.1016/j.tra.2016.03.010

Higueras-Castillo, E., Molinillo, S., Coca-Stefaniak, J. A., & Liébana-Cabanillas, F. (2020). Potential early adopters of hybrid and electric vehicles in Spain—Towards a customer profile. *Sustainability*, *12*(11), 4345. https://doi.org/10.3390/su12114345

Jeon, C., Yoo, J., & Choi, M. K. (2012). The effect of social influence on consumers' hybrid electric vehicles adoption in Korea and China. In *2012 14th international conference on advanced communication technology (ICACT)* (pp. 336–340). Global IT Research Institute (GiRI).

Knupfer, S. M., Hensley, R., Hertzke, P., & Schaufuss, P. (2017). *Electrifying insights: How automakers can drive electrified vehicle sales and profitability* (Advanced Industries). McKinsey & Company. https://www.mckinsey.com/industries/automotive-and-assembly/our-insights/electrifying-insights-how-automakers-can-drive-electrified-vehicle-sales-and-profitability

Lampo, A. (2023). *How is technology accepted? Fundamental works in user technology acceptance from diffusion of innovations to UTAUT-2*. Proceedings of the 8th International Conference on Industrial and Business Engineering (ICIBE 2022), Macau SAR, 260–266. https://doi.org/10.1145/3568834.3568903

Lampo, A., Duarte, P., & Silva, S. (forthcoming). The role of environmental concern and technology show-off on electric vehicles adoption: The case of Macau. *International Journal of Emerging Markets*. https://doi.org/10.1108/IJOEM-10-2021-1637

Lampo, A., & Silva, S. (2022). How social influence and image impact on the intention to use a technology: A study from the battery electric vehicle domain. In *Proceedings of the 13th international conference on e-business, management and economics* (pp. 142–146). Association for Computing Machinery (ACM).

Madigan, R., Louw, T., Wilbrink, M., Schieben, A., & Merat, N. (2017). What influences the decision to use automated public transport? Using UTAUT to understand public acceptance of

automated road transport systems. *Transportation Research Part F: Traffic Psychology and Behaviour*, *50*, 55–64. https://doi.org/10.1016/j.trf.2017.07.007
McKinsey. (2021). *Why the automotive future is electric*. https://www.mckinsey.com/industries/automotive-and-assembly/our-insights/why-the-automotive-future-is-electric
Meneses, J. (2021, September 19). *Special report – Hazy problem*. Macau Business. https://www.macaubusiness.com/special-report-hazy-problem/
Mordor Intelligence. (2023). *EV parts and components market size and share analysis – Industry research report – Growth trends*. https://www.mordorintelligence.com/industry-reports/electric-vehicle-parts-and-components-market
Parasuraman, A., & Colby, C. L. (2015). An updated and streamlined technology readiness index: TRI 2.0. *Journal of Service Research*, *18*(1), 59–74. https://doi.org/10.1177/1094670514539730
Plötz, P., Schneider, U., Globisch, J., & Dütschke, E. (2014). Who will buy electric vehicles? Identifying early adopters in Germany. *Transportation Research Part A: Policy and Practice*, *67*, 96–109. https://doi.org/10.1016/j.tra.2014.06.006
Rezvani, Z., Jansson, J., & Bodin, J. (2015). Advances in consumer electric vehicle adoption research: A review and research agenda. *Transportation Research Part D: Transport and Environment*, *34*, 122–136.
Rogers, E. M. (2003). *Diffusion of innovations* (5th ed.). Free Press.
Sagapova, N., Dušek, R., & Pártlová, P. (2022). Marketing communication and reputation building of leading European oil and gas companies on Instagram. *Energies*, *15*(22), 8683.
Son, J. (2018). Back translation as a documentation tool. *Translation & Interpreting*, *10*(2), 89–100. https://doi.org/10.12807/t&i.v10i2.765
Singh, P. P., Wen, F., Palu, I., Sachan, S., & Deb, S. (2022). Electric vehicles charging infrastructure demand and deployment: Challenges and solutions. *Energies*, *16*(1), 7.
Stauffer, N. (2020, November 25). *China's transition to electric vehicles*. MIT Energy Initiative https://energy.mit.edu/news/chinas-transition-to-electric-vehicles/
Tamilmani, K., Rana, N. P., Wamba, S. F., & Dwivedi, R. (2021). *The extended Unified Theory of Acceptance and Use of Technology (UTAUT2): A systematic literature review and theory evaluation*. https://bradscholars.brad.ac.uk/handle/10454/18159
Tanaka, A., Kawakami, T., & Omoto, Y. (2018). *Battery wars: Japan and South Korea battle China for future of EVs*. Nikkei Asia. https://asia.nikkei.com/Spotlight/The-Big-Story/Battery-wars-Japan-and-South-Korea-battle-China-for-future-of-EVs
Venkatesh, V. (2022). Adoption and use of AI tools: A research agenda grounded in UTAUT. *Annals of Operations Research*, *308*(1), 641–652. https://doi.org/10.1007/s10479-020-03918-9
Venkatesh, V., Morris, M. G., Davis, G. B., & Davis, F. D. (2003). User acceptance of information technology: Toward a unified view. *MIS Quarterly*, *27*(3), 425. https://doi.org/10.2307/30036540
Venkatesh, V., Thong, J. Y. L., & Xu, X. (2012). Consumer acceptance and use of information technology: Extending the Unified Theory of Acceptance and Use of Technology. *MIS Quarterly*, *36*(1), 157–178. https://doi.org/10.2307/41410412
Venkatesh, V., & Zhang, X. (2010). Unified Theory of Acceptance and Use of Technology: U.S. vs. China. *Journal of Global Information Technology Management*, *13*, 5–27. https://doi.org/10.1080/1097198X.2010.10856507
Wu, J., & Wu, Z. (2014). Local and international knowledge search and product innovation: The moderating role of technology boundary spanning. *International Business Review*, *23*(3), 542–551.
Zakari, A., Khan, I., Tan, D., Alvarado, R., & Dagar, V. (2022). Energy efficiency and sustainable development goals (SDGs). *Energy*, *239*, 122365.

CHAPTER 13

DISCUSSING INSIGHTS FROM THE CARBON PERFORMANCE LITERATURE FOR IB: THE INSTITUTIONAL COMPLEXITY OF CLIMATE CHANGE POLICY

Gunnar Leymann and Anna Kehl

University of Bremen, Germany

ABSTRACT

Multinational enterprises (MNEs) own and control technological resources and capabilities that make them critical actors in accelerating the transition toward net zero. Even beyond the energy sector, stakeholders are putting increasing pressure on MNEs to reduce the carbon intensity of their operations, that is, to improve their carbon performance. While there is unambiguous evidence that national climate policy is a critical catalyst for long-term carbon performance improvements, there is limited research on how MNEs' carbon strategies react to climate policies. This chapter reviews the concepts, drivers, and strategies connected to carbon performance in the broader sustainability and management literature to clarify potential complementarities to international business (IB). The authors then highlight how MNEs will face increasing institutional complexity along two dimensions: (1) the structural diversity of institutional environments and (2) institutional dynamism, primarily reflected by public policy. The proposed conceptual framework maps these two dimensions to national and subnational levels, and the authors present two data sources that allow the

Walking the Talk? MNEs Transitioning Towards a Sustainable World
Progress in International Business Research, Volume 18, 237–261

Published under exclusive licence by Emerald Publishing Limited
ISSN: 1745-8862/doi:10.1108/S1745-886220240000018017

quantitative analysis of country differences in the diversity and dynamism of national climate policy. The authors conclude that there are ample opportunities for IB researchers to explore MNEs' strategic reactions to climate policy and to inform policymakers about the consequences of national climate policy in the global economy.

Keywords: Carbon performance; climate change; climate policy; institutions; multinational enterprises; international business

INTRODUCTION

Climate change has become an existential challenge to humanity. Avoiding its most serious consequences requires a united effort to rapidly reduce the emission of greenhouse gasses, of which 76% are carbon dioxide. Yet, as we write, the combination of a stalemate in international agreements, increasing energy demand in emerging markets, and a slow transition of some of the most polluting sectors causes reason for concern. Data show that the carbon intensity of the global economy has not decreased at a rate that would meet the 1.5° C target (McKinsey, 2023), and MNEs are among the biggest polluters. In 2016, their international operations accounted for 17% of global carbon emissions (Zhang et al., 2020).

MNEs are under significant pressure from numerous stakeholders, including international organizations and governments, and have sometimes been criticized for inaction. And indeed, while MNEs were found to have higher carbon performance than domestic firms, recent evidence suggests that MNEs still shift operations with high emissions into countries with weak environmental regulation (Ben-David et al., 2021). Moreover, a survey of large MNEs discovered strategic plans to cut out suppliers that will not meet their emission goals rather than investing in the greening of their operations (Standard Chartered, 2021). Even if carbon emissions from MNEs have declined in recent years, a good part of this may be attributed to economic crises and deglobalization (Zhang et al., 2020). As Daron Acemoglu (2023) recently put it: "[L]eft to its own devices, the market will continue to channel investment to fossil fuels. Taxation, regulation, and societal pressure are all necessary …."

The energy transition can be defined as a long-term process (Bass & Grøgaard, 2021), and a shift away from fossil fuels after the oil crises of the 1970s has been underway for decades. However, as the visible consequences of climate change increase – with material impacts on the global economy (Linnenluecke et al., 2013) – we are likely to see even higher non-market pressures on MNEs to improve their carbon performance. Programs like the European Green Deal are a case in point that policymakers will further compress the temporal dimension of the transition in the coming years. This will lead to institutional pressures with significant temporal dynamism, partly driven by the necessity to avoid catastrophic tipping points with unpredictable consequences (Rockström et al., 2009), and partly driven by a new policy paradigm that pushes "mission-oriented" government intervention in technology development and market co-creation (Mazzucato, 2016).

In this chapter, we argue that the IB literature could complement research on carbon performance that, until now, has paid limited attention to the consequences of variations in corporate multinationality. Specifically, we highlight how government intervention in the energy transition will lead to increasing institutional diversity and dynamism, whereby the latter is driven by public policy. We suggest that the resulting institutional complexity raises questions about how MNEs can strategically react to increasing stakeholder pressures to improve their carbon performance. Our proposed conceptual framework acknowledges the spatial diversity and temporal dynamism of institutional environments in the context of MNEs carbon performance.

This chapter is divided into two parts. In the first part, we provide an overview of the concepts, measurements, stakeholder drivers, and corporate strategies related to the discussion about carbon performance in the broader corporate sustainability and management literature. In our review, we selected studies that either directly focus on carbon performance or offer us insights into how issues of carbon performance differ from the overarching concept of environmental performance. Thus, our review goes beyond the IB domain and is interdisciplinary. This part of the chapter makes three points. First, we identify significant uncertainty concerning the competitive viability of carbon strategies, potentially undermining strategic commitment to carbon performance improvements. Second, existing evidence highlights government policy as the most effective tool to resolve some of these uncertainties. However, for MNEs, this involves the complex management task of navigating a diverse set of climate change policies across their home and host economies. Third, we argue that there is a lack of dedicated IB research to unravel how the interplay of national climate policy and institutional diversity affects MNEs strategic commitment to improvements in carbon performance.

In the second part, we focus on the institutional complexity that results from the MNEs exposure to multiple, diverse institutional environments and dynamic trajectories. We discuss how the increasing interest of IB scholars in the comparative capitalism (CC) approach and policy studies can provide the foundation for a conceptual framework that allows a more granular view of climate change policy. In particular, we propose that the institutional complexity of climate change can be divided into two dimensions, namely structural diversity and process-related dynamism. Based on two policy data sources, we visualize the reality of cross-country differences in climate policy diversity and dynamism faced by MNEs. We conclude with a call for future research that explores the institutional complexities of climate change policy outside the energy industry and on national and subnational levels of analysis.

ENVIRONMENTAL AND CARBON PERFORMANCE: DIFFERENCES, DRIVERS AND STRATEGIES

According to Bass and Grøgaard (2021), two emerging strands in the IB literature focus on energy issues, namely (1) the growing concern with MNEs as renewable energy producers, and (2) the role of MNEs in transitioning toward low carbon production. Here, we focus more on the second area, specifically on the non-market drivers of carbon performance and the associated carbon mitigation strategies. We review studies from the environmental and carbon performance literature to highlight differences and understand the peculiarities of carbon performance as a construct.

Concepts and Measurement

In today's global economy, corporations are increasingly held accountable for their environmental conduct by stakeholders such as customers, investors, and regulators. Corporate environmental performance is defined in the International Standards Organization (ISO) 140001 standard as "measurable results of an organization's management of its environmental aspects" (Perotto et al., 2008, p. 517), which already indicates significant breadth in conceptualization and operationalization (Trumpp et al., 2015). And indeed, the large body of literature concerned with the green business transition applies various firm-level performance concepts, such as environmental performance (Haque & Ntim, 2018; Ramanathan et al., 2014; Villiers et al., 2011; Zameer et al., 2021), environmental innovation (Yalabik & Fairchild, 2011), the adoption of sustainable operations (Jiao et al., 2020), and environmental responsiveness (Henriques & Sadorsky, 1996), sometimes with unclear boundaries.

Even when focusing on environmental performance, one can find operationalizations that range from multidimensional indicators (Villiers et al., 2011) to individual measures of greenhouse gas emissions (Haque & Ntim, 2018). These unclear boundaries are problematic because particular dimensions of environmental performance might be difficult to compare incomparable (Iwata & Okada, 2011). Indeed, one of the factors that sets climate change mitigation apart from other sustainability issues is the necessity for large-scale investments in technology and energy supply throughout the entire value chain, not all of which are cost-competitive (Fujii et al., 2013).

The carbon performance literature provides a narrower discussion of how well a corporation reduces its carbon footprint over time. Measuring carbon performance involves measuring a corporation's carbon emissions to output ratio and benchmarking it against competitors or industry-specific net-zero pathways (Dietz et al., 2021). However, some studies also include process-based carbon performance measures that reflect the extent of managerial and strategic responses (Busch & Hoffmann, 2011). These have been criticized for not reflecting substantive change (Luo & Tang, 2022), which is why we separate carbon performance (outcome) from carbon strategies (process).

Even if the concept of carbon performance is narrower than environmental performance, there is still potential for inconsistencies in the use of concepts and measurements. Hoffmann and Busch (2008) suggested four such areas: the measurement perspective, indicator specification, emissions scope, and type of reporting scheme. Carbon performance can be measured from carbon inputs or outputs. Carbon inputs refer to the amount of carbon or carbon-containing materials used or consumed by the corporation. Carbon outputs are the direct and indirect emissions that result from the firm's activities and can be measured in absolute or relative terms. Absolute emissions help to understand the scale of carbon performance improvements, but they do not explain how much a business depends on carbon emissions. For example, MNEs could outsource production to reduce both output and total emissions, even though there has been no change in net emissions per unit output (Busch, 2010).

In terms of emission scope, there is usually a separation between direct Scope 1 emissions, indirect Scope 2 emissions from energy consumption, and indirect Scope 3 emissions from the entire supply chain. Scope 3 emissions are often challenging to estimate, and it may be specifically troublesome for MNEs with a geographically distributed supply chain (Boukherroub et al., 2017; Zhang et al., 2020). Only recently have there been more studies that aim to understand the holistic carbon footprint of MNEs, but these still focus on attributing emissions to the MNE's FDI footprint (Zhang et al., 2020). While identifying the ownership structures of MNEs is difficult, measuring the emissions of arm's length suppliers is likely to be highly complex.

Finally, the reporting scheme can affect the reliability of the carbon performance measure. Emission data can result from voluntary self-reporting or structured reporting under a framework like the European Emissions Trading System. The former causes concern that the measurements may contain bias, and the Carbon Disclosure Project, which is the data source for many empirical studies, makes this explicit. Besides the possibility of false or inaccurate disclosure (Callery & Perkins, 2021; Clark & Crawford, 2012), there is a self-selection effect that leads to an overrepresentation of high-performing firms in voluntarily disclosed data (Busch & Lewandowski, 2018; Chen & Silva Gao, 2012). The main drawback of mandatory reporting is that it often comprises only Scope 1 emissions of the regulated facility and lacks standardization across different regulatory regimes (Andrew & Cortese, 2011). Considering these variations in how carbon performance is conceived of, scholars have found that the available measures, especially those attained from third-party providers, may not always be consistent (Busch, Johnson, et al., 2022), which could affect research findings.

Stakeholder Drivers of Environmental and Carbon Performance

According to Bass and Grøgaard (2021), non-market stakeholder pressures remain key drivers in the energy transition. The literature on stakeholder drivers of environmental (carbon) performance also suffers from blurry conceptual boundaries regarding both drivers and outcomes. For instance, Ramanathan et al. (2014) considered firms' internal and external stakeholders. On the other hand, Jiao et al. (2020) differentiated between primary stakeholder pressures, including customers, competitors, suppliers, employees, and shareholders, and secondary stakeholder pressures, including communities, non-governmental organizations (NGOs), and media. Outcome measures also vary substantially, and only a few studies explicitly measure carbon performance (Damert et al., 2017; Haque, 2017; Moussa et al., 2020; Qian & Schaltegger, 2017). Here, we follow the broad distinction between internal and external stakeholders to structure our review and include studies concentrating on sustainable production, environmental performance, or carbon performance as an outcome variable.

Internal Stakeholder Pressures

Internal stakeholders that have been identified as relevant to companies' environmental performance include management, board, and shareholders. Management-specific factors have been found to influence engagement in transition-related activities significantly. For instance, greater awareness (Zameer et al., 2021) and a positive attitude among managers toward environmental issues (Jiao et al., 2020) positively impact companies' adoption of green production methods. Conversely, a negative attitude toward sustainability-related matters impedes green transitioning (Aragòn-Correa et al., 2020). Furthermore, Jiao et al. (2020) found that various stakeholders such as customers, competitors, suppliers, employees, and shareholders influence managerial perceptions of sustainable operations. Therefore, managers are critical in mediating the pressures exerted by different stakeholders on a firm's engagement in sustainable operations. Similarly, Ramanathan et al. (2014) demonstrated that pressures from stakeholders indirectly affect management, and other studies support this mediating role of managerial awareness (Zameer et al., 2021).

Another strand of research examines the characteristics of boards as drivers for environmental and carbon performance. The literature, however, provides contrasting results on the impact of board size and independence (Haque, 2017; Villiers et al., 2011; Walls et al., 2012), although empirical evidence seems to opt for more independent directors to have a more robust environmental orientation (Moussa et al., 2020). There is an ongoing debate regarding the effect of diversity factors, such as gender, age, and cultural background of board members (Post et al., 2011). Haque (2017) showed that gender diversity is associated with firm-level initiatives to improve carbon performance but not with actual emission reduction. The findings of Moussa et al. (2020) contrast these results, as they also found a direct effect of board characteristics on carbon performance.

Research on the impact of shareholders refers, in particular, to the proportion of institutional and private ownership. Institutional ownership was found to positively affect companies' environmental performance (Clò et al., 2017). It is also worth noting that sovereign wealth funds increasingly invest in activities related to Environmental, Social and Governance (ESG) criteria (Bortolotti et al., 2023; Megginson et al., 2023), indicating a shift in investment strategy toward (environmental) sustainability. Shareholders' impact on environmental performance may vary based on their motivation to pressure companies in the green transition. Shareholders often demand better carbon reporting due to their financial exposure and not necessarily because of their interest in firms' responsibility for social and environmental impact (Alsaifi et al., 2020). Although the main interest of shareholders might lie in reducing financial transition risks, these concerns can lead to environmental considerations. For example, the world's largest investment fund, BlackRock, has recently changed its strategy to consider sustainability indicators due to shareholders' concerns about financial transition risks (Arvidsson & Dumay, 2022). In addition to shareholder pressures to reduce financial transition risks, shareholder activism is more impactful. Shareholder activism mainly exists in the form of filed resolutions (Clark & Crawford, 2012) and was found to have a positive impact on firm environmental performance (M. D. P. Lee & Lounsbury, 2011; Walls et al., 2012), as well as on carbon transparency (Flammer et al., 2021).

External Stakeholder Pressures

External stakeholders can exert significant pressure on firms to adopt sustainable business practices. These pressures can stem from various sources, including communities, the media, lobby groups, NGOs, and government regulation. The latter is often separated from other pressures due to its coercive nature (González-Benito & González-Benito, 2010).

Henriques and Sadorsky (1996) found that while local communities were of moderate importance in pressuring firms to formulate environmental plans, environmental organizations, and lobby groups had low impact. A study on the effect of local community lobbying on carbon performance in large firms found no significant impact, but the authors highlight that the institutional context, namely Japan, might have affected their results (Cole et al., 2013). Similarly, Jiao et al. (2020) found that media, community, and NGOs are neither directly significant nor influential in pressuring firms to adopt sustainable operations. Zhou et al. (2022) studied the interaction effects of different stakeholders, revealing that community stakeholders can influence corporate environmental performance through their impact on policymakers. This suggests an endogenous relationship between the government and other external stakeholder pressures.

Governments are recognized as primary drivers of firms' environmental and carbon performance, mainly through regulation. A large number of studies have shown that environmental regulation strongly affects environmental performance (Ramanathan et al., 2014; Reid & Toffel, 2009; Wu et al., 2020), whereas market-based policy instruments also indirectly affect environmental outcomes by stimulating green innovations (Wu et al., 2020). However, one exception is the study of Jiao et al. (2020), who did not find that governmental pressures affected the adoption of sustainable operations, but their results were survey-based, which might not reflect the actual government regulations in place. In the specific case of carbon performance, there is more unambiguous evidence that government is among the top drivers for substantive change (Cadez & Czerny, 2016; Littlewood et al., 2018).

The increasing institutionalization of carbon disclosure practices may be an indirect channel through which governments can exert pressures. However, in addition to the problems related to differences in reporting schemes, policies that address reporting are generally voluntary rather than mandatory (Aragòn-Correa et al., 2020). As a result, higher levels of environmental disclosure may not necessarily translate into better carbon performance. Aragón-Correa et al. (2016) found that higher levels of environmental disclosure were associated with lower levels of environmental performance among the top 100 most international non-financial companies, compared to their domestic counterparts. This suggests that MNEs may use transparency as a means to improve their environmental image rather than as a way to reduce carbon emissions. Thus, policies that regulate carbon reporting have been criticized for not focusing on performance outcomes (Arvidsson & Dumay, 2022).

Research on the link between carbon transparency requirements and carbon performance has produced mixed results. Qian and Schaltegger (2017) argued that disclosure can motivate firms to improve their carbon performance. However, this may again require active government intervention. Guenther et al. (2016)

demonstrated an association between various stakeholders and firms' disclosure practices and that government intervention moderates the link between transparency and carbon performance. The authors find that improved disclosure is more likely to result in higher carbon performance in countries where governments have established more stringent climate policies.

Our review suggests that environmental (carbon) performance is often not directly related to stakeholder drivers. However, stakeholders may push firms to adopt green initiatives, carbon reporting, or environmental strategies that have the potential to affect performance outcomes. Managers are essential mediators of whether pressures translate into a substantive change in performance. These strategic management decisions depend on unambiguous and binding stakeholder pressures, which explains why the government remains a crucial actor in the energy transition. In other words, we argue that many stakeholder drivers depend on the formal institutional context they operate. It is necessary to note again that the issue of carbon performance and the broader environmental performance are difficult to compare. This is because a substantive increase in carbon performance is not easily achieved through the types of operational adjustments often explored in the broader sustainability literature, requiring stronger strategic commitment on behalf of firms.

Corporate Carbon Strategies and Strategic Commitment

Our review of stakeholder drivers revealed limited evidence that would support a direct link to carbon performance, but rather that such pressures result in strategic responses with unclear performance outcomes. A firm's carbon strategy, sometimes also referred to as climate strategy (Kolk & Pinkse, 2005), was defined by Damert and Baumgartner (2018) as "a set of market and non-market activities that aim to reduce and legitimize the impact of a firm's business activities on climate change" (p. 267). Hence, the authors explicitly refer to the possibility that firms use carbon strategies to legitimize rather than directly reduce their emissions.

Carbon strategies can be separated into continuum- and typology-based classifications. The former takes an evolutionary perspective on carbon strategies, arguing that strategies gradually shift from resistance with no commitment toward proactive strategizing with high commitment (Damert & Baumgartner, 2018; S.-Y. Lee, 2012). Typology-based approaches acknowledge that carbon strategies entail different configurations of individual strategic instruments and that there are multiple ways to increase carbon performance effectively. Empirical studies usually capture both dimensions, which we label strategic commitment (to represent the continuum approach) and scope (to represent the diversity of strategic instruments).

In terms of scope, studies have identified many individual strategic instruments, such as process and product improvements, emission reduction targets, supply chain optimization, market and business model development, external relationship development, environmental certifications, and regulatory engagement. To gain a better understanding of the most common carbon strategies and how their level of commitment varies, scholars have measured the commitment to strategic instruments and grouped them into similar clusters (Backman et al., 2017; Kolk & Pinkse, 2005; S.-Y. Lee, 2012; Sprengel & Busch, 2011; Weinhofer & Hoffmann, 2010).

Kolk and Pinkse (2005) analyzed the carbon strategies of 136 corporations from the *Financial Times* (FT) 500. They found that 31% of the corporations were in a careful exploration stage, and 36% were in a planning stage. Neither group had decided to commit scarce resources to carbon performance improvements. S.-Y. Lee (2012) clustered the strategic carbon instruments of South Korean corporations and found that most of them fell into a wait-and-see category with limited levels of overall commitment. Less than 10% of their sample consisted of corporations strongly committed to multiple strategic instruments.

Backman et al. (2017) proposed a continuum framework for assessing carbon strategies, arguing that firms first engage in Walk I, mainly related to non-market issues of governance and information management, and then Walk II, referring to investments in new technology and processes. They found that high commitment strategies, including an active engagement in Walk II, characterized 28% of the identified firms. While an encouraging 46% of firms were in a transition phase, these still had weak climate change governance and no significant resource commitments. Even inside the automobile industry, which has been under strong stakeholder pressures for years, Damert and Baumgartner (2018) found strategic laggards, albeit largely constrained to emerging market firms.

The evidence suggests that bold resource commitments to carbon strategies are not commonplace. Hypothetically, this explains why studies have found a limited substantive effect of these strategies on carbon performance (Cadez & Czerny, 2016; Damert et al., 2017; Doda et al., 2016). Some researchers focused on individual strategic instruments, such as emission targets, that were found to be ineffective, except in their most strict (but also most rare) formulation (Dahlmann et al., 2019). Corporations also altered their accounting practices to conceal a lack of substantive commitment, leading to carbon strategies that are "purely symbolic and dovetail with greenwashing practices" (Velte, 2021, p. 51). Longitudinal case studies of Australian corporations revealed a tendency to diffuse ambitious carbon strategies over time, ultimately reverting to business-as-usual practices (Wright & Nyberg, 2017).

MNEs are not an exception here. In a recent survey by Standard Chartered, 67% of MNEs prioritized reducing supplier emissions over their own, but 78% planned to eliminate suppliers who did not align with their carbon strategies rather than considering supplier upgrading (Standard Chartered, 2021). More than half of the respondents reported a lack of knowledge and data transparency concerning their suppliers, particularly in emerging markets, which could hinder investments in suppliers' emission mitigation efforts. Recent emission data also indicate that MNEs locate high-emission activities in emerging markets and developing countries (Ben-David et al., 2021; Zhang et al., 2020), perhaps to evade the costs associated with stronger regulatory pressures at home. Duan and Jiang (2021) have reached a similar conclusion, arguing that MNEs' emission reductions are driven by the recent deglobalization trends rather than firm-specific investments in technology.

The lack of bold strategic commitment by many corporations, including MNEs, may result from high uncertainty coupled with the irreversibility of investing in cleaner production and renewable energy systems. Several factors contribute to this uncertainty, including difficulties in measuring emissions in complex value chains, the unclear formulation of climate change policies, and, crucially, the competitive viability of carbon strategies. Companies must make irreversible technology investments and costly energy-sourcing decisions to increase strategic commitment. These strategic choices can affect a corporation's profitability years into the future, and given the uncertainty considering their competitive viability, corporations are more likely to prioritize flexibility over bold strategic commitment (S.-Y. Lee & Klassen, 2016; Rugman & Verbeke, 1998; Trigeorgis & Reuer, 2017).

In line with this, the evidence for a business case for carbon performance improvements is unclear. Busch and Lewandowski (2018) discovered a positive relationship between carbon and financial performance in a meta-study, but only for relative carbon performance measures. They speculated that corporations are mainly rewarded for marginal efficiency improvements. Delmas et al. (2015) found that carbon performance positively impacts Tobin's Q as a long-term performance measure but that short-term performance implications were negative. In a replication study with an expanded time frame, Busch, Bassen, et al. (2022) found a negative relationship between carbon and financial performance both in the short and long terms. The authors further discuss corroborating evidence from event studies that show adverse market reactions to carbon performance improvements, contending that "investors do not seem to perceive a firm's lower carbon emissions as value driving" (Busch, Bassen, et al., 2022, pp. 166–167). They conclude that government policy is essential to reduce the uncertainty surrounding the competitive viability of carbon strategies. In line with this, Littlewood et al. (2018) identified government and regulatory stakeholders as the most impactful on a firm's strategic commitment to an effective carbon strategy.

The Role of MNEs in Carbon Performance Research

The attention devoted to the peculiarities of MNEs in carbon performance research remains limited. In terms of the measurement and disclosure of carbon performance in MNEs, we identified data availability problems. Even if Scope 3 data are much more accessible these days, it still relies heavily on self-disclosure, which can be unreliable for two reasons. On the one hand, attributing carbon emissions to individual MNEs is difficult due to carbon transfer (carbon leakage) through foreign investment and trade channels (López et al., 2019; Zhang et al., 2020) and complicated ownership structures. On the other hand, it may be that multinationality affects the ability or willingness of MNEs to disclose accurate data (Aragón-Correa et al., 2016), especially concerning suppliers in the Global South (Standard Chartered, 2021).

Stakeholder drivers can also be affected by multinationality, and there is a long-standing argument that MNEs are more exposed to stakeholder pressures than national firms (Christmann, 2004). More multinational firms were found to be better able to develop green dynamic capabilities (Maksimov et al., 2022), possibly enabling them to navigate and satisfy stakeholder demands more effectively. The counterargument is that the international diversification of the MNE's network increases the complexity of stakeholder pressures to the point of managerial bottlenecks (Ahmadova et al., 2022; Aragòn-Correa et al., 2020), resulting in higher strategic uncertainty. We also reviewed the carbon strategy literature, which suggests that firms tend to retain strategic flexibility in the face of uncertainty, for example, by satisfying stakeholder demands without committing scarce resources.

We found two traditionally competing narratives concerning the role of cross-border differences in regulation. First, MNEs exploit poor regulatory stringency in host countries to shift emissions abroad (Ben-David et al., 2021), especially in high-pollution sectors (Cai et al., 2008; Eskeland & Harrison, 2002; Javorcik & Wei, 2003). Second, MNEs absorb the regulatory stringency from various locations and translate it into a single highest environmental standard that is applied throughout their value chains through self-regulation (Christmann, 2004; Christmann & Taylor, 2001; Di Ubaldo et al., 2022; Long et al., 2022). The reality of many MNEs is possibly more complex and dependent on strategic decisions (Rugman & Verbeke, 1998). For instance, Nippa et al. (2021) found that MNEs' carbon performance improvements as a response to regulation are highest in their home country, but foreign affiliates perform less well. Moreover, MNEs with higher international diversification were found to have higher environmental performance until overexposed to diverse institutional pressures (Ahmadova et al., 2022).

In summary, we suggest that government intervention remains the most critical stakeholder driver of carbon performance improvements. Specifically, government pressures were found to be more effective at increasing managers' strategic commitment compared to other stakeholders (Littlewood et al., 2018), which is possibly explained by its role in low-carbon market creation that generates a business case for carbon performance (Busch, Bassen, et al., 2022; Mazzucato, 2016). However, greater degrees of multinationality also increase MNEs' exposure to the "unforeseen complexities" (Rugman & Verbeke, 1998, p. 373) resulting from diverse regulations and dynamic policies on the national level.

THE TWO-DIMENSIONAL COMPLEXITY OF INSTITUTIONAL ENVIRONMENTS IN THE NET ZERO TRANSITION

While the continuing enactment of national climate regulation will allow MNEs to more reliably price in the costs from their environmental externalities, we argue that a focus on regulatory stringency fails to capture the full institutional complexity that results from national climate policy. The concept of institutional complexity was first described by Kostova and Zaheer (1999) based on organizational institutionalism and understood as the multiplicity and heterogeneity of institutional domains and task environments fragmented across the MNE network (Marano & Kostova, 2016). Here, we approach the concept of institutional complexity from the angle of the economic strand of institutional research in IB, focusing on complementing institutions' diversity (and interdependence) with policy-centered concepts that explain their dynamism.

The Diversity and Dynamism of Institutional Environments

In a recent paper, Casson and Li (2022) unpacked the concept of complexity into two dimensions: "Structure refers to the range and diversity of the elements in a system and the 'tapestry' of the connections between them. Process, on the other hand, refers to the behaviour of the system over time ..." (p. 2039). In the context of institutional change, Leymann and Lundan (2023) similarly distinguish structural from transition effects, whereby the former arise from the steady state of institutional structure, that is, "the tapestry," and the latter reflect how the process of institutional change, that is, its dynamism while changing from A to B, alters the behavior of MNEs. Hence, we propose that institutional complexity is reflected in the diversity and dynamism of institutional environments.

Fully capturing diversity and dynamism presupposes a fine-grained framework of an institutional environment that goes beyond acknowledging the stringency of individual institutional pressures (Aguilera & Grøgaard, 2019). It requires concepts reflecting the diversity of the structure of institutional systems that produce specific patterns of interaction, and it also requires concepts that are attuned to the procedural characteristics of how these systems change. There are two complementary developments in the IB literature that offer such a perspective, namely the CC view (Jackson & Deeg, 2008a) and the increasing interest in public policy as a separate but interconnected explanatory factor (Clegg, 2019; Lundan, 2018; Utesch-Xiong et al., 2022).

The CC approach was formulated to critique the IB discipline's narrow conception of institutions as indicators representing individual institutions (Jackson & Deeg, 2008a). This view was contrasted with an understanding of institutional environments as "complex bundles" (Aguilera & Grøgaard, 2019, p. 29) emerging from complementarities and contradictions between institutions that enable the strategic agency of economic actors. The CC view builds on Hall and Soskice's (2001) distinction between liberal market economies (LMEs) and coordinated market economies (CMEs). CMEs are characterized by higher degrees of coordination, referring to the extent to which interaction between societal actors, including firms, organized labor, and government, is institutionalized rather than organized through market mechanisms. Hence, there are differences in the degree of interdependence between agents, which results in varying levels of structural complexity.

For example, CMEs such as Germany and Austria are characterized by a consensus-based decision-making process (Hall & Soskice, 2001) and promote cooperation between economic actors and other stakeholders (Redding, 2005).

Germany's success in developing its renewable energy industry can be partly attributed to the coordinated efforts of government, industry, finance, science, and society (Ćetković & Buzogány, 2016). However, consensus-based policymaking can also hinder progress when political debates are dominated by stakeholders that are unwilling to compromise (Jackson & Deeg, 2008b). In Germany, MNEs actively engaged in this coordinated policy process were more committed to climate change mitigation (Eberlein & Matten, 2009), implying the importance of institutional embeddedness in CMEs.

The renewable energy policies in LMEs tend to be somewhat reactive, as in the UK, where industry development was driven mainly by external pressures such as European climate policy targets (Ćetković & Buzogány, 2016). Moreover, LMEs can prioritize cost-effectiveness over other goals, such as job creation and innovation, which was seen in the UK support scheme for renewable electricity generation ("Renewables Obligation") implemented in 2002. The scheme mandated energy suppliers to increase the share of renewable energy sources. Still, its initial market-based design offered little motivation to adopt less mature and more costly renewable technologies (Ćetković & Buzogány, 2016). This lack of central guidance by the government generates diverse non-market environments with many subnational policy initiatives and voluntary efforts implemented by corporations (Eberlein & Matten, 2009), prioritizing dynamic competition for best practice over long-term planning with stakeholders.

This also implies a variation in the dynamism of LMEs and CMEs. More coordinated economies tend to be characterized by long-term orientation (Redding, 2005), which could benefit the strategic commitment necessary in the LTE. This aligns with recent findings of a positive influence of national cultural values – including those that emphasize long-term focus and uncertainty avoidance – on firm-level carbon performance (Luo & Tang, 2022). Interestingly, some policies with such a long-term focus have been adopted by LMEs, too. For example, the German feed-in-tariff for renewable energies has stimulated the adoption of feed-in-tariffs in the USA, showing potential cross-border demonstration effects and convergence on best practices between countries (Meckling & Jenner, 2016). Not only do such demonstration effects lead to unpredictable policy dynamism, but they also highlight that the institutional diversity of the CC approach may not fully reflect the diversity and dynamism of the underlying policy environments.

Therefore, we argue that the CC approach in IB would benefit from incorporating concepts from the policy literature that explicitly acknowledge the dynamism of public policy within the broader structures of national institutional environments. This seems critical to understand better how MNEs strategically position themselves toward the green transition. As Clegg (2019) put it: "Institutions shape behavior, but it is policy that changes behavior" (p. 111). For example, the credibility and consistency of public policy were found to positively affect corporations' green innovation decisions (Greco et al., 2022; Rogge & Schleich, 2018), and this raises questions about how MNEs are impacted by the institutional complexity along their value chains, which possibly introduces new uncertainties as the transition progresses.

The policy literature has proposed concepts to capture how policy affects corporate decision-making. Rogge and Reichardt (2016) have classified policy environments into three core dimensions: policy process, policy instrument mix, and policy characteristics. The policy process dimension takes a procedural perspective on a policy environment's overall stringency and goals and its evolution over time. The policy instrument mix analyzes the evolution, configuration, and complementarity of the individual policy tools governments implement. Policy characteristics refer to the outcome conditions of policy change that might affect corporate responses, such as the dimensions of credibility and consistency.

In Fig. 13.1, we propose a conceptual framework that complements the structural diversity of institutional environments with a policy-centered perspective on institutional dynamism. We argue that the energy transition produces structural institutional pressures that drive the emergence of low-carbon markets but that the policy necessary to get to this new structure introduces procedural complexities. The MNE-specific exposure to this institutional complexity may depend on managerial awareness, the multinationality of the network, and available resources and capabilities (Ahmadova et al., 2022). Our framework offers granularity in terms of the constitutive elements of institutional environments relevant for MNEs' (carbon) strategies, and it provides a starting point to conceptualize subnational diversity in municipal or even city-level policy initiatives that do not always align with the national institutional environment (Beugelsdijk & Mudambi, 2013; Pisani et al., 2019).

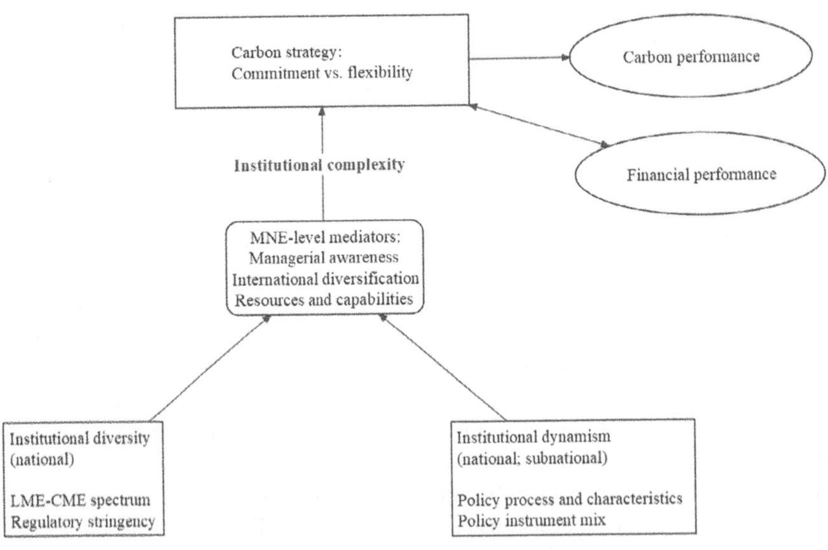

Fig. 13.1. Conceptual Framework.

Data on Climate Policy Diversity and Dynamism

Does the reality of international climate policy support our claim of increasing diversity and dynamism? To illustrate cross-country differences in the policy process, we draw from the OECD Environmental Policy Stringency Index (OECD EPS), which captures the evolution of market-, non-market- and technology-support-policy stringency for 40 countries over time (Kruse et al., 2022). The OECD EPS does not capture data on individual policies but instead evaluates the overall stringency of the policy environment.

Fig. 13.2 illustrates the average stringency of market- and non-market-policy plotted against the mean-corrected standard deviation from 1990 to 2019. Laggard countries seem more likely to be unstable, whereas leading countries have more stable trajectories, possibly because their transition efforts reached temporary saturation. Nevertheless, there are still differences in policy dynamism within higher-performing countries. For example, Austria is remarkably stable compared to other countries, and it is a typical case of a CME with higher stakeholder institutionalization and long-term focus. By contrast, the LME Australia has a higher variation in policy stringency over time, hypothetically reflecting more competition between economic actors resulting in policy dynamism.

While prior studies on pro-market reforms have found such dynamism can exhibit positive effects on firms' performance and internationalization activity, there is qualifying evidence that more general forms of institutional dynamism may lead to costly learning processes that cause a reduction of MNEs' long-term capital commitments (Leymann & Lundan, 2023; Putzhammer et al., 2020).

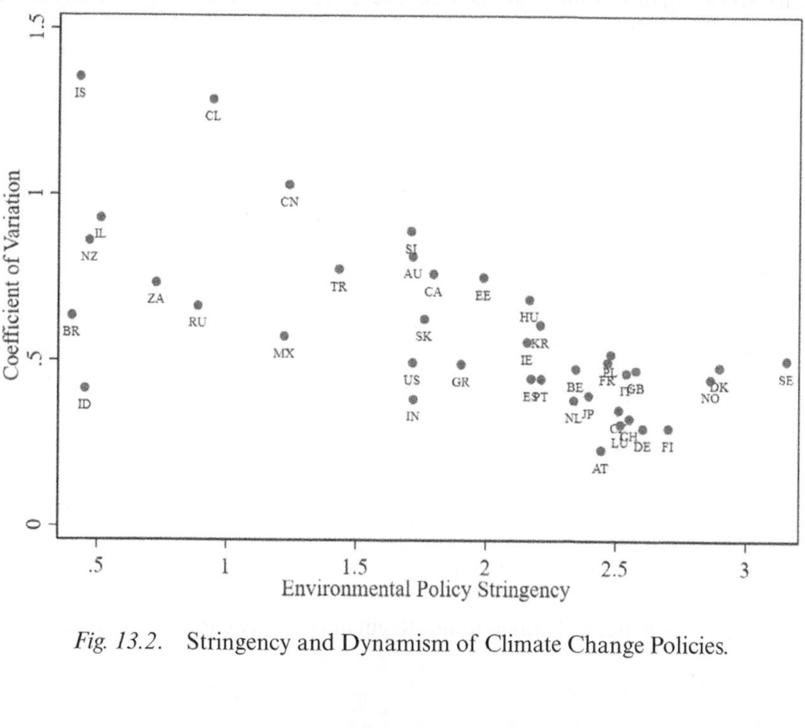

Fig. 13.2. Stringency and Dynamism of Climate Change Policies.

Especially when the policy goal is not market liberalization (external to the firm) but an increase in the firm's carbon performance (an internal change), we would think that high levels of variation and unpredictability of the policy process can negatively impact long-term investment decisions.

Rogge and Reichardt (2016) also highlighted that there can be national differences in the policy instrument mix. For the sake of brevity, we focus on some stylized facts that highlight the dynamism of economic, regulatory, and research and development (R&D)-focused climate policy instruments. The following data are based on the New Climate Institute's Climate Policy Database, which collects policy data for 198 countries (Nascimento et al., 2022). However, we constrain our analysis to the 40 countries represented in the OECD EPS and the economic, regulatory, and R&D dimensions of policy instruments.

Fig. 13.3 shows the number of additionally implemented economic and regulatory policy instruments in our sample of 40 countries from 1990 to 2019. We can see a steep increase until 2008 and then less of an upward trend with more significant fluctuations. The growth in economic instruments seems to have outpaced regulatory instruments in the 2010s. These policies allow firms to retain more flexibility than direct regulatory measures, which may support MNEs in maintaining strategic flexibility regarding their carbon strategy (Kolk & Pinkse, 2005). Since regulation might be the most effective way to bring about more strategic commitment, we may see another growth spurt in regulatory instruments.

Besides the significant number of new economic and regulatory policy instruments introduced each year, the data show that R&D-focused policy instruments are far less frequent (Fig. 13.4). In 2019, 90 new regulatory and economic instruments stood against four that focused exclusively on technology deployment,

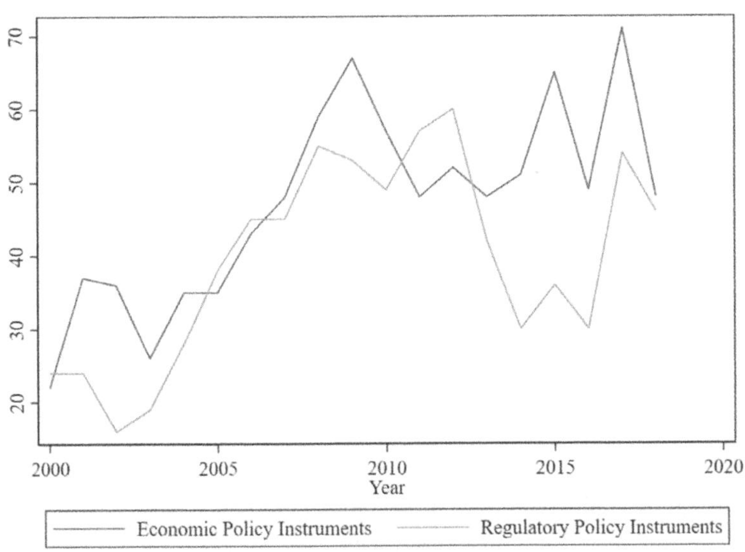

Fig. 13.3. Change in Economic and Regulatory Climate Policy Instruments.

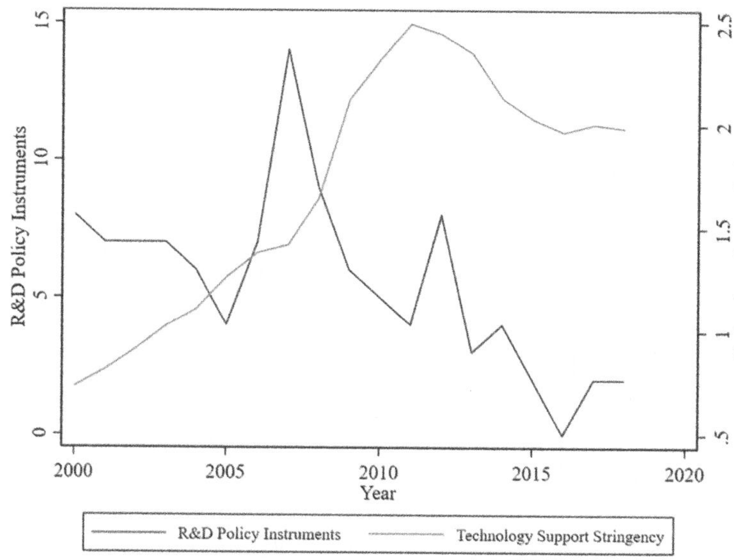

Fig. 13.4. Change and Stringency of R&D Climate Policy Instruments.

diffusion, or development. This is unsurprising as these possibly reflect the "mission-oriented"-type innovation policies in which the state takes an active entrepreneurial role through extensive resource commitments (Mazzucato, 2016). What is surprising, however, is that there is a downward trend in newly introduced R&D policy instruments and in the stringency of technology-support policies across the OECD EPS countries since 2011. This observation goes hand in hand with the fact that the volume of newly registered green patents also decreased over these years (Kruse et al., 2022).

Finally, MNEs are faced with dynamism in the quantities of new policy instruments across countries and varying instrument diversity between countries. Fig. 13.5 visualizes a policy diversity index, namely Shannon entropy. If a country has a high Shannon entropy value, it means that it has a larger number of different policy instruments that are present in more similar proportions, resulting in a more diverse policy system. Conversely, a low Shannon entropy value means that the institutional environment is dominated by few policy instruments and, thus, less diverse. Fig. 13.5 also visualizes the coordination index calculated by Witt and Jackson (2016) to represent the countries' tendencies toward the LME or CME variant. We find a weak but significant negative correlation (Pearson: −0.38; $P < 0.001$) between policy diversity and the degree of coordination, suggesting that high LMEs are slightly more associated with diverse policy instrument mixes. It is possible that institutional context matters here, since the greater diversity of elements in the policy mix might be counterbalanced in some cases by less coordination between relevant stakeholders. This raises the question of how the strategic commitment to carbon performance may differ in such environments.

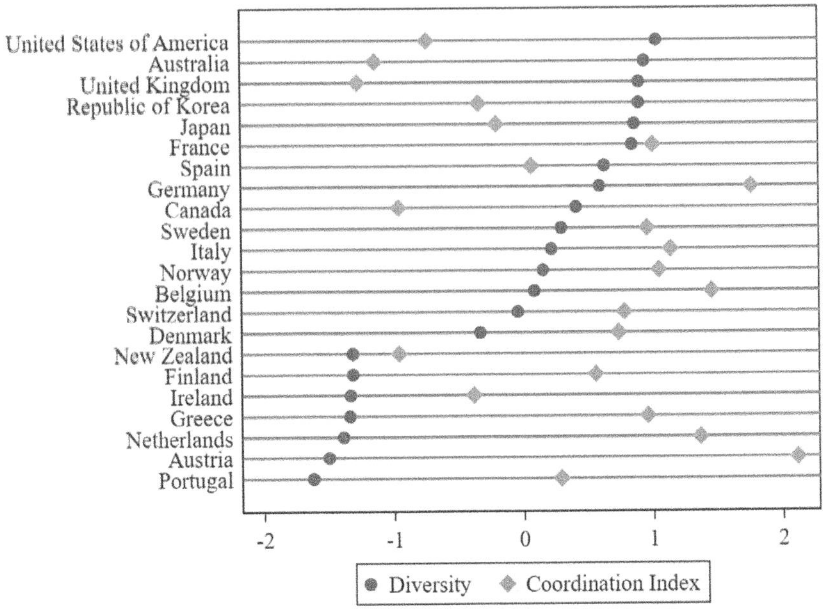

Fig. 13.5. Policy Instrument Diversity and Institutional Coordination Indices of Selected Countries.

In summary, we argue that the policy data corroborate our claim that MNEs face institutional complexity arising from the climate policy environment and that IB researchers have yet to explore how MNE networks react to variations in policy over time. Moreover, we find that the climate policy mix in OECD EPS countries concerning technology-support measures has become weaker in recent years, which seems an exciting research gap to explore how MNEs react to such policy failures, for example, by possibly optimizing their network exposure to locations with strong technology-support policies.

CONCLUSION

In this chapter, we have argued that the energy transition is associated with increased institutional complexity for MNEs and that institutional complexity might interfere with the overarching goal of improving carbon performance throughout the global economy. Based on a literature review of corporate environmental and carbon performance, we identified national governments as essential stakeholders. We developed the argument that the weak business case for carbon performance improvements creates significant uncertainty for firms and that government intervention effectively reduces such strategic uncertainty on a national level. From the perspective of MNEs, the impact is not as clear. Increasing government intervention may come with institutional complexity for MNEs that are exposed to multiple national institutional environments and policy trajectories.

In the second part of our chapter, we unpacked institutional complexity into two dimensions: structural diversity and dynamism. While extant research captures the structural complexity of multiple, possibly contradicting institutional pressures in the MNE network, we argued that the dynamism dimension, reflected mainly by public policy is another important facet of institutional complexity in the climate change context. We proposed a framework that can map the diversity and dynamism of institutional complexity to national and potentially subnational levels. Furthermore, we have presented and explored data sources that allow the development of institutional complexity measures and support the recent call for more large-sample quantitative studies on MNEs' reactions to policy (Nippa et al., 2021).

We identify two core implications for future IB research. First, the energy transition provides a unique context to explore the call for "thickness" in institutional IB research (Aguilera & Grøgaard, 2019; Jackson & Deeg, 2008a). Taking the perspective of the economic strand of institutionalism in IB, we argue that the role of institutions is no longer constrained to a "thin" background condition as resolving (or causing) frictions in market exchange. The "thick" conception of institutions views them as active shapers of how societies approach grand challenges such as climate change and if solutions to such challenges are rewarded in the marketplace (Mazzucato, 2018). The former implies that the underlying variations in national institutions and policy may strongly affect the effectiveness of the non-governmental stakeholder drivers of carbon performance. The latter means that regulatory, market, and innovation policy is increasingly at the heart of market co-creation in the energy transition (Mazzucato, 2016), raising questions about how MNEs react to the tighter involvement of public actors.

MNEs have strategic options both to avoid institutional complexities and to structure their network in a way to benefit from complementarities in policy environments, but both options are likely to involve the development of non-market capabilities. A recently published book on the aviation industry illustrates the complex interrelationship between managers, strategy, and government policy, highlighting the importance of industry-level studies in further uncovering the complexity of MNEs' non-market environment (Walls & Wittmer, 2022). Here, we focused on the institutional environment as an exogenous factor. Future research might take up the challenge of understanding how MNEs utilize corporate political activities to control the institutional complexity of their operations (Doh et al., 2021). Finally, we may see greater policy harmonization across countries (Meckling & Jenner, 2016), especially in difficult-to-abate sectors such as shipping and aviation. Some bilateral agreements have emerged in this area, known as green corridors that connect (air)ports through a route with more stringent environmental standards (Joerss et al., 2023). This type of policymaking might spur new competition between countries regarding who gets to set standards and decide upon technological paths, with significant effects on the competitiveness of individual MNEs. We hope that our chapter stimulates research on the relationship between institutional diversity, dynamism, and the carbon strategies of MNEs.

REFERENCES

Acemoglu, D. (2023, April 7). *In search of a new political economy*. Project Syndicate. https://www.project-syndicate.org/onpoint/shift-from-market-fundamentalism-to-economy-for-social-benefit-by-daron-acemoglu-2023-04

Aguilera, R. V., & Grøgaard, B. (2019). The dubious role of institutions in international business: A road forward. *Journal of International Business Studies*, *50*(1), 20–35. https://doi.org/10.1057/S41267-018-0201-5/TABLES/2

Ahmadova, G., Bueno García, M., Delgado-Márquez, B., & Pedauga, L. (2022). Firm- and country-specific advantages: Towards a better understanding of MNEs' environmental performance in the international arena. *Organization & Environment*, 36(3), 468–497. https://doi.org/10.1177/10860266221129699

Alsaifi, K., Elnahass, M., & Salama, A. (2020). Carbon disclosure and financial performance: UK environmental policy. *Business Strategy and the Environment*, *29*(2), 711–726. https://doi.org/10.1002/BSE.2426

Andrew, J., & Cortese, C. (2011). Accounting for climate change and the self-regulation of carbon disclosures. *Accounting Forum*, *35*(3), 130–138. https://doi.org/10.1016/j.accfor.2011.06.006

Aragón-Correa, J. A., Marcus, A., & Hurtado-Torres, N. (2016). The natural environmental strategies of international firms: Old controversies and new evidence on performance and disclosure. *Academy of Management Perspectives*, *30*(1), 24–39. https://doi.org/10.5465/amp.2014.0043

Aragòn-Correa, J. A., Marcus, A. A., & Vogel, D. (2020). The effects of mandatory and voluntary regulatory pressures on firms' environmental strategies: A review and recommendations for future research. *Academy of Management Annals*, *14*(1), 339–365. https://doi.org/10.5465/annals.2018.0014

Arvidsson, S., & Dumay, J. (2022). Corporate ESG reporting quantity, quality and performance: Where to now for environmental policy and practice? *Business Strategy and the Environment*, *31*(3), 1091–1110. https://doi.org/10.1002/BSE.2937

Backman, C. A., Verbeke, A., & Schulz, R. A. (2017). The drivers of corporate climate change strategies and public policy: A new resource-based view perspective. *Business & Society*, *56*(4), 545–575. https://doi.org/10.1177/0007650315578450

Bass, A. E., & Grøgaard, B. (2021). The long-term energy transition: Drivers, outcomes, and the role of the multinational enterprise. *Journal of International Business Studies*, *52*(5), 807–823. https://doi.org/10.1057/s41267-021-00432-3

Ben-David, I., Jang, Y., Kleimeier, S., & Viehs, M. (2021). Exporting pollution: Where do multinational firms emit CO_2? *Economic Policy*, *36*(107), 377–437. https://doi.org/10.1093/epolic/eiab009

Beugelsdijk, S., & Mudambi, R. (2013). MNEs as border-crossing multi-location enterprises: The role of discontinuities in geographic space. *Journal of International Business Studies*, *44*(5), 413–426. https://doi.org/10.1057/jibs.2013.23

Bortolotti, B., Loss, G., & van Zwieten, R. W. (2023). The times are they a-changin'? Tracking sovereign wealth funds' sustainable investing. *Journal of International Business Policy*, *6*, 276–305. http://hdl.handle.net/2318/1895226

Boukherroub, T., Bouchery, Y., Corbett, C. J., Fransoo, J. C., & Tan, T. (2017). Carbon footprinting in supply chains. In Y. Bouchery, C. J. Corbett, J. C. Fransoo, & T. Tan (Eds.), *Sustainable supply chains: A research-based textbook on operations and strategy* (pp. 43–64). Springer International Publishing. https://doi.org/10.1007/978-3-319-29791-0_3

Busch, T. (2010). Corporate carbon performance indicators revisited. *Journal of Industrial Ecology*, *14*(3), 374–377. https://doi.org/10.1111/j.1530-9290.2010.00239.x

Busch, T., Bassen, A., Lewandowski, S., & Sump, F. (2022). Corporate carbon and financial performance revisited. *Organization & Environment*, *35*(1), 154–171. https://doi.org/10.1177/1086026620935638

Busch, T., & Hoffmann, V. H. (2011). How hot is your bottom line? Linking carbon and financial performance. *Business & Society*, *50*(2), 233–265. https://doi.org/10.1177/0007650311398780

Busch, T., Johnson, M., & Pioch, T. (2022). Corporate carbon performance data: Quo vadis? *Journal of Industrial Ecology*, *26*(1), 350–363. https://doi.org/10.1111/jiec.13008

Busch, T., & Lewandowski, S. (2018). Corporate carbon and financial performance: A meta-analysis. *Journal of Industrial Ecology*, *22*(4), 745–759. https://doi.org/10.1111/jiec.12591

Cadez, S., & Czerny, A. (2016). Climate change mitigation strategies in carbon-intensive firms. *Journal of Cleaner Production*, *112*, 4132–4143. https://doi.org/10.1016/j.jclepro.2015.07.099

Cai, S., Souza, R. D., Goh, M., Li, W., Lu, Q., & Sundarakani, B. (2008). The adoption of green supply chain strategy: An institutional perspective. In *Proceedings of the 4th IEEE international conference on management of innovation and technology, ICMIT* (pp. 1044–1049). IEEE. https://doi.org/10.1109/ICMIT.2008.4654512

Callery, P. J., & Perkins, J. (2021). Detecting false accounts in intermediated voluntary disclosure. *Academy of Management Discoveries*, *7*(1), 40–56. https://doi.org/10.5465/amd.2018.0229

Casson, M., & Li, Y. (2022). Complexity in international business: The implications for theory. *Journal of International Business Studies*, *53*(9), 2037–2049. https://doi.org/10.1057/s41267-021-00495-2]

Ćetković, S., & Buzogány, A. (2016). Varieties of capitalism and clean energy transitions in the European Union: When renewable energy hits different economic logics. *Climate Policy*, *16*(5), 642–657. https://doi.org/10.1080/14693062.2015.1135778

Chen, L. H., & Silva Gao, L. (2012). The pricing of climate risk. *Journal of Financial and Economic Practice*, *12*(2), 115–131. https://doi.org/10.2139/ssrn.1940727

Christmann, P. (2004). Multinational companies and the natural environment: Determinants of global environmental policy. *Academy of Management Journal*, *47*(5), 747–760. https://doi.org/10.5465/20159616

Christmann, P., & Taylor, G. (2001). Globalization and the environment: Determinants of firm self-regulation in China. *Journal of International Business Studies*, *32*(3), 439–458. https://doi.org/10.1057/PALGRAVE.JIBS.8490976

Clark, C. E., & Crawford, E. P. (2012). Influencing climate change policy: The effect of shareholder pressure and firm environmental performance. *Business & Society*, *51*(1), 148–175. https://doi.org/10.1177/0007650311427594

Clegg, J. (2019). From the editor: International business policy: What it is, and what it is not. *Journal of International Business Policy*, *2*(2), 111–118. https://doi.org/10.1057/s42214-019-00025-w

Clò, S., Ferraris, M., & Florio, M. (2017). Ownership and environmental regulation: Evidence from the European electricity industry. *Energy Economics*, *61*, 298–312. https://doi.org/10.1016/j.eneco.2016.12.001

Cole, M. A., Elliott, R. J. R., Okubo, T., & Zhou, Y. (2013). The carbon dioxide emissions of firms: A spatial analysis. *Journal of Environmental Economics and Management*, *65*(2), 290–309. https://doi.org/10.1016/j.jeem.2012.07.002

Dahlmann, F., Branicki, L., & Brammer, S. (2019). Managing carbon aspirations: The influence of corporate climate change targets on environmental performance. *Journal of Business Ethics*, *158*(1), 1–24. https://doi.org/10.1007/s10551-017-3731-z

Damert, M., & Baumgartner, R. J. (2018). Intra-sectoral differences in climate change strategies: Evidence from the global automotive industry. *Business Strategy and the Environment*, *27*(3), 265–281. https://doi.org/10.1002/bse.1968

Damert, M., Paul, A., & Baumgartner, R. J. (2017). Exploring the determinants and long-term performance outcomes of corporate carbon strategies. *Journal of Cleaner Production*, *160*, 123–138. https://doi.org/10.1016/j.jclepro.2017.03.206

Delmas, M. A., Nairn-Birch, N., & Lim, J. (2015). Dynamics of environmental and financial performance: The case of greenhouse gas emissions. *Organization & Environment*, *28*(4), 374–393. https://doi.org/10.1177/1086026615620238

Di Ubaldo, M., McGuire, S., & Shirodkar, V. (2022). Voluntary programs and emissions revisited: What is the effect of EU trade agreements with environmental provisions? *Journal of International Business Policy*, *5*(3), 1–23. https://doi.org/10.1057/s42214-022-00133-0

Dietz, S., Bienkowska, B., Hastreiter, N., Komar, V., Scheer, A., & Sullivan, R. (2021). *TPI's methodology report: Management quality and carbon performance*. Technical Report, Transition Pathway Initiative. https://www.transitionpathwayinitiative.org/publications/90.pdf

Doda, B., Gennaioli, C., Gouldson, A., Grover, D., & Sullivan, R. (2016). Are corporate carbon management practices reducing corporate carbon emissions? *Corporate Social Responsibility and Environmental Management*, *23*(5), 257–270. https://doi.org/10.1002/csr.1369

Doh, J., Budhwar, P., & Wood, G. (2021). Long-term energy transitions and international business: Concepts, theory, methods, and a research agenda. *Journal of International Business Studies*, *52*(5), 951–970. https://doi.org/10.1057/S41267-021-00405-6/TABLES/2

Duan, Y., & Jiang, X. (2021). Pollution haven or pollution halo? A re-evaluation on the role of multinational enterprises in global CO_2 emissions. *Energy Economics*, *97*, 105181. https://doi.org/10.1016/j.eneco.2021.105181

Eberlein, B., & Matten, D. (2009). Business responses to climate change regulation in Canada and Germany: Lessons for MNCs from emerging economies. *Journal of Business Ethics*, *86*(2), 241–255. https://doi.org/10.1007/s10551-009-0194-x

Eskeland, G. S., & Harrison, A. E. (2002). *Moving to greener pastures? Multinationals and the pollution haven hypothesis* [Working Paper 8888]. NBER Working Paper Series. http://www.nber.org/papers/w8888

Flammer, C., Toffel, M. W., & Viswanathan, K. (2021). Shareholder activism and firms' voluntary disclosure of climate change risks. *Strategic Management Journal*, *42*(10), 1850–1879. https://doi.org/10.1002/smj.3313

Fujii, H., Iwata, K., Kaneko, S., & Managi, S. (2013). Corporate environmental and economic performance of Japanese manufacturing firms: Empirical study for sustainable development. *Business Strategy and the Environment*, *22*(3), 187–201. https://doi.org/10.1002/bse.1747

González-Benito, J., & González-Benito, Ó. (2010). A study of determinant factors of stakeholder environmental pressure perceived by industrial companies. *Business Strategy and the Environment*, *19*(3), 164–181. https://doi.org/10.1002/bse.631

Greco, M., Germani, F., Grimaldi, M., & Radicic, D. (2022). Policy mix or policy mess? Effects of cross-instrumental policy mix on eco-innovation in German firms. *Technovation*, *117*, 102194. https://doi.org/10.1016/j.technovation.2020.102194

Guenther, E., Guenther, T., Schiemann, F., & Weber, G. (2016). Stakeholder relevance for reporting. *Business & Society*, *55*(3), 361–397. https://doi.org/10.1177/0007650315575119

Hall, P. A., & Soskice, D. (2001). An introduction to varieties of capitalism. In P. A. Hall & D. Soskice (Eds.), *Varieties of capitalism: The institutional foundations of comparative advantage* (pp. 1–68). Oxford University Press.

Haque, F. (2017). The effects of board characteristics and sustainable compensation policy on carbon performance of UK firms. *The British Accounting Review*, *49*(3), 347–364. https://doi.org/10.1016/j.bar.2017.01.001

Haque, F., & Ntim, C. G. (2018). Environmental policy, sustainable development, governance mechanisms and environmental performance. *Business Strategy and the Environment*, *27*(3), 415–435. https://doi.org/10.1002/BSE.2007

Henriques, I., & Sadorsky, P. (1996). The determinants of an environmentally responsive firm: An empirical approach. *Journal of Environmental Economics and Management*, *30*(3), 381–395. https://doi.org/10.1006/JEEM.1996.0026

Hoffmann, V. H., & Busch, T. (2008). Corporate carbon performance indicators. *Journal of Industrial Ecology*, *12*(4), 505–520. https://doi.org/10.1111/j.1530-9290.2008.00066.x

Iwata, H., & Okada, K. (2011). How does environmental performance affect financial performance? Evidence from Japanese manufacturing firms. *Ecological Economics*, *70*(9), 1691–1700. https://doi.org/10.1016/j.ecolecon.2011.05.010

Jackson, G., & Deeg, R. (2008a). Comparing capitalisms: Understanding institutional diversity and its implications for international business. *Journal of International Business Studies*, *39*(4), 540–561. https://doi.org/10.1057/palgrave.jibs.8400375

Jackson, G., & Deeg, R. (2008b). From comparing capitalisms to the politics of institutional change. *Review of International Political Economy*, *15*(4), 680–709. https://doi.org/10.1080/09692290802260704

Javorcik, B. S., & Wei, S.-J. (2003). Pollution havens and foreign direct investment: Dirty secret or popular myth? *Contributions in Economic Analysis & Policy*, *3*(2), 1–32. https://doi.org/10.2202/1538-0645.1244

Jiao, J., Liu, C. G., & Xu, Y. (2020). Effects of stakeholder pressure, managerial perceptions, and resource availability on sustainable operations adoption. *Business Strategy and the Environment*, *29*(8), 3246–3260. https://doi.org/10.1002/BSE.2569

Joerss, M., Kersing, A., Kramer, A., Stone, M., & Mohr, D. (2023). *Green corridors: A lane for zero-carbon shipping*. McKinsey Sustainability. https://www.mckinsey.com/capabilities/sustainability/our-insights/green-corridors-a-lane-for-zero-carbon-shipping

Kolk, A., & Pinkse, J. (2005). Business responses to climate change: Identifying emergent strategies. *California Management Review*, *47*(3), 6–20. https://doi.org/10.2307/41166304

Kostova, T., & Zaheer, S. (1999). Organizational legitimacy under conditions of complexity: The Case of the multinational enterprise. *The Academy of Management Review*, *24*(1), 64–81. https://doi.org/10.2307/259037

Kruse, T., Dechezleprêtre, A., Saffar, R., & Robert, L. (2022). *Measuring environmental policy stringency in OECD countries: An update of the OECD composite EPS indicator*. OECD. https://doi.org/10.1787/90ab82e8-en

Lee, M. D. P., & Lounsbury, M. (2011). Domesticating radical rant and rage: An exploration of the consequences of environmental shareholder resolutions on corporate environmental performance. *Business and Society*, *50*(1), 155–188. https://doi.org/10.1177/0007650310394640

Lee, S.-Y. (2012). Corporate carbon strategies in responding to climate change. *Business Strategy and the Environment*, *21*(1), 33–48. https://doi.org/10.1002/bse.711

Lee, S.-Y., & Klassen, R. D. (2016). Firms' response to climate change: The interplay of business uncertainty and organizational capabilities. *Business Strategy and the Environment*, *25*(8), 577–592. https://doi.org/10.1002/bse.1890

Leymann, G., & Lundan, S. (2023). From structural to transition effects: Institutional dynamism as a deterrent to long-term investments by MNEs. *International Business Review*, *32*(3), 102070. https://doi.org/10.1016/j.ibusrev.2022.102070

Linnenluecke, M. K., Griffiths, A., & Winn, M. I. (2013). Firm and industry adaptation to climate change: A review of climate adaptation studies in the business and management field. *WIREs Climate Change*, *4*(5), 397–416. https://doi.org/10.1002/wcc.214

Littlewood, D., Decelis, R., Hillenbrand, C., & Holt, D. (2018). Examining the drivers and outcomes of corporate commitment to climate change action in European high emitting industry. *Business Strategy and the Environment*, *27*(8), 1437–1449. https://doi.org/10.1002/bse.2194

Long, W., Luo, L., Sun, H., & Zhong, Q. (2022). Does going abroad lead to going green? Firm outward foreign direct investment and domestic environmental performance. *Business Strategy and the Environment*, *32*(1), 484–498. https://doi.org/10.1002/bse.3156

López, L.-A., Cadarso, M.-Á., Zafrilla, J., & Arce, G. (2019). The carbon footprint of the U.S. multinationals' foreign affiliates. *Nature Communications*, *10*(1), Article 1. https://doi.org/10.1038/s41467-019-09473-7

Lundan, S. M. (2018). From the editor: Engaging international business scholars with public policy issues. *Journal of International Business Policy*, *1*(1), 1–11. https://doi.org/10.1057/s42214-018-0007-y

Luo, L., & Tang, Q. (2022). National culture and corporate carbon performance. *Australian Journal of Management*, *47*(3), 503–538. https://doi.org/10.1177/03128962211038664

Maksimov, V., Wang, S. L., & Yan, S. (2022). Global connectedness and dynamic green capabilities in MNEs. *Journal of International Business Studies*, *53*(4), 723–740. https://doi.org/10.1057/s41267-019-00275-z

Marano, V., & Kostova, T. (2016). Unpacking the institutional complexity in adoption of CSR practices in multinational enterprises. *Journal of Management Studies*, *53*(1), 28–54. https://doi.org/10.1111/joms.12124

Mazzucato, M. (2016). From market fixing to market-creating: A new framework for innovation policy. *Industry and Innovation*, *23*(2), 140–156. https://doi.org/10.1080/13662716.2016.1146124

Mazzucato, M. (2018). *The value of everything: Making and taking in the global economy*. Allen Lane, an imprint of Penguin Books.

McKinsey. (2023). *The global energy landscape to 2050*. https://www.mckinsey.com/industries/oil-and-gas/our-insights/charting-the-global-energy-landscape-to-2050-emissions

Meckling, J., & Jenner, S. (2016). Varieties of market-based policy: Instrument choice in climate policy. *Environmental Politics*, *25*(5), 853–874. https://doi.org/10.1080/09644016.2016.1168062

Megginson, W. L., Malik, A. I., & Zhou, X. Y. (2023). Sovereign wealth funds in the post-pandemic era. *Journal of International Business Policy*, *6*(1), 1–23. https://doi.org/10.1057/S42214-023-00155-2

Moussa, T., Allam, A., Elbanna, S., & Bani-Mustafa, A. (2020). Can board environmental orientation improve U.S. firms' carbon performance? The mediating role of carbon strategy. *Business Strategy and the Environment*, *29*(1), 72–86. https://doi.org/10.1002/bse.2351

Nascimento, L., Kuramochi, T., Iacobuta, G., den Elzen, M., Fekete, H., Weishaupt, M., van Soest, H. L., Roelfsema, M., Vivero-Serrano, G. D., Lui, S., Hans, F., Jose de Villafranca Casas, M., & Höhne, N. (2022). Twenty years of climate policy: G20 coverage and gaps. *Climate Policy*, *22*(2), 158–174. https://doi.org/10.1080/14693062.2021.1993776

Nippa, M., Patnaik, S., & Taussig, M. (2021). MNE responses to carbon pricing regulations: Theory and evidence. *Journal of International Business Studies*, *52*(5), 904–929. https://doi.org/10.1057/s41267-021-00403-8

Perotto, E., Canziani, R., Marchesi, R., & Butelli, P. (2008). Environmental performance, indicators and measurement uncertainty in EMS context: A case study. *Journal of Cleaner Production*, *16*(4), 517–530. https://doi.org/10.1016/j.jclepro.2007.01.004

Pisani, N., Kolk, A., Ocelík, V., & Wu, G. (2019). Does it pay for cities to be green? An investigation of FDI inflows and environmental sustainability. *Journal of International Business Policy*, *2*(1), 62–85. https://doi.org/10.1057/s42214-018-00017-2

Post, C., Rahman, N., & Rubow, E. (2011). Green governance: Boards of directors' composition and environmental corporate social responsibility. *Business & Society*, *50*(1), 189–223. https://doi.org/10.1177/0007650310394642

Putzhammer, M., Slangen, A., Puck, J., & Lindner, T. (2020). Multinational firms' pace of expansion within host countries: How high rates of pro-market reform hamper the local exploitation of foreign expansion knowledge. *Journal of International Management*, *26*(1). https://ideas.repec.org//a/eee/intman/v26y2020i1s1075425318303156.html

Qian, W., & Schaltegger, S. (2017). Revisiting carbon disclosure and performance: Legitimacy and management views. *The British Accounting Review*, *49*(4), 365–379. https://doi.org/10.1016/j.bar.2017.05.005

Ramanathan, R., Poomkaew, B., & Nath, P. (2014). The impact of organizational pressures on environmental performance of firms. *Business Ethics: A European Review*, *23*(2), 169–182. https://doi.org/10.1111/BEER.12042

Redding, G. (2005). The thick description and comparison of societal systems of capitalism. *Journal of International Business Studies*, *36*(2), 123–155. https://doi.org/10.1057/palgrave.jibs.8400129

Reid, E. M., & Toffel, M. W. (2009). Responding to public and private politics: Corporate disclosure of climate change strategies. *Strategic Management Journal*, *30*(11), 1157–1178. https://doi.org/10.1002/SMJ.796

Rockström, J., Steffen, W., Noone, K., Persson, Å., Chapin, F. S., Lambin, E., Lenton, T. M., Scheffer, M., Folke, C., Schellnhuber, H. J., Nykvist, B., de Wit, C. A., Hughes, T., van der Leeuw, S., Rodhe, H., Sörlin, S., Snyder, P. K., Costanza, R., Svedin, U., … Foley, J. (2009). Planetary boundaries: Exploring the safe operating space for humanity. *Ecology and Society*, *14*(2). https://www.jstor.org/stable/26268316

Rogge, K. S., & Reichardt, K. (2016). Policy mixes for sustainability transitions: An extended concept and framework for analysis. *Research Policy*, *45*(8), 1620–1635. https://doi.org/10.1016/J.RESPOL.2016.04.004

Rogge, K. S., & Schleich, J. (2018). Do policy mix characteristics matter for low-carbon innovation? A survey-based exploration of renewable power generation technologies in Germany. *Research Policy*, *47*(9), 1639–1654. https://doi.org/10.1016/j.respol.2018.05.011

Rugman, A. M., & Verbeke, A. (1998). Corporate strategies and environmental regulations: An organizing framework. *Strategic Management Journal*, *19*(4), 363–375. https://doi.org/10.1002/(SICI)1097-0266(199804)19:4<363::AID-SMJ974>3.0.CO;2-H

Sprengel, D. C., & Busch, T. (2011). Stakeholder engagement and environmental strategy – The case of climate change. *Business Strategy and the Environment*, *20*(6), 351–364. https://doi.org/10.1002/bse.684

Standard Chartered. (2021). *Carbon dated*. Standard Chartered. https://www.sc.com/en/insights/carbon-dated/

Trigeorgis, L., & Reuer, J. J. (2017). Real options theory in strategic management. *Strategic Management Journal*, *38*(1), 42–63. https://doi.org/10.1002/smj.2593

Trumpp, C., Endrikat, J., Zopf, C., & Guenther, E. (2015). Definition, conceptualization, and measurement of corporate environmental performance: A critical examination of a multidimensional construct. *Journal of Business Ethics*, *126*(2), 185–204. https://doi.org/10.1007/s10551-013-1931-8

Utesch-Xiong, F., Leymann, G., & Lundan, S. M. (2022). More policy is not always effective policy: How policy coherence affects firm internationalization. *International Business Review*, 102082. https://doi.org/10.1016/j.ibusrev.2022.102082

Velte, P. (2021). Environmental performance, carbon performance and earnings management: Empirical evidence for the European capital market. *Corporate Social Responsibility and Environmental Management*, *28*(1), 42–53. https://doi.org/10.1002/csr.2030

Villiers, C. de, Naiker, V., & van Staden, C. J. (2011). The effect of board characteristics on firm environmental performance. *Journal of Management*, *37*(6), 1636–1663. https://doi.org/10.1177/0149206311411506

Walls, J. L., Berrone, P., & Phan, P. H. (2012). Corporate governance and environmental performance: Is there really a link? *Strategic Management Journal*, *33*(8), 885–913. https://doi.org/10.1002/SMJ.1952

Walls, J. L., & Wittmer, A. (Eds.). (2022). *Sustainable aviation: A management perspective*. Springer International Publishing. https://doi.org/10.1007/978-3-030-90895-9

Weinhofer, G., & Hoffmann, V. H. (2010). Mitigating climate change – How do corporate strategies differ? *Business Strategy and the Environment*, *19*(2), 77–89. https://doi.org/10.1002/bse.618

Witt, M. A., & Jackson, G. (2016). Varieties of capitalism and institutional comparative advantage: A test and reinterpretation. *Journal of International Business Studies*, *47*(7), 778–806. https://doi.org/10.1057/s41267-016-0001-8

Wright, C., & Nyberg, D. (2017). An inconvenient truth: How organizations translate climate change into business as usual. *Academy of Management Journal*, *60*(5), 1633–1661. https://doi.org/10.5465/amj.2015.0718

Wu, J., Yang, J., & Zhou, Z. (2020). How does environmental regulation affect environmental performance? A case study of China's regional energy efficiency. *Expert Systems*, *37*(3), e12326. https://doi.org/10.1111/EXSY.12326

Yalabik, B., & Fairchild, R. J. (2011). Customer, regulatory, and competitive pressure as drivers of environmental innovation. *International Journal of Production Economics*, *131*(2), 519–527. https://doi.org/10.1016/J.IJPE.2011.01.020

Zameer, H., Wang, Y., & Saeed, M. R. (2021). Net-zero emission targets and the role of managerial environmental awareness, customer pressure, and regulatory control toward environmental performance. *Business Strategy and the Environment*, *30*(8), 4223–4236. https://doi.org/10.1002/BSE.2866

Zhang, Z., Guan, D., Wang, R., Meng, J., Zheng, H., Zhu, K., & Du, H. (2020). Embodied carbon emissions in the supply chains of multinational enterprises. *Nature Climate Change*, *10*(12), 1096–1101. https://doi.org/10.1038/s41558-020-0895-9

Zhou, Y., Luo, L., & Shen, H. (2022). Community pressure, regulatory pressure and corporate environmental performance. *Australian Journal of Management*, *47*(2), 368–392. https://doi.org/10.1177/03128962211017172

PART IV

WALKING THE TALK IN GLOBAL VALUE CHAINS

CHAPTER 14

HOW INSTITUTIONS INFLUENCE FIRMS' CLIMATE CHANGE STRATEGIES: EXTENDING THE PERSPECTIVES OF INTERNATIONAL BUSINESS AND GLOBAL VALUE CHAINS WITH BUSINESS SYSTEMS

Mohammad B. Rana[a] and Matthew M. C. Allen[b]

[a]*Aalborg University, Denmark*
[b]*Manchester Metropolitan University, UK*

ABSTRACT

The changing roles of the United Nations (UN) and national institutions have made addressing climate change a critical concern for many multinational enterprises' (MNEs) survival and growth. This chapter discusses how such institutions, which vary in their nature and characteristics, shape firm strategies for climate change adaptation. Exploring different versions of institutional theory, the chapter demonstrates how and why institutional characteristics affect typical patterns of firm ownership, governance, and capabilities. These, in turn, influence companies' internationalisation and climate-change strategies. Climate change poses challenges to how we understand firms' strategic decisions from both an international business (IB) (HQ–subsidiary relations) and global value chains

Walking the Talk? MNEs Transitioning Towards a Sustainable World
Progress in International Business Research, Volume 18, 265–294

ISSN: 1745-8862/doi:10.1108/S1745-886220240000018018

(GVC) (buyer–supplier relations) perspective. However, climate change also provides opportunities for companies to gain competitive advantages – if firms can reconfigure and adapt faster than their competitors. Existing IB and GVC research tends to downplay the importance of climate change strategies and the ways in which coherent or dysfunctional institutions affect firms' reconfiguration and adaptation strategies in a globally dispersed network of value creation. This chapter presents a perspective on the institutional conditions that affect firms' climate change strategies regarding ownership, location, and internalisation (OLI), and GVCs, with 'investment' and 'emerging standards' playing a significant role. The authors illustrate the discussion using several examples from the Global South (i.e. Bangladesh) and the Global North (i.e. Denmark, Sweden, and Germany) with a special emphasis on the garment industry. The aim is to encourage future research to examine how a 'business systems', or varieties of capitalism, institutional perspective can complement the analysis of sustainability and climate change strategies in IB and GVC studies.

Keywords: Climate change; strategy; institutions; international business; global value chain

INTRODUCTION

Issues concerning sustainability, particularly those around man-made changes in average land and sea temperatures as well as the earth's atmosphere, commonly referred to as climate change, have been a critical concern for firms' survival and growth in the IB and GVCs literatures. Studies have explored how climate change issues affect MNEs' firm-specific advantages (FSAs), especially the knowledge and competencies that underpin their product differentiation and process reconfigurational capabilities to reduce environmental pollution (Kolk & Pinkse, 2008). However, significant variation exists in the climate change risk profiles of different regions (Romilly, 2007); these variations can include institutional voids, affecting the ability of MNEs to focus on the challenges of climate change (Pinkse & Kolk, 2012). Previous studies have tended to view climate change strategy through a unitary lens of economic imperatives, highlighting, for instance, how MNEs' climate change mitigation strategies influence their sales, profits, market share, or, indeed, survival (Chakrabarty & Wang, 2013; Kolk & Pinkse, 2008).

The IB and GVC literatures have both discussed the dynamics of, and influences on, MNEs' strategies across globally dispersed value chains, yet neither adopts a comprehensive perspective on institutions to examine MNEs' climate change strategies. In particular, both tend to downplay how climate change strategies and the ways in which coherent or dysfunctional institutions affect firms' reconfiguration and adaptation strategies in globally dispersed networks of value creation. By addressing these issues, we draw on the 'business systems' perspective, and seek to explain how institutional conditions affect firms' climate-change strategies regarding OLI, and GVCs, with 'investment' and 'emerging standards' playing a significant role.

While institutions present a dual perspective, that is, constraints and opportunities, affecting failure and success for MNEs (Rana & Sørensen, 2021), previous studies have demonstrated a partial view of institutions to explain climate change issues that firms address. Thus, there is a need for a theoretical discussion of institutions in relation to climate-change issues as to how and why variations in institutional systems shape firm capabilities, and how complementarities (or a lack of complementarities) and path dependency in national institutional systems not only constrain firms' efforts to mitigate climate change, but also enable them to harness opportunities from policies and discourses on climate change.

This perspective chapter, in line with the Sustainable Development Goal (SDG) on Climate Action, therefore, seeks to explain how institutions shape firms' strategies for climate change adaptation. It discusses the key questions of how institutions and firms, as powerful actors, interact and evolve socio-politically manifested climate change agendas and priorities, and how this process shapes firms' capabilities, structures, and strategies in a globally interconnected business network.

Climate change issues, such as CO_2 emissions, water depletion, sustainable energy, resource conservation, green transportation, and green manufacturing, raise key strategic choices for firms and national and international organisations (Boje & Rana, 2021; Kolk, 2016; Lund-Thomsen & Lindgreen, 2014; Ripple et al., 2024). To be ecologically and humanely (i.e. ethically) sustainable, a society must adopt 'a new order of things' (Dovers, 2001, p. 3). Climate change and sustainability represent profound challenges for both businesses and people (Ripple et al., 2024); tackling them involves a re-evaluation of the basic assumptions and structures of societal institutions.

To de-carbonise the economy by 2050, the European Union (EU) has introduced the carbon border adjustment mechanism to reduce the risk of carbon leakage and level the 'playing field' for European industries towards the de-carbonisation of production processes. In a similar vein, carbon taxes and emissions trading systems have been introduced worldwide, in particular for high-emitting industries, such as steel, cement, aluminium, fertiliser/chemical, and energy production (European Commission, 2021). Not all countries in the world have or implement similar policies to reduce their carbon footprints; therefore, the EU's new rules and policies may create arbitrage opportunities in the production and sale of some goods from certain countries. A higher carbon price (via a carbon tax, tradable certificates, or simply stricter regulation) will make goods relatively more expensive for countries implementing this policy, pushing carbon-intensive production into countries that have weaker institutional systems characterised by less stringent regulations or lower carbon prices, and increasing the emissions from these latter countries.

The new policies and rules on climate change will not affect all industries directly (Ripple et al., 2024). The apparel industry, for example, will not be subjected to as much direct regulation even though it is a heavily polluting industry. Nonetheless, this industry will experience an indirect impact as it uses aluminium, chemicals, and energy across its activities, such as textile production, dyeing, fabrication of metal accessories, garment manufacturing, washing, packaging, and transportation. All of this activity takes place within globally dispersed value chains. The European Environment Agency has estimated that the amount of clothing bought per person in the EU increased by 40 per cent between 1996 and 2012, and that over half of all clothing is not recycled, resulting in significant water consumption, pollution, greenhouse gas (GHG) emissions, and landfill use. As a result, in 2021, the European Parliament adopted a resolution on the new Circular Economy Action Plan, initiating additional measures to achieve a carbon-neutral, environmentally sustainable, toxic-free, and fully circular economy by 2050. To achieve this ambition, the EU has implemented tighter recycling rules and compulsory targets for materials use and consumption by 2030 (News European Parliament, 2021).

The new rules will not only affect apparel brands from the Global North (i.e. advanced economies), but also suppliers from emerging economies in the Global South with weak institutional settings (Rana & Allen, 2021b). Institutional contexts and their variations between countries, therefore, are critical factors for firms when determining their strategies to address climate change issues (Rana et al., 2021).

Critical press reports and media stories have highlighted the overproduction of, and waste associated with, clothes and textiles, particularly in the Global North, where the fast fashion trend has resulted in short-term usage to such an extent that some consumers may use clothing items only once or twice before throwing them away. As a result, the global textile industry has been under scrutiny due to its negative impact on the climate, particularly through CO_2 emissions related to various aspects of the industry's GVC. This is relevant in the context of the so-called science-based targets, which help companies by showing how much and how quickly they can reduce their GHG emissions and prevent the worst effects of climate change. Meanwhile, the industry's impacts on global CO_2 emissions are also distributed highly unequally, as are the risks that producer and importer countries face due to climate change. For instance, Bangladesh, the second-largest apparel exporter in the world, contributes very little to global climate change, but is likely to be among the countries worst affected by it, with 17 per cent of the country's land projected to be underwater by 2100 (Paul & Das, 2022).

In the context of this globalising industry and its widespread effect on climate change, a central question arises as to how producer and importer countries can adapt and mitigate their environmental impacts. In this chapter, we use an institutional theory lens – the business systems approach – to address these questions and focus on how typical IB and GVC studies can draw on the business systems framework to analyse and explain better firms' climate change strategies. We seek, then, to show how institutional theory – in the form of the business systems approach – can help to conceptualise how transnational and national institutions interact with transnational industries and firms in the face of the global climate crisis (Agrawal et al., 2008).

How national institutions affect climate change mitigation and adaptation agendas and how firms respond to climate-change imperatives are subjects of contention that have not yet been comprehensively addressed in typical IB or GVC studies. These issues have been widespread in recent years, and each has been approached differently by academic fields investigating globalising business and sustainability issues. Hence, both literatures have relevance for the analysis of climate change mitigation and adaptation for MNEs.

The Swiss Re firm reports that climate change could reduce the global economy by up to 18 per cent of gross domestic product (GDP) by 2050 if global temperatures rise by 3.2°C. The impact of climate change is forecast to be the hardest for Asian economies, with a 5.5 per cent hit to GDP in the best-case scenario and a 26.5 per cent hit in the worst case. Meanwhile, the World Economic Forum's Global Risks Report 2021 identifies the failure to take climate action as the most impactful and second-most likely long-term risk the world is facing at a time when populations are struggling to mitigate the consequences of the COVID-19 pandemic (Marchant, 2021).

The remainder of this chapter is organised as follows. First, we briefly review different and important institutional perspectives, and how they can be applied in relation to climate change strategies. Second, we discuss the key dimensions of the IB and GVC frameworks to demonstrate the firms' climate-change strategies that they employ and the role played by institutions. We explain how these strategies are relevant and how they are applied with reference to examples from transnational industries, particularly the apparel industry, as one of the most polluting industries. Finally, we present conclusions and argue that an institutional approach is crucial in explaining the dynamics and strategies for climate-change issues, and can help to complement typical IB and GVC analyses. We hope this chapter will encourage future IB and GVC studies to consider the institutional perspective to enhance analytical rigour in examining sustainability and climate-change strategies.

INSTITUTIONAL PERSPECTIVES AND THEIR RELEVANCE TO CLIMATE-CHANGE ADAPTATION STRATEGIES

In this section, we explore various institutional perspectives. We focus on national-level institutions, as this is the level at which many key changes are manifested and applied, and they help to explain how different actors – including states, firms, and civil society organisations – contribute to the pursuit of climate-change mitigation and adaptation in transnational industries, such as clothing and textiles. Our first entry point into this discussion is early institutional theory and new institutional theory in economics and neo-institutionalism in organisation studies, followed by a national business systems perspective. We also explore how the institutional perspective can be employed in the context of climate change strategies and sustainability.

Institutional Perspectives

The concept of institutionalism, particularly early institutional theory, originates from Max Weber's ideas of 'rationalization' and the 'iron cage' (i.e. structures) and Émile Durkheim's functional structuralism. Weber and Durkheim explain that rules, norms, values, and beliefs collectively institute 'modes of behaviour' that rationalise forms of transactions, relationships, and social orders (Durkheim, 1974). In the context of this chapter, the focus is on the rules, norms, values, and beliefs of dominant actors, such as the state, firms, and civil society actors (e.g. non-governmental organisations, NGOs), regarding climate change mitigation and adaptation. Different actors' interpretations of, and responses to, climate change challenges are, therefore, associated with pressures to conform, power struggles, and identity reconstruction based on their internal resources, knowledge, and cognitive frames. Consequently, the focus of the strategising process is not on objectively valid knowledge about 'climate change', but rather on how actors in organisations produce and reproduce knowledge about climate change through their interactions with the state, other businesses, and civil society actors.

Based on the theoretical foundations laid by early institutional researchers, new-institutional economics (Coase, 1998; North, 1991; Williamson, 1975), neo-institutionalism in organisational studies (Scott, 2008), and business systems (Whitley, 1992; Morgan, 2001) have worked to assess how institutional characteristics influence actor behaviour (Whitley, 1999).

North (1991) defined institutions as 'humanly devised constraints that structure political, economic and social interactions'. Constraints include both formal rules (constitutions, laws, and property rights) and informal norms (sanctions, taboos, customs, traditions, and codes of conduct) which perpetuate order and safety within a market or society. In the context of climate change challenges, formal rules and regulations can limit the ability of firms (and other actors) to pollute the environment, particularly through CO_2 emissions and water consumption. However, informal norms clearly have significant implications as well, particularly for the question of whether climate change can be addressed as a priority issue and how a society (or a firm) would like to address a specific aspect of climate change. For instance, a critical issue is whether actors – in terms of their norms and values – accept that climate change is 'real' and thus prioritise it; this is subsequently manifested by the actions of the state, firms, and civil society actors across the country.

The degree to which institutions are effective in setting constraints can be reduced by factors such as a government's limited coercive force, the lack of an organised state, or the presence of strong religious precepts. In the case of addressing climate change challenges, we should expect that governments' willingness and ability to act to reduce CO_2 emissions within their borders will differ substantially.

North (1991) argues that the way institutions are organised in a country shapes the extent of property rights, information asymmetry, agency (control), and transaction costs experienced by a firm operating in that country. The difficulty in monitoring the degree to which suppliers in weak institutional contexts carry out climate change mitigation strategies in GVCs has resulted in lead firms and other organisations creating private regulations and self-reporting-based indexes (Rana et al., 2019). Examples of such regulations are the 'Accord' and the 'Alliance' initiatives established by European and American firms purchasing garments from Bangladesh that suppliers to global brands are obliged to follow (Kazi & Rahman, 2021).

Scott (2001, 2008, p. 48) summarises the neo-institutionalism approach to the study of organisations, highlighting how 'institutions are social structures that have attained a high degree of resilience'. Institutions are composed of cultural-cognitive, normative, and regulative elements which, together with associated activities and resources, provide stability and meaning to social life.

For instance, tackling climate change requires supportive regulations and social norms to which firms should conform in formulating strategies. However, in a country where most firms are family owned and not listed on stock markets, owners tend to have higher influence in corporate decision making because they usually sit on company boards. This is the case in Bangladesh. Therefore, if business owners do not have a positive mindset towards addressing climate change developed in line with institutional norms and rules, firms will see climate change as an extra cost or, at best, an opportunity for instant economic benefit, instead of responding to climate change issues to achieve long-term outcomes.

The national business systems approach builds on early institutionalism, providing a comprehensive operational framework for studying institutions (Morgan et al., 2010; Whitley, 1987, 1992). The characteristics that make this framework distinct are, first, the degree to which it focusses on institutions at the national level and the idea of a coherent and relatively systematic national institutional framework that impacts firms; and second, the degree of agency that each firm is assumed to possess, and therefore the degree of importance given to the strategies and organisational structures of firms and the variety of ways in which they respond to institutional and market pressures in a global context (Rana & Allen, 2018; Rana & Morgan, 2019). This approach posits that institutions are constructed and reconstructed through their continuous interactions with firms and other actors, leading to a coherent system that shapes companies' structures, strategies, motivations, and capabilities to adapt to, and compete in, changing conditions (Rana, 2015).

For example, the UN's climate change agenda has contributed to change in many national markets. Variations in the readiness and ability to create an enabling capacity among national institutions will make the national market and firms either more or less competitive. For instance, in a collaborative business system, such as Denmark's, where state institutions are dynamic and highly efficient, firms are likely to be more efficient in responding to the climate change agenda in terms of product development, process reconfiguration, and marketing as compared to firms based in incoherent and fragmented business systems, such as Bangladesh's. In the latter type of business system, firms do not receive support from institutions of similar quality to those present in collaborative business systems. Thus, Danish firms are likely to have a competitive advantage over Bangladeshi companies. For example, Danish energy firms are more competitive in renewable and green energy production because national institutions, over the last 40 years, have provided supportive policies and incentives to boost research, education, and incentives for firms to catch up with the cutting-edge of renewable and green energy technologies.

The business systems approach emphasises how globalisation and internationalisation may change firms and institutions. Thus, it considers how firms which exist in both national contexts and global networks are influenced by international phenomena, and collectively implement creative responses to changing markets and institutions (Rana & Morgan, 2019). It takes the variable nature of firms into consideration, which the GVC literature does not tend to, and considers how institutions co-constitute firms, which the IB literature does not tend to. Often, rules are not developed at all by state actors in the Global South: firms there often apply climate change adaptation strategies in response to the requirements of Global North buyers. This eventually influences states to form new rules due to pressures from global actors and local suppliers to international brands. Thus, local firms, in support of their Western business partners and multilateral institutions, including donor agencies, are the initiators of climate-change adaptation. This is the case for the practice of the circular economy in the Bangladeshi garment industry, which was initiated by the P4G (Pioneering green partnership) and Global Fashion Agenda, with the government not having designed the policies and regulations as of the time of writing.

Table 14.1 sets out the proximate/formal and background/informal institutions; it illustrates the business system characteristics of firms that future studies on sustainability and climate change can consider in their analyses (see Rana, 2015; Whitely, 1992).

Our rationale for adopting a business systems perspective is its focus on how firms are shaped by national-level institutions, as the institutions at this level tend to be strongest and have the most significant effect on the overall economic outcomes of societies (Allen et al., 2021). Nevertheless, this perspective also recognises that there can be regional or sectoral differences, depending on the state's policies and structures (see Whitley, 2005). For instance, in a unitary state system (e.g. Denmark), rules and incentive structures apply equally for all firms in the country. However, countries with a federal state system, such as China, may have different incentive structures for different industries in different provinces. This can encourage some industries to reconfigure more quickly than others by taking advantage of the differentiated rules and incentive structures at the regional/provincial level. The circular economy districts established in various regions in China are good examples of regional institutional differences that affect climate change strategies; these include Guangzhou Huadu in Guangdong Province in Southern China and Suzhou New District in Jiangsu Province of Eastern China.

The business systems perspective argues that the interactions between firms and institutions give rise to a particular firm capability, with institutions acting as systemic influences on firms, enabling them to pursue some strategies more readily than other strategies (Redding, 2005; Whitley, 1992, 1999, 2010).

Climate change issues primarily affect the following three characteristics of firms seeking to gain competitive advantage in a business system (Rana, 2015, 2014) (see firm characteristics in Table 14.1):

(i) the nature of firms' ownership structure and governance,
(ii) the nature of inter-firm collaboration and relationships, and
(iii) the nature of management in organisations, including the vision, aim, and commitment to change.

Table 14.1. Institutional Features Conditioning Firm Characteristics on Climate Change and Sustainability Strategies.

Institutional Features and Firm Characteristics	Description
Institutional features	
Norms governing trust and authority relationships	• Trust in formal institutions and interaction-based relationships to support climate change strategy adoption • Importance of social capital in accessing resources, information, and justice for achieving sustainability goals • Authority hierarchy governing the degree of autonomy and power given to subordinates for independent decision making
Rationales and logics affecting decisions and justifications	• Both material and ideational logics affecting decisions over costs, technology features, production and operation features, interaction features, and sustainability/climate change features
State structures and policies	• Degree of dominance and directive role of the state in response to adapting and mitigating climate change issues across industries • Degree of complementarity and collaboration between state organs and between firms/business groups and state organs/policies • State encouragement of intermediary organisations in developing and implementing economic policies on sustainability and climate change-related initiatives and research
Financial system	• Size, liquidity, and significance of capital markets and their roles in reconfiguring and adopting long-term technological capacities and science-based targets on climate change mitigation • Dominance and support of bank-based financing versus capital market-based financing (e.g. the presence of a large pension fund can make a difference in facilitating long-term investment for sustainability-focussed transformation by firms)
Labour system	• Effectiveness of the public skills development and certification system on sustainability and climate change knowledge and skills • Quality of education and research system • Strength of employer and labour federations and their role in coordinating bargaining power and employee rights in the context of green transformation and reskilling
Firm characteristics	
Nature of ownership and governance of firms	• The degree to which family-owned, state-owned, foundation-owned, or public-limited companies are dominant and how they integrate sustainability agendas in their corporate strategies (i.e. both environmental, social, and economic agenda) • The degree to which private managerial hierarchies (salaried managers) coordinate economic activities as opposed to owners, and how the degree of autonomy delegated to managers allows them to take risks to undertake innovative and long-term strategies for meeting the climate-change agenda • The degree to which firms are networked with foreign firms (e.g. HQ–subsidiary relationships, joint ventures, licencing, and arm's length relationships in the GVC), and how the agency (power) and nature of control by the lead firms/HQ in the network plays a role in adopting the climate change agenda in the management and operation of firms. • The degree to which the specialisation of managerial capabilities and activities within authority hierarchies contribute to resource and operation reconfiguration towards the sustainability agenda • The degree to which risks are managed through mutual dependence with business partners, complementary institutions/intermediaries, and employees in transforming the operation and structure of firms towards achieving sustainability goals

(*Continued*)

Table 14.1. (*Continued*)

Institutional Features and Firm Characteristics	Description
	• The degree to which large conglomerates (domestic and foreign) are dominant and proactive in bringing new knowledge and resources for achieving sustainability capability, and how their agency co-develops new policies to mitigate the challenges of climate change
Nature of the networks and relationships between firms	• The extent of long-term cooperative relations between firms within and between sectors/industries, providing resource dependency opportunities and complementarity in achieving sustainability goals • The significance of intermediaries/supporting firms and institutions in the coordination of market transactions and efficiency in the business model reconfiguration • Stability, integration, and efficiency of business groups/associations in formulating coherent and indiscriminate policies for all types of firms to gain sustainability capabilities • Dependence of cooperative relations on personal ties and trust; and/or the degree of dependency on social capital for accessing compatible resources for achieving sustainability goals
The internal management dynamics (i.e. the nature of management in organisations)	• The degree to which the formalisation of authority and subordination relationships in firms contributes to the implementation of new routines and strategies for achieving sustainability goals • Task, skill, and role specialisation and individualisation suitable for promoting adaptation to climate change • The degree to which the differentiation of authority roles and the decentralisation of operational control contribute to the development and implementation of innovative strategies for adapting to climate change and social sustainability agenda • The degree to which the distance and superiority of managers affect the voices, new ideas, and decisions from the bottom level of operations (supervisors and workers) in developing new strategies for environmental and social sustainability • The extent to which employer–employee commitment (i.e. long-term/short-term or the nature of the incentive structure) contributes to transforming the organisational structure to achieve sustainability goals

Gaining a competitive advantage or improving sustainability will, therefore, depend on how coherent the state policy and regulatory supports are for adaptation and reconfiguration; how complementary the financial institutions are for the reconfiguration of R&D, production, technology, and marketing; and how efficient and supportive the labour institutions (including the education system) are in renewing and restructuring the skill development and employment systems.

Climate change issues present challenges for garment suppliers in Bangladesh, which they need to address if they are to remain competitive; for example, buyers from Western countries may choose not to purchase products from Bangladeshi suppliers if they fail to maintain or meet the required climate standards. Conversely, a climate change perspective can also be a source of competitive advantage if companies can obtain or create the relevant knowledge and technology and can reconfigure their organisational capabilities in pursuit of climate change issues faster than their competitors. This ability of the firm will reflect how it can implement green production processes to make more environment-friendly products, have an efficient cost structure and supply chain, and develop differentiation strategies on sustainability (Ivang & Rana, 2019).

Institutionalism in Practice

There are four key conditions in an institutional system that, when present, can hinder firms' ability to adapt to climate change: (i) when state agencies cannot develop consistent policies and regulations with the same speed with which the global rules are framed and enacted; (ii) when the state cannot enforce policies and regulations efficiently or cannot create or reform complementary institutions (such as financial and labour institutions); (iii) when state institutions fail to adequately collaborate with firms and support them to reconfigure themselves for climate change mitigation and adaptation; and (iv) when there is a lack of socially and environmentally conscious consumers.

The final condition requires a high-quality education system and background cultural institutions that encourage such an institutional norm in society. Through these institutions, the state influences consumers to develop the norms to consume and legitimise sustainable products. Unless consumers' minds are ready to consume environmentally sustainable products, and they have the ability and willingness to pay for them, firms will find it difficult to develop sustainable products or reconfigure their processes so that they are less environmentally damaging. In such a context, gaining a competitive advantage in the global market would be difficult as reconfiguring product design (i.e. R&D), technology, manufacturing, and supply chains would be too expensive and cumbrous. For example, environmental policy in Denmark provides tax waivers to house owners to renovate and improve the energy efficiency of their houses by changing roofs and windows. As this policy encourages house owners to buy energy-efficient products, it creates a national market for companies working in renewable energy and sustainable roofs and windows, helping Danish companies to compete in the global market.

A strong and coherent institutional system provides innovation opportunities for firms through research budgets; strong collaboration with firms to tackle the climate change challenges they face; a high level of trust in collaboration between organisations, universities, and state institutions; and high-level managerial and scientific knowledge produced at universities (Whitley, 2007). Trust and commitment in organisations regarding change and adaptation are, therefore, created through institutional policies, incentives, and a credible education and research system. For example, Bangladeshi apparel companies do not tend to collaborate with universities for product, process, and capacity development and innovation due to low trust. The sole exception to this is task-specific consultancy work. This general lack of inter-organisational collaboration, arguably, hampers firms' innovation capabilities in relation to climate change mitigation (Rana & Allen, 2021b).

Such collaboration is unlikely to improve unless state policy provides research grants that focus on the challenges that companies face, and encourage firms and universities to collaborate. Collaboration between industry and academia also requires universities to have high-class research and innovation capabilities that firms are aware of and can access. Firms in business systems that lack such collaboration are more likely to experience difficulties in reconfiguring their capabilities for quick adaptation to climate change and are less able to develop local solutions using local knowledge (Witt & Redding, 2014).

A business systems perspective looks for multiple logics and rationalities in the relationships between institutional systems and firm behaviour. Studies focussing on IB and GVC that seek to explain organisations' climate change strategies can consider this perspective to gain greater insights. In comparison, the new institutionalism in economics based on North's (1991) work primarily focusses on the 'constraint' dimension of institutions and overlooks how firms can actively change or complement institutional voids or weak institutional systems (Jackson & Deeg, 2008; Rana & Allen, 2021b; Whitley, 2010).

Future studies on climate change can identify certain key institutions and their characteristics that shape resources for firms and the opportunities and constraints under which firms operate – for example, the state, the financial system, the skills and training system, and labour market regulations and trade union formation (e.g. monopolies or oligopolies, restrictions on labour supply, or weak trade unions that are politically motivated and fragmented). The background cultural system (i.e. norms, authority, identity, and rationales) is also highly significant (see Table 14.1).

For example, it is important to consider whether cultural relations are predominantly individualistic, based on opportunism and economic incentives, or collective, relying on shared notions of community, trust, and social capital. Where non-market relations are predominant, studies can ask what roles the three main actors – the state, employers, and labour – play in shaping institutions and how they affect climate change strategies at the national and firm levels. Where the state, employers, and labour and, in some cases, NGOs or trade unions, which may act on behalf of labour (e.g. in Northern European economies), work together, the outcome is a form of corporatist capitalism (Whitley, 1999). This may be highly inclusive and network-based (i.e. aiming to treat all workers and firms in the economy in the same way), as in Denmark, which resembles a collaborative network-based business system, or a collaborative, but slightly less inclusive collaborative business system, such as Germany that privileges export-sector employees and large conglomerates that have a significant influence on policy outcomes (Rana & Morgan, 2019). Compared to those in Germany, institutions in Denmark tend to enable labour representatives to have a greater influence on policy; consequently, climate change strategies tend to include labour perspectives more (e.g. health and safety, and skill development), and technological adaptation and research have greater emphasis in educational institutions (Morgan et al., 2005).

Where labour is absent from the dominant coalition, three broad climate-change strategies are possible. First, the state is dominant and encourages particular forms of business to emerge, often based on the exchange of political favours (e.g. South Korea and India). Consequently, climate change policies from the state dominate firms' work to reconfigure themselves to adapt to climate change. The second is where the state and businesses tend to be equally powerful and negotiate cooperatively among themselves (e.g. Japan) (Rana & Morgan, 2019). In this context, firms and the state co-evolve climate change policies and strategies. The labour perspective is reflected in climate-change policies, while the education system focusses on high-level research and development on climate change issues, from which firms can benefit.

The third possibility is the one currently seen in Bangladesh. Here, institutional systems become incoherent and fragmented, and the country fails to develop an effective business system. In such a context, skill certification is weak and poorly organised, and thus the wages for specific skills are not well defined or consistent. Labour unions are politically oriented with low bargaining power in many areas, such as labour welfare and development, and thus NGOs step in to push for improved labour welfare. Workers and lower-level managers do not have a voice in strategic decisions, leading to poor labour management and top-down decision making on social sustainability and climate-change strategies. Although specific policies on tax incentives and low-cost capital are offered for climate change adaptation, access to these schemes often depends on social capital and affiliation with political executives. In such a situation, accessing resources and government schemes requires employers to have strong ties to political elites. Consequently, firms' growth and ability to reconfigure for climate-change adaptation is dependent upon these relationships. Educational institutions rarely offer high-level market-oriented skills and research knowledge to catch up with the cutting-edge technology required for climate-change adaptation (Rana & Allen, 2021b). Therefore, firms in such business systems view climate-change strategies from the perspectives of cost, pressure, and legitimacy, and not as an opportunity to develop long-term sustainable competitive advantage.

In fragmented business systems, firms' responses to climate change in IB and GVC depend on how companies build their strategies in line with national and global institutional systems and how they act as institutional entrepreneurs to change or complement the institutional gaps to grow and sustain in business. However, firms need to fulfil the requirements of Scope 3 emissions, as defined by the UN Global Compact to adapt to climate change issues. Scope 3 emissions are the result of activities from assets that are not owned or controlled by the reporting organisation; however, those activities outside the organisation in its GVC indirectly impact emissions. Scope 3 emissions include all sources not within Scopes 1 and 2 emissions; these latter two cover the GHG emissions that a company makes directly (Scope 1), such as its combustion-engine vehicles or gas-powered heating, as well as those it makes indirectly (Scope 2), such as the electricity it buys to run its computers or cool its buildings. Scope 3 emissions arise up and down the company's value chain and relate to the emissions that its suppliers and buyers make. Scope 3 emissions, therefore, often arise because of MNEs' internationalisation strategies and their role within GVCs.

How companies respond to the need to reduce Scope 3 emissions reflects the type of capitalism that they operate within. In particular, there are two critical issues that firms should consider when addressing their climate change impact.

- First, a firm needs to look at the GVC to map out the critical-value generating points that emit the most CO_2 and take the necessary steps to reduce or eliminate these emissions. Global firms that have internalised their value chains (Dunning, 1998) are likely to be able to implement such efforts more quickly by adopting new technology and reconfiguring the process of value creation, value delivery, and value proposition across the GVC. This may be relatively easy and less time consuming because although the value chain for such a firm is internationally dispersed, it is contained within the firm's organisational boundaries. However, firms may not be able to internalise some value-creating activities, such as transportation and distribution.

Conversely, firms that tend to outsource production and supply are likely to find it more difficult to reduce their environmental impacts because all suppliers across borders may not have equal capabilities and resources. They are, furthermore, embedded in institutional contexts that may lack the necessary complementarity to reconfigure operations in pursuit of GHG emissions reduction.

- Second, firms should prioritise a circular economy that builds on regenerative and restorative mechanisms using sold products and/or waste as the input for the further production of new products. This can significantly reduce CO_2/GHG emissions, water consumption, and 'fugitive emissions', such as unintentional and the undesirable emission, leakage, or discharge of toxic gases (Ivang & Rana, 2019). As a comprehensive strategy, firms can also develop their own methods or collaborate with relevant firms for carbon capture (e.g. Power2X business) and storage using a set of technologies in power plants, transportation, and production systems. The stored carbon can be then traded in an emissions trading scheme, which can offset firms' CO_2 emission balances in the GVC and provide a competitive advantage by creating a new source of revenue, provided firms have the requisite knowledge, technology, and managerial capability.

For example, taking advantage of institutional supports and incentives, Power2X, in collaboration with researchers from Aalborg University, Denmark, and local industries, built a complete industrial-scale electricity-to-methanol plant in Aalborg. The plant makes eMethanol from CO_2 (which is emitted by various industries in the area) and hydrogen using renewable electricity. This technology can make heavy transport greener, because heavy transport in general (ships, lorries, trains, and aircraft) accounts for more than half of CO_2 emissions from the entire transport sector. At the same time, it provides a complementary advantage to other firms in reusing captured and stored CO_2, which contributes to green transportation that the possibility for this kind of strategy emerges from an institutional system with high complementarity (i.e. effective complementarity and collaboration between a technology company, research institution, and local firms, AAU News, 2021). This sort of complementarity is insufficient or absent in a fragmented institutional system, such as that found in Bangladesh.

Both strategies require institutional complementarity and path dependency to accomplish the corporate goal regarding climate change. Firms' institutional settings fundamentally shape firms' opportunities in terms of how they use capital, labour, and technology for climate change adaptation. The business systems and institutionalist research imply that there is a limited number of ways in which institutions cohere, to a greater or lesser extent, into a 'national business systems'. This can be explained in many ways, but generally relies on two key concepts: institutional complementarity and path dependency.

The first concept, institutional complementarity (Deeg, 2005; Morgan, 2007; Whitley, 2005), draws attention to how the benefits of one institution are greater if other institutions encourage the same type of actor behaviour. For instance, combining a labour system that is based on long-term employment with a financial market that is highly volatile and subjects senior managers to short-term financial pressures is likely to be difficult. The ability to expand or contract the labour force is a major way in which managers can respond to short-term financial pressures, and can manage how changes in the economy influence their firm's share price. Therefore, if companies are subjected to short-term financial performance goals, offering lifetime employment for most employees in the firm will be extremely difficult. Long-term employment policies and investment for sustainable product and process development require long-term commitment and finance (Rana & Morgan, 2019).

Similarly, firms in institutional systems characterised by a lack of trust between managers and workers will face severe challenges in sustaining long-term employment policies. Firms will, as a result, be shaped by their home-country institutional setting; this is true for both firm strategy in general, and climate change, in particular.

The following case goes some way towards explaining the concept of institutional complementarity in firms' strategy on climate change and sustainability.

When the extent of the climate crisis first became apparent, Denmark undertook consistent policies to support firms in reconfiguring their production processes, energy usage, and products to adapt to climate change challenges. In 2014, it developed long-term policies and strategies, such as the 'Danish Nature Policy' (Naturplan Danmark) and 'Global Climate Action Strategy: A Green and Sustainable World (klimastrategi)', aiming to reduce emissions by 70 per cent by 2030 and achieve climate neutrality by 2050. To strengthen the focus on adaptation and sustainable development, state organisations cooperated with the private sector on green solutions. This was complemented by two subsequent policies, 'Green Growth in Denmark' and 'Green Future Fund', aiming to generate up to EUR 15 billion for financing green companies and projects for climate change adaptation and reconfiguration (State of Green, 2021). Such a highly ambitious set of goals and policies could never be achieved without a skilled workforce and research outputs; in pursuing this ambition, Danish research grant authorities and universities have prioritised sustainability and the climate change agenda in their education and research to complement firms' restructuring and adaptation strategies. As a result, Danish firms can now reduce GHG emissions, store and trade CO_2 using local innovative technology, and use green energy throughout the value chain – practices for which Denmark already has earned a strong reputation.

In contrast, although Bangladesh has developed some policies and strategies for transforming its energy sector, it currently relies on fossil fuels, which generate more than two thirds of the total energy produced; less than 2 per cent of electricity comes from renewable sources, primarily solar energy. Policy implications, however, face difficulties in achieving the goals due to a lack of complementarity and collaboration between several ministries (i.e. those responsible for developing policies and those responsible for implementing them), research institutions, business groups, and financial systems. High-level bureaucratic inefficiency, a lack of green funds and research grants, and the absence of a long-term vision and policy contribute to ineffective institutional efforts to support firms' abilities to adapt to climate change. Although some policy initiatives, such as schemes that direct banks to disburse loans to firms for climate change adaptation, have been implemented, they are often applied in a discriminatory way, depending on who owns the firm and how closely the owners are networked with political elites. Recently, Bangladesh's Infrastructure Development Company Ltd. (Idcol) received USD256.5 million from the UN's Green Climate Fund to promote private sector investment through the adoption of energy-saving technologies and equipment for the textile and ready-made garment sectors (Rahman, 2020). However, large textile firms with adequate knowledge of technology and managerial capability are likely to receive this discounted loan for climate change adaptation, creating a one-sided benefit for big players due to discriminatory practices in the institutional system.

The second key concept is path dependency (Deeg, 2005; Morgan, 2007). Once actors invest and follow existing institutional practices and norms, they are often reluctant to change because they may have strongly embedded routines. They also tend to adapt to change in an incremental way unless a major crisis occurs or pressure is exerted by a different set of institutions or the impetus for change comes from some external source, such as competition or political upheaval.

Institutional complementarities and path dependencies vary considerably, not just in the degree to which they exist, but also in how far and in what ways they are present in society. In some incoherent and dysfunctional institutional systems, such as Bangladesh's, some complementarities can be negative, that is, they reinforce weak firms and weak economic development, locking institutions and firms into a low-performance equilibrium. Equally, complementarities can be positive for some sectors in society and less so for others, creating a discriminatory or varied institutionally conditioned performance for different firms (Rana & Allen, 2021b).

This explains why some societies in the Global North may specialise in different sectors, depending on the sorts of complementarities that are strongest. For example, Danish firms are particularly strong in the green energy sector and related industries (ventilation, cooling, and pumps). Long-term finance for sustainable growth from pension funds and foundation-led investments supports these firms (Kristensen & Morgan, 2018). This example illustrates how firms' strategies, capabilities, and structures in an institutional system evolve for endogenous reasons as much as exogenous ones – that is, the rise of new sectors or new pressures from global and national institutions such as COP26 and the UN Sustainability Goals. These external pressures/incentives can cause the decline and reorganisation of mature sectors and pose difficulties in developing supporting institutions that affect the growth and survival of the existing business models followed by firms in a particular industry (Boje & Rana, 2021). The business systems framework enables analyses to more readily include the factors discussed above than do other theories. However, combining insights from the business systems framework with typical analytical foci in IB or GVC studies will potentially provide the most cogent explanation of firms' responses to the climate emergency.

CLIMATE CHANGE STRATEGIES IN IB AND GVC RESEARCH

The IB and GVC literatures have not, so far, focussed much attention on climate change strategies. IB research focusses primarily on MNE governance, performance, and competitive advantage in cross-border contexts. Studies of GVCs often examine how lead firms in global value networks/global production networks drive value-creating activities and the upgrading of value chain actors (Kostova et al., 2016; Ponte et al., 2019). Despite these differences, there is a similarity between IB and GVC research. Both focus on the same boundary structure in which organisations interact (i.e. in global and cross-border contexts) and both aim to explain (a) how organisations are interlinked with multiple actors in global business operations to create value and (b) how actors across countries coordinate economic activities, which are embedded in various national institutional contexts and influence socio-economic development (Gereffi, 2002, 2018). However, the underlying rationales that explain how firms gain and control competitive competencies in value-creating processes across countries make these two approaches different in their application. Aware of these differences, we present a discussion of how climate change strategies from an institutional perspective can be explained in the IB and GVC domains.

Climate Change Strategies in IB

Despite the increasing attention being paid to the climate change agenda and the institutional perspective, IB studies have often overlooked how institutions in home and host contexts affect MNEs' governance and ownership to reconfigure capabilities for climate change adaptation (Allen et al., 2019; Doz & Prahalad, 1980). Although the 'institutional context' is an important dimension in IB literature (see Peng et al., 2008), IB studies have tended to downplay suppliers' perspectives on GVCs from the Global South (developing countries in Asia) that are linked with foreign subsidiaries (i.e. with buyers) from the Global North. This lack of focus on how Global South suppliers are linked to different foreign subsidiaries results, partially, from how the IB literature tends to neglect how the institutions in the lead firm's home-country influence the nature of that company and its relationship to its subsidiaries and suppliers (Rana & Allen, 2021a). In particular, IB research does not typically examine the GHG emissions of their subsidiaries and suppliers that are connected across borders.

Using the OLI paradigm, IB studies can explain climate change strategies in two types of relationship contexts: (i) HQ–subsidiary and (ii) subsidiary–supplier relationships.

Dunning (2000) argues, using an OLI approach, that MNEs' ownership and location advantages underpin their decisions to internalise or externalise their cross-border operations. Internalisation means the MNE retains control over operations; externalisation means outsourcing the work. The OLI model contends that a company must possess net firm- and/or country-specific ownership advantages over other firms from other countries in serving a particular national market. In the context of a climate change agenda, ownership advantages may stem from institutional incentives and/or institutional complementarities, such as managerial and technology competencies embedded in the local education system. Companies may be able to draw on these institutionally conditioned resources to make themselves ready for climate change adaptation. If these advantages do not already exist in the local context, firms may employ strategies to acquire them, using internationalisation (i.e. through foreign direct investment, FDI; joint ventures; licencing; or strategic alliances).

There are two scenarios that may negatively affect MNEs' abilities to integrate a climate change agenda into their operations:

- First, the financial pressures on investors/owners reduce their ability to adopt a positive mindset towards addressing climate change issues, and thus firms are not willing to accept many risks to reconfigure their operations for long-term gains.
- Second, management and employees are not motivated or willing to accept the risk arising from the reconfiguration process; this applies in contexts where managers and workers view this change as a threat to their jobs or as unimportant. Similarly, if an MNE's headquarters does not support the change process in subsidiaries or if suppliers work in less formalised institutional systems, senior managers may find it difficult to transition their organisations away from fossil fuels at all or partially. The inability to adapt to climate change is likely to occur most when the MNE's policy is not consistently evolved to reconfigure subsidiaries' organisational capabilities in terms of new skills and technology development that requires additional investment in the host context.

However, firms may be able to develop new capabilities that help them adopt more environmentally sustainable practices if corporate governance conditions enable it (Haxhi & Aguilera, 2017). For example, H&M Foundation, a part of H&M, consistent with H&M's corporate strategy entered an international strategic partnership with the Hong Kong Research Institute of Textiles and Apparel and invested millions of dollars to develop various technologies for scaling up the circular economy and climate change initiative by developing low-cost technologies, such as AI-powered Smart Garment Sorting System for automating textile recycling, bio-based leather regenerated from leather waste, sound waves that separate microplastics from water, a green machine for recycling blended textiles at scale, cellulose powder that improves cotton quality, and CO_2 capturing textiles (H&M Foundation, 2024).

The Partnership for Cleaner Textiles (PaCT), a programme funded by the International Finance Corporation (IFC), a global organisation that promotes, amongst other things, norms and institutions around sustainability, and with which fashion multinationals are networked, helps Bangladeshi garment suppliers to adopt cleaner production practices. This initiative demonstrates how institutions can constitute firms, changing their natures and capabilities. This initiative, which provides a useful example for the global supplier perspective in IB, has helped 338 factories to reduce freshwater consumption by 25 million m^3 a year and wastewater discharge by 21.08 million m^3 a year. As a result, these suppliers now can save 2.5 million MWh per year in energy and avoid CO_2 emissions of up to 489,796 tons per year. PaCT assists factories to recover chemicals, such as salt and caustic soda, from production processes, and provides low-cost green investment funds to garment suppliers from the USD200 million Green Transformation Fund, which is supported by IFC group and World Bank as part of UN sustainability green fund (PaCT, 2021). This programme supports suppliers to develop dynamic capabilities to reconfigure their competencies and comply with the new requirements that MNE buyers, such as H&M, emphasise for their suppliers. In this regard, the global climate change initiatives undertaken by global organisations in collaboration with MNEs, suppliers, local institutions, and civil society complement suppliers' reconfigurational processes (Rana & Tajuddin, 2021).

IB studies tend to overlook the role of national and transnational institutions, such as those promoted by, for example, the PaCT, in MNEs suppliers' adaptation to the effects of climate change. The supply-side actors in manufacturing industries, particularly the component suppliers and manufacturers, are often in emerging economies. They are generally seen as rule takers or rule followers and tend to be less powerful due to their limited organisational capabilities and market influence compared to their counterparts that control key resources, such as brands. This is where the tension appears, as suppliers and MNEs experience unequal institutional conditions that result in different capabilities and strategies to mitigate and adapt to climate change challenges (Okereke, 2006). Below we discuss OLI from the perspective of climate change strategies.

- *In OLI, the concept of Ownership* has been extended to different dimensions, such as asset advantage ownership (Oa) and transactional advantage ownership (Ot) (Dunning, 1977, 1983). Oa signifies the exclusive possession and use of certain kinds of income-generating assets, and Ot considers a firm's ability to coordinate separate value-adding activities across national boundaries and to reduce environmental (i.e. climate change and socio-political) and foreign exchange risks (Dunning, 1988, p. 25). Ownership can be analysed in several ways, including the cognitive perspective (ability and access to information and knowledge about the assets); the equity versus non-equity perspective (cooperative relations-based); and the tangible versus intangible asset-based perspective (intellectual and relational assets) (Eden & Dai, 2010). Further, the OLI framework also emphasises *institutional ownership advantages* (Oi), including both formal and informal institutional structures, incentives, and enforcement mechanisms (constraints) that shape firm choices and performances in relation to climate change impact (Dunning & Lundan, 2008a, 2008b, 2010).

For instance, the Swedish fashion multinational H&M has been a forerunner in adopting climate change strategies in its corporate strategies, global sourcing, and marketing. In its downstream operations, H&M has promoted sustainable fashion and consumption and has taken measures to reduce CO_2 emissions, water consumption, and chemical waste. It ensures that the operations across its upstream (designing, component sourcing, production, and shipment) and downstream (sales and distribution) value networks consistently adopt climate change mitigation strategies. The challenge in this approach lies in reorganising and governing upstream operations in line with downstream operations, primarily because its upstream operations are externalised (i.e. it sources apparel products from suppliers in China, Bangladesh, Vietnam, India, Turkey, Indonesia, etc.) (Rana & Allen, 2021a). Since supplier countries have relatively weak institutional settings that tend to undermine efforts to address fire safety, water, and air pollution issues, H&M has developed a restrictive form of governance and collaborates with suppliers to ensure employee wellbeing and climate change adaptation. It has developed a long-term plan to curb climate change challenges in supply and production. H&M has effectively implemented a circular economy model in parallel with a linear business model in global production networks by using advantages conferred by asset ownership, transaction ownership, and institutional ownership in the Swedish institutional system (Rana & Tajuddin, 2021).

In the Swedish institutional system, which is a collaborative and networked business system, family-owned firms tend to pursue long-term sustainable strategies to protect their family reputation, while large firms can source long-term finance from financial institutions that access large-volume pension funds with an aim of investing in sustainability-driven projects. H&M, therefore, has been able to invest heavily in recycling technology, developing a mechanism to collect used clothes and produce artificial cellulosic fibres, and reduce CO_2 emissions and water consumption among its suppliers. This is incentivising and encouraging suppliers to transform production processes and use environmentally sustainable components in garment manufacturing. This strategy consistently contributes to the reduction of CO_2 emissions, water consumption, and waste disposal in the long run. H&M's institutional advantages and corporate-ownership advantages in terms of knowledge, managerial skills, and finance, together with its long-term investment commitment, have allowed the company to reconfigure its internalised activities and externalised manufacturing operations including some in developing countries (i.e. its globally dispersed supplier networks) (Rana & Tajuddin, 2021). In contrast to the GVC literature, IB research has traditionally examined internalised operations, exploring how MNEs' strategies are affected by ownership advantages, location advantages, and internalisation advantages, although they are not necessarily relating climate change strategies across countries.

(a) *'L' (Location) advantages (where to invest?)* indicate that the net ownership advantages must be profitably exploited in the context of conditions (skills, knowledge, markets, raw materials, technologies, etc.) outside the home country and in the host country (Eden & Dai, 2010). Using the firm perspective in international production, John Dunning emphasises the institutional complementarity in which transaction costs (Williamson, 1975) have been the central argument of the OLI eclectic paradigm because differences in institutional contexts are associated with variation in transaction costs. Therefore, location advantages depend on institutional characteristics, and transaction costs impact how subsidiaries and their supply chain actors can adopt climate change strategies. MNEs are resource-rich, thus they may undertake several strategic options (including FDI, strategic alliances, and franchising) and market transactions when they expand globally. As a result, different factor conditions will apply as to how climate change strategies will be employed in different locations.

For example, following Facebook and Apple, the tech giant Google has invested DKK4.5 billion in renewable energy-driven data centres in Fredericia, Denmark (Jacobsen & Jensen, 2018). Its reason for internationalising its data centres is that Denmark has the locational advantage of institutional incentives for renewable energy, state-of-the-art renewable energy infrastructure, and a global reputation as a green energy country, helping tech giants to achieve global legitimacy and net-zero carbon emission goals.

(b) *'Internalisation' (I) advantages (how to invest?)* are those which represent benefits for a firm to use its ownership advantages inside its organisational structure rather than leasing or outsourcing them. Building on the resource-based view and transactional cost theory, Dunning points to the motivations (i.e. the willingness of owners and management) and capabilities of MNEs as factors that drive internalisation (I) decisions, that is, what to internalise and what to externalise. Thus, internalisation strategies and location choices are closely dependent on the sum of ownership advantages [Oa+Ot+Oi] (Eden & Dai, 2010). In explaining the internalisation concept, Buckley and Strange (2011) point to two important issues, namely the cost of governance and the risks of governance within the hierarchy of the firm and the markets (external institutional settings), which are critical to climate change strategies.

MNEs tend to have three kinds of problems when making long-term strategic decisions on climate change issues, namely the information/knowledge problem, the coordination problem, and the motivation problem (from managers', subsidiaries', or stakeholders' point of view). The way in which these problems are handled depends on the organisational norms and practices around risk, organisational capabilities, and visions as well as the institutional contexts in which the firm is embedded. The choice to either take or avoid risks depends on the comparative costs of external and internal governance, which are necessarily influenced by the internal and external transaction costs within the institutional context (Buckley, 1993).

For instance, Spanish fashion multinational Inditex internalises 53 per cent of its 8,543 supplying factories worldwide; these are in relatively close proximity to its HQ in Arteixo, A Coruña, Spain. This is done to mitigate the risk inherent in the need to supply their shops to meet fast fashion trends and to have the reconfiguration capability needed to respond to changes in the market and regulations, including those related to climate change. As it internalises such a large amount of production, Inditex can quickly implement climate change strategies in its supply chain and operations based on the ownership advantages it possesses. Thus, it has been able to reduce its energy consumption and GHG emissions more quickly through strong coordination within its internalised network in Spain. Specifically, 100 per cent of Inditex's global energy consumption in its own facilities (HQ, production, logistics, and stores) comes from clean sources. Its current focus is on transforming the energy consumption of its suppliers that are externalised in the value chain (Boje & Rana, 2021).

Inditex provides a useful contrast to H&M suppliers outside its ownership structure. This requires H&M to form collaborative relationships with externalised suppliers to support them in transforming their energy consumption practices, which takes more time and strategic effort.

Moreover, certain countries rely heavily on fossil fuels in their energy mix. For instance, 95 per cent of Bangladesh's energy comes from fossil fuels, complicating suppliers' adoption of green energy strategies in their operations. In contrast to firms based in Denmark, such suppliers in Bangladesh will, therefore, have difficulty in transforming their energy consumption from highly polluting to green energy. Climate change adaptation processes that require the reconfiguration of value-creating activities result in tension due to differences in cognition (or mindset that determines which aspect of the climate change agenda to prioritise), institutional conditions, and firm capabilities. It is true that ownership advantages give firms a competitive advantage through certain location advantages, but it may also pose some disadvantages when firms wish to change before the institutions to support that transition have developed sufficiently (Cantwell & Mudambi, 2005; Nachum & Zaheer, 2005; Nohria & Ghoshal, 1997). Therefore, in general, the IB literature would benefit from integrating key insights from the business systems framework into analyses of how firms respond to climate change. In particular, ensuring that home and host country institutions are fully included in analyses so that the context within which firms develop new capabilities is important. This will require (1) a solid understanding of how home-country institutions constitute firms, both lead firms and suppliers, and (2) an awareness of how firms' relationships with other organisations will vary as a result, affecting their abilities to achieve climate change objectives.

Climate Change Strategies in the GVC

The concept of the GVC refers to the full range of activities embedded in a variety of institutional contexts that different economic actors carry out to bring a product or service from conception to end-use and recycling (Ponte et al., 2019). This concept helps map activities that are geographically dispersed and explains the role of lead firms and formal institutions (such as national and transnational institutions) in the economic and social upgrading of actors in the value chain (Gereffi et al., 2005). These activities include design, production, processing, assembly, distribution, maintenance/repair, marketing, finance, consumer services, and disposal/recycling.

GVC studies tend to assume that lead firms from all countries outsource or externalise some activities to suppliers (Sako & Zylberberg, 2017). However, this may not always be the case: several Danish apparel SMEs have internalised their supply chains in Italy, Poland, and Romania. Therefore, studies examining the impact of climate change can examine the role of actors in a hierarchical network in the GVC or a network defined by relational governance with an arm's length relationship, which is typically externalised, for example, H&M. This neglect of how firms' actions vary stems, in part, from the GVC literature's approach to institutions: it tends to downplay how the institutions in lead firms' countries vary and how those institutions co-constitute firms, leading to differences in their strategic priorities and capabilities.

There is a clear power imbalance in the governance structure between SME brands and their suppliers, as well as between MNEs and their suppliers, due to differences in trade volumes, innovation competency, and suppliers' dependency on lead buyers (Khan et al., 2015; Lee & Gereffi, 2021). *Processes and power* in the global value network are key dimensions that climate change strategies should consider when explaining the strategic choices for MNEs or SMEs and the consequences that they can have for their suppliers' upgrading and reconfigurational capabilities (Gereffi, 2021).

There are two major issues that will affect GVC studies on climate change strategies.

- *Investment perspective on climate change strategies in GVCs:*

The reconfiguration of climate change strategies requires firms to have as much support from management and technology as they do from investment. There is a growing community of financial institutions acting and demonstrating leadership on climate change initiatives. Some organisations are allocating capital and driving financial flows towards more low-carbon and climate-resilient activities. Others are pursuing strategies to change corporate behaviour, influencing policy outcomes by building the data and tools required to embed climate-change issues into how the market functions.

For many developing countries, enhancing resilience means not only confronting the severe economic vulnerabilities stemming from a lack of export diversification and a dependence on a few industries in the GVC, but also adapting to increased exposure to environmental shocks, which can only be mitigated through organisational capability and long-term investment (Gereffi et al., 2022; Whitfield & Staritz, 2020). Many large MNEs have been planning to reshore some activities within their GVCs to enhance resilience. Reshoring is likely to reduce the existing power asymmetries between lead firms and their suppliers that moderate the efforts by governments in supply-side countries in the Global South to mitigate the effects of climate change.

The extent to which suppliers will be able to reconfigure their operations in response to climate change depends on the extent to which the institutions around them can change and support the transition process. Since Bangladesh generates most of its energy from fossil fuels, 152 large garment suppliers have recently constructed green factories and internalised the solar panel-based green energy system, partly financed by global organisations and the national climate fund. They have received Leadership in Energy and Environmental Design certification, giving them more bargaining power to receive more orders at higher prices from Western buyers in the GVC. However, small and medium-sized factories are under different resource and capability conditions since they have limitations in accessing finance from national and multilateral organisations.

An investment perspective on climate change strategy facilitates an assessment of the impact of FDI in GVCs and the roles that national and multilateral organisations play. To stimulate economic transformation through GVCs, policymakers and firms in developing countries need to better understand MNEs' business strategies, the internationalisation pathways for domestic firms, and how climate policies can create a conducive environment for both local and international firms. The United Nations Conference on Trade and Development 2020 report reveals that the global sustainability and climate change funds have reached USD1.2–1.3 trillion, including USD260 billion of green bonds, USD105 billion of social bonds, and over USD900 billion of sustainability-themed equity funds (Zhan, 2021). The global effort to mobilise investment to achieve the SDGs will change the future patterns of FDI globally, which will impact sources of financing, sectoral development, and geographical locations. However, a long-term national investment plan for technology and infrastructure development in relation to climate change will also impact the degree to which the focal sectors in a country will be able to upgrade and compete in GVCs.

- *Emergent standards affecting climate-change adaptation strategies:*

Today's GVCs and the end-to-end processes that underpin activities, such as raw material extraction and processing, consumer fulfilment, and end-of-life disposal or recycling, have been built on a paradigm of localised production nodes and globalised flows. How the Fourth Industrial Revolution and emerging global standards will impact the role of production and restructure the paradigm in connection with climate change in GVCs remains a question to be investigated. The key emergent standards today typically come from global organisations, civil-society organisations, and technological breakthroughs, each of which continues to develop new requirements and policies in relation to climate change, consumption and disposal, artificial intelligence, 3D production, global ethics, and labour rights. National governments and companies, therefore, need to rethink their policies and institutional ownership to create necessary transformations in GVCs for their industrial development and investment strategies.

Thus, studies should examine the opportunities for multi-stakeholder collaboration in national economies to build readiness and resilience in institutions that can play a role in next-generation sustainability in GVCs.

Global economic policies, especially trade and investment policies, are shifting from multilateral cooperation to regional and bilateral solutions. This is compounded by the shift in national economic policymaking through new industrial policies, including protectionism, digital governance, and standard building (i.e. compliance criteria) in relation to climate change mitigation. The aggravated systemic competition (in trade, investment, technology, etc.) between economic powers impacts firms' relationships in GVCs. For instance, the application of new technologies in global production networks by MNEs will have far-reaching consequences for the configuration of, and governance within, GVCs. These new technologies include robotics-enabled automation, AI-enhanced systems, and supply chain digitalisation (including platforms, Internet of Things, blockchain, additive manufacturing, and mass customisation) (Brun et al., 2019).

While robotics-enabled automation can reduce wastage in production and enhance efficiency and productivity, AI-enhanced systems in supply chain management can record and analyse vast quantities of data from upstream and downstream operations and enable real-time analysis of different parameters of climate change issues in relation to a range of value-adding activities across the globe. This technology suggests solutions for optimised supply chain management between buyers and suppliers in relation to resilience and sustainability. It will reduce information asymmetry and ensure data-driven customised governance and compliance in many GVCs. However, the extent to which firms can adopt such new technology will depend on their internal resource capability and the institutional contexts they are embedded and operate within.

The Higg Index is a recently introduced tool supporting the pursuit of digital governance within the global apparel and footwear industries. It presents a measure of sustainability for each factory throughout a product's lifecycle, from raw material production to end-of-life, based on detailed information on climate change and social responsibility parameters provided by suppliers on a voluntary and continuous basis. At present, the metrics created by the Higg Index are limited to the company's internal use for the evaluation and improvement of its environmental performance. However, over time, it has the potential to share information with buyers and other value chain actors to become a global standard for addressing climate change in the global garment and textile industry; this would, however, depend on dominant institutional actors granting it legitimacy.

Despite efforts such as these, the differences in approaches and standards between countries and regions on emissions targets, as well as suppliers' responsiveness, will continue to impact the governance choices for buyers' GVCs and how value chain actors can adapt to institutional changes and standards. Hence, including insights from the business systems framework on the role of wider institutions in firms' strategies to address the climate change emergency is important.

CONCLUSION

This chapter has illustrated and discussed various dimensions of climate change strategies from institutional perspectives, including those perspectives from IB and the GVC literatures. While both the IB and GVC literatures can explain some aspects of climate change strategies and sustainability, this chapter has tried to highlight relatively downplayed aspects in both, and to offer new insights to address those aspects. We argue that combining IB and GVC analyses with insights from the business systems framework will provide more cogent explanations of how firms in different countries respond strategically to the climate change emergency.

Reflecting on the internalised global network of MNEs (i.e. MNEs and their subsidiaries) and the externalised global production network of multinational buyers and their suppliers in developing countries, we have illustrated how different dynamics in institutions, highlighted in the business systems framework, such as the ownership and control of firms and the role of government, influence climate change strategies across countries. In so doing, we have presented several examples from the Global North (i.e. Europe, especially Denmark and Sweden) and the Global South (particularly from Bangladesh with reference to the global garment industry) to substantiate the insights reflected in the theoretical discussion so that future researchers can follow the theoretical reasoning with reference to the OLI and GVC frameworks.

We argue that the changing nature of institutional structures and firms' capabilities shape climate change strategies. How firms develop their climate change strategies, thus, depends on the changing nature of institutions, complementarity and path dependency in the institutional environment, and the extent to which emerging technologies, investments, standards, and market forces support firms' operations in the GVC. Therefore, we argue that the use of a further institutional perspective, namely the business systems perspective, which covers these areas, will help to ensure the analytical rigour of research on climate change strategies in IB and GVCs. By doing so, this chapter contributes to sustainable strategic management from an institutional perspective in the IB and GVC literatures.

ACKNOWLEDGEMENTS

The first author greatly acknowledges the grant for this research from the Ministry of Foreign Affairs of Denmark-DFC- 20-10-CBS. The authors are also grateful to Peter Lund-Thomsen from Copenhagen Business School and Olav Jull Sørensen from Aalborg University Business School for their insightful comments on the early draft of this chapter.

REFERENCES

AAU News. (2021). Retrieved January 30, 2021, from https://www.en.engineering.aau.dk/Cases/Power2X/

Agrawal, A., McSweeney, C., & Perrin, N. (2008). *Local institutions and climate change adaptation* [Social Development Notes No. 113]. World Bank, Washington, DC. https://www.unepfi.org/fileadmin/documents/FinancialInstitutionsTakingActionOnClimateChange.pdf

Allen, M. L., Allen, M. M. C., Cumming, D., & Johan, S. (2021). Comparative capitalisms and energy transitions: Renewable energy in the European Union. *British Journal of Management*, *32*(2), 611–629.

Allen, M., Rana, M. B., & Liu, J. (2019, July 4–6). *Comparative institutional analysis and institutional voids: The influence of firm heterogeneity and host-country institutions and contexts* [Conference presentation]. EGOS Conference, Edinburgh, United Kingdom.

Boje, D., & Rana, M. B. (2021). Defining a sustainability-driven business modeling strategy with a storytelling science approach. In S. Markovic, C. Sancha, & A. Lindgreen (Eds.), *Handbook of sustainability-driven business strategies in practice* (pp. 59–77). Edward Elgar. https://doi.org/10.4337/9781789908350.00014

Brun, L., Gereffi, G., & Zhan, J. (2019). The "lightness" of industry 4.0 lead firms: Implications for global value chains. In P. Bianchi, C. Ruíz Durán, & S. Labory (Eds.), *Transforming industrial policy for the digital age* (pp. 37–67). Edward Elgar Publishing. https://doi.org/10.4337/9781788976152.00008

Buckley, P. J. (1993). The role of management in internalisation theory. *MIR: Management International Review*, *33*(3), 197–207. http://www.jstor.org/stable/40228157

Buckley, P. J., & Strange, R. (2011). The governance of the multinational enterprise: Insights from internalization theory. *Journal of Management Studies*, *48*(2), 460–470.

Cantwell, J., & Mudambi, R. (2005). MNC competence-creating subsidiary mandates. *Strategic Management Journal, 26*(12), 1109–1128. http://dx.doi.org/10.1002/smj.497

Chakrabarty, S., & Wang, L. (2013). Climate change mitigation and internationalization: The competitiveness of multinational corporations. *Thunderbird International Business Review, 55*(6), 673–688.

Coase, R. H. (1998). The new institutional economics. *American Economic Review, 88*, 72–79.

Deeg, R. (2005). Path dependency, institutional complementarity, and change in national business systems. In G. Morgan, R. Whitley, & E. Moen (Eds.), *Changing capitalisms? Internationalization, institutional change and systems of economic organization* (pp. 21–52). Oxford University Press.

Dovers, S. (2001). Institutions for sustainability. *Tela Series: Environment, Economy and Society, 7*, 13–15. Retrieved January 25, 2022, from https://openresearch-repository.anu.edu.au/bitstream/1885/40972/3/een0101.pdf

Doz, Y., & Prahalad, C. K. (1980, March). How MNCs cope with host government intervention. *Harvard Business Review*. https://hbr.org/1980/03/how-mncs-cope-with-host-government-intervention

Dunning, J. H. (1977). Trade, location of economic activity and the MNE: A search for an eclectic approach. In B. Ohlin, P. O. Hesselborn, & P. M. Wijkman (Eds.), *The international allocation of economic activity* (pp. 395–418). MacMillan.

Dunning, J. H. (1983). Market power of the firm and international transfer of technology: A historical excursion. *International Journal of Industrial Organization, 1*(4), 333–351. https://doi.org/10.1016/0167-7187(83)90002-4

Dunning, J. H. (1988). The eclectic paradigm of international production: A restatement and some possible extensions. *Journal of International Business Studies, 19*(1), 1–31. http://www.jstor.org/stable/154984

Dunning, J. (1998). Location and the multinational enterprise: A neglected factor? *Journal of International Business Studies, 29*,45–66. https://doi.org/10.1057/palgrave.jibs.8490024

Dunning, J. H. (2000). The eclectic paradigm as an envelope for economic and business theories of MNE activity. *International Business Review, 9*(2), 163–190.

Dunning, J. H., & Lundan, S. (2008a). *Multinational enterprises and the global economy* (2nd ed.). Edward Elgar.

Dunning, J. H., & Lundan, S. (2008b). Institutions and the OLI paradigm of the multinational enterprise. Asia Pacific. *Journal of Management, 25*(4), 573–593. https://ssrn.com/abstract=1490986

Dunning, J. H., & Lundan, S. (2010). The institutional origins of dynamic capabilities in multinational enterprises. *Industrial and Corporate Change, 19*(4), 1225–1246.

Durkheim, É. (1974) [1953]. *Sociology and philosophy* (D. F. Pocock, Trans., with an introduction by J. G. Peristiany). Free Press.

Eden, L., & Dai, L. (2010). Rethinking the O in Dunning's OLI/eclectic paradigm. *Multinational Business Review, 18*(2), 13–34. https://doi.org/10.1108/1525383X201000008

European Commission. (2021). European Green Deal: Commission proposes transformation of EU economy and society to meet climate ambitions. Retrieved August 14, 2021, from https://ec.europa.eu/commission/presscorner/detail/en/ip_21_3541

Gereffi, G. (2021). Foreword. In M. B. Rana & M. Allen (Eds.), *Upgrading the global garment industry: Internationalization, capabilities, and sustainability* (New Perspectives on the Modern Corporation Series, pp. xvi–xviii). Edward Elgar Publishing.

Gereffi, G. (2018). *Global value chain and development: Redefining the contours of 21st century capitalism.* Cambridge University Press.

Gereffi, G. (2002). *The international competitiveness of Asian economies in the apparel commodity chain* [EDB Working Paper Series No. 5]. Asian Development Bank, Manila, Philippines.

Gereffi, G., Humphrey, J., & Sturgeon, T. (2005). The governance of global value chains. *Review of International Political Economy, 12*(1), 78–104.

Gereffi, G., Pananond, P., & Pedersen, T. (2022). Resilience decoded: The role of firms, global value chains, and the state in COVID-19 medical supplies. *California Management Review, 64*(2), 46–70. https://doi.org/10.1177/00081256211069420

Gereffi, G., Pananond, P., & Pedersen, T. (2022). Resilience decoded: The role of firms, global value chains, and the state in COVID-19 medical supplies. *California Management Review*, *64*(2), 46–70. https://doi.org/10.1177/00081256211069420

Greenwood, R., Sahlin-Andersson, K., Suddaby, R., & Olver, C. (Eds.). (2012). *Institutional theory in organization studies*. SAGE.

Haxhi, I., & Aguilera, R. V. (2017). An institutional configurational approach to cross-national diversity in corporate governance. *Journal of Management Studies*, *54*(3), 261–303. https://doi.org/10.1111/joms.12247

H&M Foundation. (2024). Retrieved January 1, 2024, from https://hmfoundation.com/project/planet-first/

Ivang, R., & Rana, M. B. (2019). Better world fashion: Circular economy and competitive advantage. In *IVEY teaching case 9B19A021* (pp. 1–16). Western University, Ivey Publishing. https://www.iveypublishing.ca/s/product/better-world-fashion-circular-economy-and-competitive-advantage/01t5c00000CwovCAAR

Jackson, G., & Deeg, R. (2008). From comparing capitalisms to the politics of institutional change. *Review of International Political Economy*, *15*(4), 680–709. http://www.jstor.org/stable/25261991

Jacobsen, S., & Jensen, T. (2018). *Corrected: Google invests $700 million in Danish data centre, secures green energy*. Reuters. https://www.reuters.com/article/alphabet-denmark-idINKCN1NP0GI

Kazi, M. R., & Rahman, S. (2021). Governance mechanism for the sustainability of global garment production networks: Perspectives from rule-takers. In M. B. Rana & M. M. C. Allen (Eds.), *Upgrading the global garment industry* (pp. 362–387). Edward Elgar Publishing. https://doi.org/10.4337/9781789907650.00023

Khan, Z., Lew, Y. K., & Sinkovics, R. R. (2015). The mirage of upgrading local automotive parts suppliers through the creation of vertical linkages with MNEs in developing economies. *Critical Perspectives on International Business*, *11*(3–4), 301–318.

Kolk, A. (2016). The social responsibility of international business: From ethics and the environment to CSR and sustainable development. *Journal of World Business*, *51*(1), 23–34.

Kolk, A., & Pinkse, J. (2008). A perspective on multinational enterprises and climate change: Learning from "an inconvenient truth"?. *Journal of International Business Studies*, *39*, 1359–1378. https://doi.org/10.1057/jibs.2008.61

Kostova, T., Marano, V., & Tallman, S. (2016). Headquarters–subsidiary relationships in MNCs: Fifty years of evolving research. *Journal of World Business*, *51*(1), 176–184.

Kristensen, P. H., & Morgan, G. (2018). Danish foundations and cooperatives as forms of corporate governance: Origins and impacts on firm strategies and societies. In N. Boeger & C. Villiers (Eds.), *Shaping the corporate landscape: Towards corporate reform and enterprise diversity* (pp. 271–288). Hart Publishing.

Lee, J., & Gereffi, G. (2021). Innovation, upgrading, and governance in cross-sectoral global value chains: The case of smartphones. *Industrial and Corporate Change*, *30*(1), 215–231.

Lund-Thomsen, P., & Lindgreen, A. (2014). Corporate social responsibility in global value chains: Where are we now and where are we going?. *Journal of Business Ethics*, *123*(1), 11–22.

Marchant, N. (2021). *This is how climate change could impact the global economy*. World Economic Forum. Retrieved September 23, 2021, from https://www.weforum.org/agenda/2021/06/impact-climate-change-global-gdp/

Morgan, G. (2001). Transnational communities and business systems. *Global Networks*, *1*(2), 113–130.

Morgan, G. (2007). National business system research: Progress and prospects. *Scandinavian Journal of Management*, *23*(2), 127–145.

Morgan, G., Campbell, J. L., Crouch, C., Pedersen, O. K., & Whitley, R. (Eds.). (2010). *The Oxford handbook of comparative institutional analysis*. Oxford University Press.

Morgan, G., Whitley, R., & Moen, E. (2005). Introduction: Changing capitalisms? In G. Morgan, R. Whitley, & E. Moen (Eds.), *Changing capitalisms? Internationalisation, institutional change and systems of economic organisation* (pp. 1–20). Oxford University Press.

Nachum, L., & Zaheer, S. (2005). The persistence of distance? The impact of technology on MNE investment motivations. *Strategic Management Journal*, *26*(8), 747–767. http://dx.doi.org/10.1002/smj.472

Nohria, N., & Ghoshal, S. (1997). The differentiated network: Organizing multinational corporations for value creation. *Journal of International Business Studies*, *29*(2), 429–431.

North, D. (1991). Institutions. *Journal of Economic Perspectives*, *5*(1), 97–112. https://doi.org/10.1257/jep.5.1.97

News European Parliament. (2021). Retrieved January 01, 2021, from https://www.europarl.europa.eu/news/en/headlines/society/20201208STO93327/the-impact-of-textile-production-and-waste-on-the-environment-infographic

Okereke, C. (2006). Global environmental sustainability: Intragenerational equity and conceptions of justice in multilateral environmental regimes. *Geoforum*, *37*(5), 725–738.

Paul, R., & Das, K. N. (2022). *Bangladesh garment export growth seen slowing to "normal" 15% this year*. Reuters. https://www.reuters.com/markets/asia/bangladesh-garment-export-growth-seen-slowing-normal-15-this-year-2022-08-10/

Peng, M. W., Wang, D. Y. L., & Jiang, Y. (2008). An institution-based view of international business strategy: A focus on emerging economies. *Journal of International Business Studies*, *39*(5), 920–936. http://www.jstor.org/stable/25483310

PaCT. (2021). Retrieved February 06, 2022 from https://www.textilepact.net/

Pinkse, J., & Kolk, A. (2012). Multinational enterprises and climate change: Exploring institutional failures and embeddedness. *Journal of International Business Studies*, *43*, 332–341. https://doi.org/10.1057/jibs.2011.56

Ponte, S., Gereffi, G., &Raj-Reichert, G. (Eds.). (2019). *Handbook on global value chains*. Edward Elgar Publishing.

Romilly, P. (2007). Business and climate change risk: A regional time series analysis. *Journal of International Business Studies*, *38*, 474–480. https://doi.org/10.1057/palgrave.jibs.8400266

Rahman, F. M. (2020, November 18). Bangladesh gets $256.5m Green Climate Fund. *The Daily Star*, 1.

Rana, M. B. (2014). *Rethinking business system theory from the perspective of civil society, transnational community, and legitimacy: Strategies of European MNCs in Bangladesh* [Unpublished Ph.d. thesis]. Department of Business and Management, Aalborg University, Denmark.

Rana, M. B. (2015). Tri-space framework for understanding MNC behaviour and strategies: An institutionalism and business system perspective. In S. T. Marinova (Ed.), *Institutional impacts on firm internationalization* (pp. 299–333). Palgrave Macmillan.

Rana, M. B., & Allen, M. (2018). Business systems perspective on entrepreneurship. In R. Turcan & N. Fraser (Eds.), *The Palgrave handbook of multidisciplinary perspectives on entrepreneurship* (pp. 271–291). Palgrave Macmillan. https://doi.org/10.1007/978-3-319-91611-8_13

Rana, M. B., & Allen, M. (2021a). Upgrading the global garment industry: Internationalisation, capabilities and sustainability. In M. B. Rana & M. Allen (Eds.), *Upgrading the global garment industry: Internationalization, capabilities and sustainability* (p. 1). New Perspectives on the Modern Corporation series. Edward Elgar Publishing.

Rana, M. B., & Allen, M. (2021b). Why apparel suppliers are locked into the upgrading ladder in Bangladesh: A institutional and business systems perspective. In M. B. Rana & M. Allen (Eds.), *Upgrading the global garment industry: Internationalization, capabilities and sustainability* (pp. 138–161). New Perspectives on the Modern Corporation Series. Edward Elgar Publishing.

Rana, M. B., Allen, M., & Servais, P. (2021). Supplier's internationalization in global apparel value chain from Bangladesh to Ethiopia: Buyer's business model, institutions, and entrepreneurial capability. In M. B. Rana & M. Allen (Eds.), *Upgrading the global garment industry: Internationalization, capabilities, and sustainability* (pp. 13–45). New Perspectives on the Modern Corporation series. Edward Elgar Publishing.

Rana, M. B., Hoque, I., & Allen, M. (2019). *MNE's business model, governance & relational condition, and suppliers' capability development: A comparative study on cross border dyads* [Abstract]. European International Business Academy, Leeds, United Kingdom.

Rana, M. B., & Morgan, G. (2019). Twenty-five years of business systems research and lessons for international business studies. *International Business Review*, *28*(3), 513–532.

Rana, M. B., & Sørensen, O. J. (2021). Levels of legitimacy development in internationalization: Multinational enterprise and civil society interplay in institutional void. *Global Strategy Journal*, *11*(2), 269–303. https://doi.org/10.1002/gsj.1371

Rana, M. B., & Tajuddin, S. A. (2021). Circular economy and sustainability capability: The case of H&M. In M. B. Rana & M. Allen (Eds.), *Upgrading the global garment industry: Internationalization, capabilities and sustainability* (New Perspectives on the Modern Corporation Series, pp. 253–282). Edward Elgar Publishing. https://doi.org/10.4337/9781789907650.00019

Redding, G. (2005). The thick description and comparison of societal systems of capitalism. *Journal of International Business Studies*, *36*(2), 123–155.

Ripple, W. J., Wolf, C., van Vuuren, D. P., Gregg, J. W., & Lenzen, M. (2024). An environmental and socially just climate mitigation pathway for a planet in peril. *Environmental Research Letters*, *19*, 021001. https://doi.org/10.1088/1748-9326/ad059e

Sako, M., & Zylberberg, E. (2017). Supplier strategy in global value chains: Shaping governance and profiting from upgrading. *Socio-Economic Review*, *17*(3), 687–707.

Scott, W. R. (2001). *Institutions and organizations* (2nd ed.). SAGE.

Scott, W. R. (2008). *Institutions and organizations: Ideas and interests*. (3rd ed.). SAGE.

State of Green. (2021). Retrieved December 07, 2021 from https://stateofgreen.com/en/

Whitfield, L., & Staritz, C. (2020). The learning trap in late industrialisation: Local firms and capability building in Ethiopia's apparel export industry. *The Journal of Development Studies*, *57*(6), 980–1000.

Whitley, R. (1987). Taking firms seriously as economic actors: Towards a sociology of firm behaviour. *Organization Studies*, *8*(2), 125–147.

Whitley, R. (Ed.). (1992). *European business systems: Firms and markets in their national contexts*. SAGE.

Whitley, R. (1999). *Divergent capitalisms: The social structuring and change of business systems*. Oxford University Press.

Whitley, R. (2005). How national are business systems? The role of states and complementary institutions in standardizing systems of economic coordination and control at the national level. In G. Morgan, R. Whitley, & E. Moen (Eds.), *Changing capitalisms? Internationalization, institutional change, and systems of economic organization* (pp. 190–231). Oxford University Press.

Whitley, R. (2007). *Business systems and organisational capabilities: The institutional structuring of competitive competences*. Oxford University Press.

Whitley, R. (2010). The institutional construction of firms. In G. Morgan, J. L. Campbell, C. Crouch, O. K. Pedersen, & R. Whitley (Eds.), *The Oxford handbook of comparative institutional analysis* (pp. 422–453). Oxford University Press.

Williamson, O. E. (1975). *Markets and hierarchies, analysis and antitrust implications: A study in the economics of internal organization*. Free Press.

Witt, M. A., & Redding, G. (2014). *The Oxford handbook of Asian business systems*. Oxford University Press.

Zhan, J. X. (2021). GVC transformation and a new investment landscape in the 2020s: Driving forces, directions, and a forward-looking research and policy agenda. *Journal of International Business Policy*, *4*(15), 206–220. https://doi.org/10.1057/s42214-020-00088-0

CHAPTER 15

COME CLOSER! ON TRANSACTION COSTS AND SPATIAL CHOICES IN A CIRCULAR ECONOMY

Corina Fehlner

Department of Strategy and Entrepreneurship, BI Norwegian Business School, Oslo, Norway

ABSTRACT

This chapter analyzes the efficiency levels of a circular economy (CE) with an emphasis on transaction costs. It examines the governance aspect of CE activities in comparison to the predominant linear value creation. Extant CE research in business studies tends to be descriptive and lacks a theoretical foundation, particularly in understanding CE management. Transaction cost theory explains efficiency in economic organizing, lending itself to the study of arrangements that maximize resource efficiency at continued economic virtue. The conceptualization proposes that CE transaction costs are greater than those within the linear economy (LE), primarily due to the uncertainties about reciprocal dependencies, looping material complexities, exchanging novel information, and increased contracting efforts. Geographically bounded and institutionally homogeneous CE initiatives may curb these rising costs. By bringing efficiency concerns into CE analysis, the chapter demonstrates the applicability of transaction cost theory and highlights CE relevance to international business by pointing out spatial choice implications.

Keywords: Circular economy; transaction cost theory; multinational enterprise; spatial choice; international value chains

---

Walking the Talk? MNEs Transitioning Towards a Sustainable World
Progress in International Business Research, Volume 18, 295–318

ISSN: 1745-8862/doi:10.1108/S1745-886220240000018019

INTRODUCTION

The CE is a sustainable economic model in which recycled materials and products reduce the extraction and processing of virgin resources; this significantly reduces emissions, making CE implementation essential in addressing the climate emergency (Ellen MacArthur Foundation, 2019; Ghisellini et al., 2016; United Nations Environment Programme, 2019). The CE differs substantially from the predominant LE by prioritizing systemic resource waste reduction over a mere production focus, fundamentally transforming economic interactions (Ranta & Saari, 2019). Therefore, conditions for economic efficiencies in circular value loops may differ from what the theoretical insights accumulated from an LE context suggest.

In addition to the waste and emission reduction benefit of the CE, and the mitigation of the costs of our impact on the climate, it may also help economies benefit from growth due to additional value creation and macroeconomic cost savings in materials and energy. For example, US$4.5 trillion of additional value is at stake by 2030, and up to US$630 billion in annual cost savings are projected for the European Union (EU) manufacturing sector alone (Ellen MacArthur Foundation, 2013; Lacy et al., 2020; Tuladhar et al., 2016). However, a more comprehensive understanding of the CE efficiency effects also requires taking into account the organizational costs involved in such an economic model. A crucial barrier to CE implementation is the failure of market players to adapt their value chain interactions, or transactions (Kirchherr et al., 2018). Therefore, research points to transaction cost theory as a valuable perspective for addressing organizational aspects of the CE (Lahti et al., 2018; Liu et al., 2018).

The purpose of this examination is to investigate how the CE transition impacts organizational costs relative to LE and how these altered costs may be strategically addressed. While case-based studies provide interesting insights, there is a need to investigate the CE phenomenon more systematically and ground the analysis more firmly in theory-based perspectives (Korhonen et al., 2018; Lahti et al., 2018).

Multinational enterprises (MNEs) hold powerful levers to change entire value chains (Gereffi et al., 2005; Mitra & Datta, 2014; Parmigiani et al., 2011), and transitioning to CE involves significant transformative and adaptive changes in their business systems (Benito & Fehlner, 2022; Di Stefano et al., 2023). Yet, there is a surprising lack of international business publications about climate change and CE influences on MNEs (Geissdoerfer et al., 2017; Ghauri et al., 2021; Montiel et al., 2021), particularly given the fact that international business contributes disproportionately to climate change (Yu et al., 2023).

Transitioning to CE means establishing loops between actors that do not currently exist, enabling products and materials to rejoin the system after their initial use to slow wastage and close the resource loop, maximize their lifetime instead of generating ever greater volumes of non-reusable waste (Cleveland & Ruth, 1997; Ellen MacArthur Foundation, 2013; Ranta & Saari, 2019). This core underlying difference may affect theoretical predictions derived from an LE context and thus makes it vital to reconceptualize transactions and their costs in CE.

The complexity of circularity and the larger number of participating actors mean that transactions are no longer simple, dyadic, linear, and instant events between two parties, as in traditional global value chains. Instead, they become complex, multitudinous, and reciprocal, and can influence future interactions (Mathews, 2014; Stuchtey et al., 2016). Therefore, established transaction cost theory should be tested for applicability and potentially broadened and adapted for the new CE context (Ketokivi & Mahoney, 2017; Nygaard, 2019).

This conceptual analysis integrates the CE model and transaction cost theory to examine CE efficiency effects from an organizational perspective. Specifically, analyzing transaction costs assumes that all other cost parameters remain equal (e.g., resources, labor, production costs, and revenues). This study proposes that the CE-induced changes along the value loops increase transaction costs relative to LE, and suggests a model to explain how circular value loops change organizational costs and why geographic and institutional proximity enables more efficient CE transactions, thereby assuming consequences for MNEs' strategic spatial decisions. While various strategies to counter high transaction costs have been found to be feasible, anecdotally, pioneer CE loops are observed in locations with limited spatial distances, which motivates this research focus (Chertow, 2008).

This study offers a governance perspective on challenges around CE implementation and thereby makes multiple contributions. First, it aims to enrich circularity research by bringing efficiency concerns into the analysis; specifically, I analyze organizational costs of principal CE value loop activities and explain the underlying phenomenon of rising transaction costs that may hinder CE adoption. Second, the chapter is theoretically important as it propounds a reconsideration of the applicability of theories developed in an LE context to predict corporate behavior in the CE context. Finally, it develops the position of the CE model in international business literature, particularly by potentially impacting MNEs' spatial choices.

THEORY

Circular Value Creation

The CE is an overarching economic model designed to minimize environmental degradation, waste, and emissions by increasing the usage efficiency of resources and energy in design and production processes, as well as during consumption, by narrowing, closing, and slowing resource flows (Bocken et al., 2016; Esposito et al., 2018; Geissdoerfer et al., 2020; Kirchherr et al., 2017).

Even though the transitions necessary to reach a sustainable economic system are recognized, their implementation is mostly unresolved; one illustration of this is the fact that the global circularity level of material usage is only 7.2%, and shrinking (Circle Economy, 2023). Management research is also lagging; while private sector initiatives play a critical role in combating climate change (Ansari et al., 2013; Wittneben et al., 2012), management research has been slow to focus on this important area (George et al., 2015; Nyberg & Wright, 2020; Snihur & Bocken, 2022) to the point where there is minimal CE research in top international business and strategy journals (Geissdoerfer et al., 2017). Only recently, management studies have begun to focus on CE (Blackburn et al., 2022; Ciulli et al., 2020; Di Stefano et al., 2023; Esposito et al., 2018; Hofstetter et al., 2021). CE benefits include decreased production and usage costs, reduced resource requirements, and increased productivity (Montiel et al., 2021; Stuchtey et al., 2016). It also minimizes macroeconomic loss by regarding waste as a secondary resource (Esposito et al., 2018).

The predominant LE extracts natural resources and transforms them via production processes and often singular consumption into degraded waste output (Esposito et al., 2018; Korhonen et al., 2018; Murray et al., 2017). It neither recognizes the limits of irreversible and unidirectional material flows and accumulated waste output (Cleveland & Ruth, 1997) nor factors in the negative environmental impacts of waste. Instead, it sees a direct relationship between economic activity and the consumption of natural resources, accepting their finite, relatively scarce nature while implying that they cannot be substituted by artificial means (Ranta & Saari, 2019). The linear production model reveals two approaches that may facilitate a transformation into a more sustainable version; minimizing the degree of non-reusable waste and optimizing the resource usage until it eventually reaches the state of terminal waste (Murray et al., 2017).

The CE proposes that economic growth be decoupled from natural resource depletion and environmental degradation. In contrast to LE, the CE does not have a "net effect on the environment; rather it restores any damage done in resource acquisition while ensuring little waste is generated throughout the production process and in the life history of the product" (Murray et al., 2017, p. 371). The Ellen MacArthur Foundation (2013) developed a prominent model that incorporates the CE's three core principles: preserving natural capital, optimizing resource yields, and fostering system effectiveness. The first addresses controlling the extraction of finite natural resources to protect and save them and managing renewable resources in a balanced manner. The second principle targets the production and consumption phases of economic activity in biological and technical cycles and aims to optimize product, component, and material circulation while maximizing utilization, thereby encompassing the concept of "industrial symbiosis" (Murray et al., 2017, p. 371), product design adaptations (Esposito et al., 2018; McDonough & Braungart, 2010; Murray et al., 2017), and consumer behavior (e.g., Ciulli et al., 2020). The third principle is about minimizing systemic leakage and negative externalities; in other words, limiting and slowing material release into biochemical cycles to ensure that the planetary ecosystem remains within healthy boundaries.

Efficiency improvements in production and recovery processes can limit environmental impact. This study focuses on efficiencies in material circulation between economic actors along the value chains of manufacturing MNEs. The CE's goal is to increase resource and energy usage efficiency in production processes, consumption, and recovering waste as secondary raw materials (Esposito et al., 2018). Core corporate CE strategies include narrowing, slowing, and closing resource flows, which are achievable by adapting product designs and business models, all of which have implications for several activities along the traditional linear value chain (Bocken et al., 2016; Geissdoerfer et al., 2017).

Narrowing resource flows contributes to CE efficiency by reducing resource usage. Products can be designed to require fewer resources per unit and consumer needs are addressed with fewer products through, for example, encouraging sharing strategies to displace one-time product sales. The overall reduction of resources is distinct from the closing and slowing strategies that target resource circulation.

Slowing resource loops means extending product usage periods by building in longevity and enabling life extensions through maintenance, repairs, refurbishments, and upgrades. Reusing the same product many times is the smallest loop, embodying the most sustainable form of circulation in the waste hierarchy, but it does demand novel loops between users or service providers (Kirchherr et al., 2017; Reike et al., 2018).

Closing resource flows is about remanufacturing or repurposing products or components, as well as recycling expired material. It depends on establishing a novel link between post-use and inbound logistics, currently unknown in the LE, to recover and recycle material and circulate resources. A core concept for closing the loop is industrial symbiosis, which involves minimizing waste and emissions while creating mutual benefits for all the economic actors involved (Kokoulina et al., 2019). Industrial symbiosis advances if "traditionally separate industries [engage] in a collective approach to competitive advantage involving physical exchange of materials, energy, water and by-products" (Chertow, 2000, p. 314). In this symbiotic process, one company's waste becomes a secondary resource for another, thereby closing the systemic loop (Esposito et al., 2018; Stuchtey et al., 2016).

A broadly accepted generic concept of the value chain was developed by Porter (1985). While this view has been advanced and nuanced, i.a., by Stabell and Fjeldstad (1998), it is a basic structure for analyzing main firm activities, despite building on an LE. Porter differentiates between primary and support activities. Primary activities are directly associated with a product's physical creation, sales, and after-sales, are of direct value to the customer, and include five core activities.

Inbound logistics describes the management of inputs and materials at the production site and implies collaboration with suppliers; operations are the actual manufacturing activities, while outbound logistics relate to the distribution of the final product; marketing and sales include all efforts to advertise and sell the product; finally, after-sales services include repairs and the supply of replacement parts to maintain the product value.

These five core activities can only operate effectively when they are supported by secondary activities. These include procurement, which is organizing and negotiating the purchase of material inputs from suppliers; technology development, which comprises designing, engineering, and improving products and processes; human resource (HR) management embodies the hiring, training, compensation, and, where necessary, the firing of personnel; corporate infrastructure covers general strategic, financial, quality, legal, and lobbying activities that support value creation.

Porter points out that these activities are interdependent so that choices and changes influence cost or performance along an MNE's value chain and those of its partners, demanding coordination. Notably, the value chain logic neither considers nor discusses the product's end-of-life stage, as is typical for the LE.

MNEs' activities and transactions characteristically spread across national boundaries. Consequently, their value chains are globally organized based on the advantages created from cross-border differences in terms of costs, competencies, and resource availability. Much of their activities are carried out outside the traditional enterprise, requiring the MNE to coordinate value chain activities between independent network partners (Buckley & Ghauri, 2004; Gereffi et al., 2005). This chapter adopts the perspective of multinational lead firms in the manufacturing sector to understand CE transition.

Based on the distinct CE characteristics, it is evident that economic activities must be adapted to maximize utilization and circulation, and the upstream and downstream value chains must be redesigned. These adaptations will likely provoke differences in economic organization efficiencies, which will be analyzed in the following sections.

Transaction Cost Theory

Transaction cost theory is concerned with the economic organization of transactions between market actors, thereby corresponding to the CE goal of reducing waste (Ketokivi & Mahoney, 2017). Transaction costs arise at trading in the market. The focal unit of analysis is the individual transaction, defined as an exchange agreement for goods or services between actors from two stages within the value chain (Macher & Richman, 2008; Shelanski & Klein, 1995). Transaction costs are different from production costs (Ketokivi & Mahoney, 2017; Williamson, 1975, 1979, 1985) and can be categorized as either direct or indirect costs, and opportunity costs in the case of inferior economic decisions. Direct transaction costs include transportation and tariffs, and costs of switching to another partner. Adjustment costs can be regarded as a "subcategory of transaction costs," with their differences not clearly conceptually separated (Argyres et al., 2019, p. 365). Indirect transaction costs arise, for example, when negotiating and drafting contracts between business partners, searching for information, or encountering a power imbalance due to information being withheld; they also include communication charges, or potentially paying lawyers to follow up on a breached agreement (Rindfleisch & Heide, 1997). Some transaction costs occur before an organizational choice is made (e.g., information gathering and decision-making), and others after the decision has been taken (e.g., monitoring, coordinating, or enforcement costs) (Coase, 1937).

The efficient organization of transactions can lower their costs, particularly in complex environments. Making appropriate choices around vertical integration, diversification, or internationalization, for example, can minimize the cost of transactions (Cuypers et al., 2020). According to Williamson (1975, 1979), there are three critical dimensions of a transaction that can help navigate an organizational decision; the specificity of the underlying asset being transacted, the uncertainty of the transaction, and the potential frequency of its recurrence.

In terms of specificity, the asset can be standardized, mixed, or specifically adapted for, and unique to, the transaction, in which case its value as compared to an alternative transaction becomes considerably lower, meaning that high asset specificity drives transaction costs unless organized under common ownership. Specific assets may include a production site, human capital development, or a tailored physical good that cannot easily be transferred (Williamson, 1991).

Naturally, transactions that are certain create little in the way of extra transaction costs, but handling costs, for example, increase in line with the degree of uncertainty. Uncertainties become expensive, especially for complex transactions and asset-specific investments.

Finally, transactions can be isolated, one-off events, or they can be occasional and in small volume, but more often, they happen more than once, or they recur regularly.

In addition to the contextual factors, human nature can also generate transaction costs. The theory builds on two behavioral assumptions.

First, actors can be affected by serving their self-interest ("opportunism") and they may exploit relationships to their favor. Identifying this behavior and finding trustworthy partners or monitoring their conduct comes at a cost. The concept of opportunism has been extended in internalization theory – a branch of transaction cost theory with widespread use in international business – to "bounded reliability," which denotes people's proclivity to reliably fulfill their commitments, but only to a limited extent so that, sometimes, promises remain unfulfilled, either intentionally or to counter over-commitment. This behavior results in a loss in terms of efficiency and effectiveness and therefore demands costly safeguarding (Verbeke & Greidanus, 2012). The second behavioral assumption is that people frequently demonstrate limited rational judgment and a constrained cognitive capability ("bounded rationality"), a condition that reduces the ability to make optimal decisions and raises information-gathering costs (Williamson, 1975). The latter is particularly problematic for uncertain transactions (Rindfleisch & Heide, 1997).

The application of transaction cost theory allows for identifying levels of organizational efficiency in CE which, in turn, can be seen as cost drivers that may initially delay the transition to CE. In initial case studies, Seuring (2004) finds that increased coordination efforts between the focal company and its suppliers demand the integration of transaction costs when analyzing ecological life cycles. Werning and Spinler (2020) find that an intensified interconnectedness between teams, as well as a need for such teams to be large and interdisciplinary, result in greater transaction costs in CE.

Yet, conventional transaction cost theory has been developed within the LE paradigm and assumes single, dyadic transactions in linear flows. The following section examines the theory's applicability in CE, in which long-term, reciprocal relationships develop over time.

TRANSACTION COST ANALYSIS AND CE VALUE LOOPS

The CE differs from the LE because it aims to narrow, slow, and close resource cycles, a goal that can only be reached through extended coordination between value chain participants and their activities. An early empirical example from Werning and Spinler (2020) of applying transaction cost theory in the CE context suggests that initial transaction cost surge as a result of the need for complex and costly activities by large interdisciplinary teams in striving to overcome barriers in interconnected exchange and supply loops.

Value loop adaptations focus on the early- and late-stage activities that are required to establish new closing loops within and between each other, as well as slowing strategies that significantly transform marketing and sales activities. The corresponding transactions differ from those in linear systems, where they are characteristically one-off, unidirectional, and isolated, and this demands a critical review of the applicability of transaction cost theory in the CE. Specifically, positive transaction costs in a CE context may have multiple stimuli, subsequently discussed, that rely on the traditional value chain activities in the manufacturing sector (Porter, 1985).

As suggested by transaction cost economics, these costs are influenced by the choice of adequate governance modes (Williamson, 1975). So, the absolute level of the costs in the subsequent analysis differs depending on their organizational contexts. Transaction costs in CE value loops established with the central orchestration of MNEs may be relatively lower than in settings with more dispersed transactional partners. However, this conjecture may depend on how the advancement of CE develops and the nature of the principal activities being performed. First, hierarchical CE organization may be more economical until uncertainties, informational asymmetries, behavioral challenges and contractual challenges are resolved and market governance becomes advantageous. Second, CE value loops that focus on activities close to the core of the product, such as remanufacturing or repairs, may be less costly to perform within the centrally coordinating brand-owning MNE, than activities at the periphery of manufacturer's influence and expertise, such as recycling or energy recovery at the end of (first) life.

Inbound Logistics and Procurement

The transition to CE impacts the physical management of inbound logistics and procurement. Physical resource input differs fundamentally from LE because it is cycled, not virgin material. Therefore, the flow and management of secondary resource inputs demand additional coordination with the end-of-life activities, either internally within the company or with suppliers or other partners (Ranta & Saari, 2019).

Recognizing alternative material sources requires the establishment of reverse supply chains and industrial symbioses, neither of which are a consideration in the LE. Reverse supply chains enable products, components, and materials to be recovered and fed back as part of the slowing and closing strategies (Govindan et al., 2015; Guide & Van Wassenhove, 2009). Analyses typically take a product perspective and assume a return to, and re-usage by, the original manufacturer (Cannella et al., 2016). Yet, broader material exchanges and circulation require industrial symbioses between various network actors from traditionally separate industries (Chertow, 2000). Tools to drive industrial symbioses include input–output matching, stakeholder support, material flow tracking, synergistic collaboration planning, and consideration of new and existing operations (Chertow, 2000).

The additional loops add complexity to the economic system because a larger number of actors are required to organize, i.a., waste collection, reverse logistics, and remanufacturing (Guide & Van Wassenhove, 2009). The absolute number of transactions also rises, leading to higher total transaction costs. Furthermore, industrial symbiosis requires a higher level of coordination effort than in an LE context as the forward and reverse chains increase interdependencies between the value chain activities of the focal company and its secondary resource providers (Chertow, 2000). Instead of a series of one-way, linear transactions between two parties, the entire circle of multiple reciprocal interactions must be considered (Mathews, 2014; Stuchtey et al., 2016).

In the following, I analyze how coordination issues affect transaction costs. Identifying adequate secondary resource suppliers results in relatively high search costs as the CE is most efficient if it embraces diverse industrial players for the industrial symbioses (Chertow, 2000), and information on the practices of previously unrelated industries must be gathered from scratch. Another central cost driver in a CE context is negotiating and setting up contracts with secondary resource suppliers that underpin new, reverse links with other value chain activities and new symbiotic industrial players. The resulting contractual costs may be disproportionately high as the material loops are unprecedented and the processes complicated because they need to respect the timing, quantity, and quality of material supplies. Then, such collaborative agreements throughout the circular system that enable the collection of products and exchange of materials create additional bureaucratic efforts for the procurement team, thereby evoking additional costs (Williamson, 1985). There is also likely to be an increase in effort and cost regarding inbound logistics and inventory management to conform with the new input sources. Once loops are established, it is crucial to exchange information frequently and transparently (Cannella et al., 2016), otherwise, information asymmetries and power imbalances arise between reciprocally dependent actors, increasing transaction costs even further (Guide & Van Wassenhove, 2009).

The concept of adjustment costs can be seen as a subset of transaction costs, as Argyres et al. (2019) pointed out, yet the difference between these two concepts is not always clearly defined. In the context of a transition to the CE, the focal company may need to invest in novel value chain links and knowledge of the new economic system. These adjustments can be costly, presenting a barrier to the transition (Kirchherr et al., 2018). Other partners in the value chain may also need to make adaptations, but their reluctance to invest may necessitate contractual safeguards for the lead company and thereby increase costs. Thus, the challenges associated with adjustment costs can impact both the lead company and its partners in the value chain, highlighting the importance of carefully managing this aspect of a CE transition.

In symbiotic industrial processes, waste is a secondary resource that may have to be tailored to the buyer's needs, unless it is a standardized raw material. This negotiation will bring additional costs to CE transactions. Once the material, a physical asset, is specified for the CE partner, flexibility to respond to changes in the market is reduced, and dependency is increased, demanding further contractual safeguards that raise transaction costs. Operationally, the recollected material must be tested, sorted, and potentially repaired by inbound logistics personnel to ensure specific value for the operational activities (Guide & Van Wassenhove, 2009). Changing from primary to secondary resources induces uncertainties, a primary driver for transaction costs. Steady resource availability is (initially) less certain than in traditional supply chains, particularly when there are standardized quality expectations to be met. Potential supply irregularities, or even scarcity, may be an issue to be factored into transaction costs. Although Cannella et al. (2016) suggest that resource stability improves as circular supply loops grow in quantity, reducing uncertainty over time, the initial uncertainty implies an intensified need for quality control as part of inbound logistics, so monitoring and potentially contractual enforcement costs may be higher than in the LE as resource suppliers are likely to be more dispersed and novel.

Participating in the "green economy" is increasingly expected from a societal perspective, and this implies a risk that CE partners may take advantage of such expectations and behave opportunistically. For example, contractors and sub-contractors may exploit the lead firm's willingness to pay extra to create a green image by reducing its CO_2 footprint by sourcing cycled materials (Nygaard, 2019). Potential bounded reliability behaviors on the part of cycled material suppliers may play a more significant role in a CE if the material is specifically tailored for a specific manufacturing process and not readily available from alternative sources. Such instances of unfulfilled commitments may be more frequent and problematic in a CE as a novel economic system demands increased commitment by decision-makers.

Finally, adding reverse supply chains to an economy doubles transportation occurrences and presumably transportation costs. However, engaging in symbioses with industrial partners and cycling existing materials, instead of excavating and processing natural resources, might mitigate these increases in transportation. On the other hand, in a sustainable economy that is designed to reduce the ecological footprints of the participants, emissions might be subject to heavier taxation, and fuel-based transportation costs might rise.

In sum, closing the resource loop with reverse supply chains and industrial symbiosis intensifies coordination demands and, therefore, transaction costs (Domenech et al., 2019). The procurement and management of cycled, secondary resources appear to demand greater efforts and come at a higher organizational cost than do linear resource inputs. Higher material loop-closing costs at the early stages of the value chain (inbound logistics and resource procurement) may be mitigated by slowing and narrowing CE strategies since slowing product deterioration demands increased activity at the sales and service stages. If products can be redesigned to reduce total resource use by the technological development team, then transaction costs will naturally decrease at the inbound stage.

Operations

Reducing resources and avoiding waste through optimized production processes have been central to the cleaner production movement since the 1990s. In accordance with these roots, the CE principle of narrowing contributes to efficiency goals by reducing overall resource usage. In the adaptation to a CE at the operations stage in the value chain, compared to resource-efficient linear operations, any potential transaction cost differences are assumed to be negligible for the purposes of this analysis.

Outbound Logistics

Outbound logistics are concerned with the distribution of the final product and may need to be slightly adapted to incorporate reverse supply chains; an amended contract with wholesalers that accounts for product or component re-integration could include the final pickup. Logistically, used goods should be collected by the same vehicle that delivers new products; indeed, additional transportation demands in a cycled economy may motivate haulage businesses and their clients to optimize their distribution and collection routes to reduce transaction costs. With increasing fuel prices in a LE, these aspects do not appear to differ significantly in CE.

Marketing, Sales, and Service

Adapting one-time sales business models that rely on limited use by a single customer is critical for effective CE transformation (Geissdoerfer et al., 2017). Systemic CE business models maximize usage intensity while maintaining the resources as long as possible or dematerialize usage altogether (Bocken et al., 2016; Geissdoerfer et al., 2020). This novelty affects sales and service activities and creates interrelations, which demands a joint analysis of these activities.

The "slowing" strategy requires the creation of novel loops between users sharing the same physical product in intensity and over time, thus maximizing resource usage and reducing the total number of products needed (Kirchherr et al., 2017). The manufacturer's business model changes from selling product ownership to providing consumers access to product utilization, thus intensifying the use of under-utilized products (Hamari et al., 2016). In this "product-as-a-service" (PaaS) scenario, product ownership and responsibility remain with the manufacturer. Marketing activity to promote new product usage becomes necessary and implies CE adjustment costs for MNEs in transition. From a sustainability perspective, providing easy access to products may provoke over-usage, a distraction from public alternatives, or even incentivize purchasing, so sharing CE business models must be carefully designed (Ciulli & Kolk, 2019). Providing access instead of ownership also demands changes in the contractual setup between producer and customer. Instead of a one-time sales contract, PaaS means more transactions and more complex and novel contracts, the crafting of which may need to be learned, driving transaction costs (Lahti et al., 2018). The increased frequency of transactions adds to the transaction costs, over and above those in the LE, and information on current product availability will have to be constantly updated to stay relevant as a sharing service provider. Potential customer misbehavior during usage is another transaction cost driver; when the customer does not own the product, this may lead to mistreatment, damage, or destruction. Therefore, the contracts need to safeguard the product owner against potential "user opportunism" (Ter Huurne et al., 2017), and further costs will arise in the event of needing to enforce the contract or otherwise control consumer behavior.

An alternative CE business model to the sharing economy incentivizes a traditional owner to extend their product's life (Bocken et al., 2016), but again, encouraging a change in consumer behavior stipulates additional and adjusted marketing activity. Furthermore, continuous re-usage of products results in additional demand for components and upgrades, spiking transactions between customer and manufacturer and adding importance to a well-functioning after-sales and service business. The significance of these traditional value chain activities rises, and their development needs to be accounted for in terms of CE knowledge creation. With an extended customer relationship, complexity increases, adding transaction costs to the service side of the value chain.

Sharing and life-extending CE business models enable manufacturers to regain the product, effectively as a bundle of resources, at the end of its (first) life, allowing them to maintain, repair, or potentially remanufacture it and integrate it into a new customer relationship loop. Consumer upgrades are sometimes possible without replacing an entire product, thereby connecting service and sales activities. However, like any change designed to reduce resource usage, adequate overview and control are required to create planning certainty. Depending on the resource quality, reverse supply chains back to the inbound logistics and procurement activities might become necessary (Guide & Van Wassenhove, 2009).

Whether products are shared between users or product life is extended by manufacturer upgrades, transportation and logistics efforts will rise, meaning that direct transaction costs for shared and regularly upgraded products in the CE are higher than in linear business models.

New business models that are designed to slow down resource usage and close material loops go hand-in-hand with new methods to create value and generate profit (Geissdoerfer et al., 2020). This transition has implications for internal processes and external supplier and customer relations, and it might even demand new partnerships (Diaz Lopez et al., 2019). Any change in a business model carries uncertainties, which increases transaction costs. Marketing, sales, and service activities need to relate to new value propositions and intensified customer relations. To counter these uncertainties, safeguarding contracts become necessary, and knowledge of these novel relationships must be built up. The complexity of ongoing transactions, instead of one-time sales, increases transaction costs for marketing, sales, and service activities in a CE, relative to the LE.

Human Resources

The core activities of hiring, training, and compensating competent personnel do not significantly change in a CE. However, HR must work more closely with internal colleagues along the value loop, their counterparts in external partner firms (enabling industrial symbiosis), and consumers (supporting new business models). HR activities need to facilitate internal and external multi-actor collaboration to a higher degree than in the LE, and this includes creating adequate working processes and infrastructure for sharing information and collaboration (Fjeldstad et al., 2012). Of course, intensified collaboration creates a complexity that may lead to higher transaction costs than in a LE, while new ways of working also imply costs in building up knowledge and learning to adapt, again demanding collaboration along the value loop. These necessary and costly structural changes may impede the CE transition (Domenech et al., 2019).

Technical Development

The CE strategies of narrowing, closing, and slowing resource use demand adaptations in the design of products (to make them durable, long-lasting, and repairable) and processes (Bocken et al., 2016; Esposito et al., 2018; McDonough & Braungart, 2010; Murray et al., 2017). Engineering products that incorporate sustainable material choices, upgradability, and modularity are critical to CE implementation, and service design enables the transition from sales to service business models.

These product design adaptations come with adjustment costs that firms transitioning from an LE to a CE must carry (Argyres et al., 2019) in addition to search costs to build up a new knowledge base around, for example, alternative material from secondary resources, or durable product design for alternative use cases, such as sharing services. Finally, it is worth noting that some products may be so heavily customized for novel CE business models that a return to an LE paradigm may be difficult, and the uncertainty related to asset specificity may lead to additional transaction costs arising from the technical development in a CE.

Infrastructure

A company's transition toward a circular value creation model demands its supporting infrastructure be adapted correspondingly.

The CE necessitates the development of adapted or new partnerships with various entities including suppliers, symbiotic industry partners, and loop operators such as material collectors, public service entities, or digital platform providers (Bressanelli et al., 2018; Hansen & Revellio, 2020). The firm's strategy and legal departments need to find suitable partners, negotiate collaboration conditions, and set up contracts. As with all new partnerships, uncertainties and potential misbehavior must be considered and factored in, further adding to transaction costs. In contractual breaches, the legal team is responsible for contractual enforcement. Besides new business-to-business relationships, the legal contracts for access-based customer relations must be drafted and enforced.

Faced with rising transaction costs and other transition barriers, MNEs are likely to ramp up their lobbying activities (which carry their own transaction costs evoked by information gathering or relationship uncertainties) to obtain policy support for the CE transition and embed their CE projects into a broader ecosystem.

Increased Transaction Costs in CE

The above transaction cost analysis for CE value creation suggests increasing organizational costs for almost all activities, as compared to the LE; in particular, the CE strategies of closing and slowing resource flows increase transaction complexity and uncertainty. Porter (1985, p. 74) states that LE value chain activities are interdependent through "vertical linkages"; in CE, reciprocal dependencies are even higher. First, closing the resource supply loop depends on inbound logistics and procurement to gain access to secondary resources that can rejoin the loop through reverse supply chains, linking to sales and service activities, or collaborating with other industries through industrial symbiosis. Second, based on innovations by the technical development activities, sales and service activities engage in adapted and novel customer relationships to slow down resource flows with access-based business models. In sum, shifting the focus from one linear product life controlled by a lead firm to increased cooperation between multiple actors from different industries and value chain sections to slow and close the resource loop seems to increase transaction costs. This leads to the first proposition (see also Fig. 15.1):

Proposition 1. Transaction costs in a CE are relatively higher than in a LE for most value chain activities.

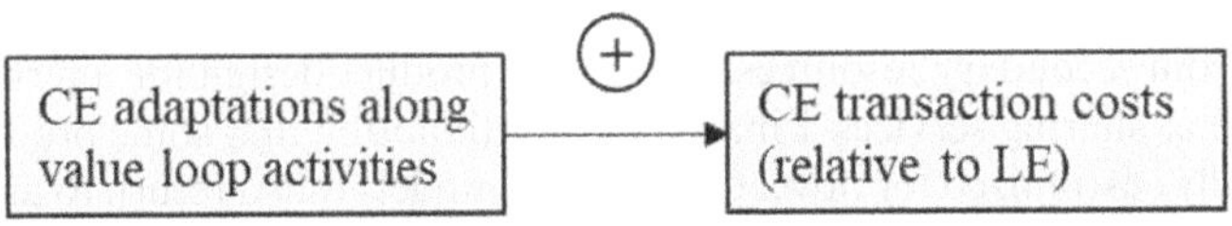

Fig. 15.1. Transaction Costs for CE Activities Relative to LE.

SPATIAL CLOSENESS AS AN ENABLER OF EFFICIENT CE ADVANCEMENT

Transaction costs can be mitigated by making efficient strategic choices in areas such as firm boundaries in terms of vertical integration and product diversification, inter-firm relationships such as alliances or supply chain connections, organization-internal structures, and internationalization decisions around market entry and operation modes (Cuypers et al., 2020). Location choices might be activity-specific and bound to certain circumstances and can either cross national boundaries or be concentrated in geographically bounded settings. In this context, location choices are about defining the spatial distances between value activities, and the choices lie on a continuum between those locations that are *far away* to those *co-located*, instead of concrete *places*, as is typically the case in international business. Williamson (1996) states that geographically co-located economic partners have chosen "site specificity," a distinct type of asset specificity designed to reduce transportation and inventory costs.

The globalization project has recently stagnated (Gygli et al., 2019; OECD, 2020), and MNEs are now considering reconfiguring their value chains to achieve efficiencies in these changing circumstances, for example, by de-internationalizing, defined as actions "that reduce a company's engagement in or exposure to current cross-border activities" (Benito & Welch, 1997, p. 9). While international business research mainly engages with market entry and expansion analysis, the globalization of supply chains may now be going into reverse, with an increased focus on nationalism and sustainability predicted to cause higher levels of geographical concentration (Herczeg et al., 2018; UNCTAD, 2020). The effect of reshoring and localizing value chains on environmental sustainability is as yet undetermined (Benito et al., 2022; Gupta et al., 2021).

Anecdotally, most CE pilots appear in geographically bounded areas such as Chinese industrial parks or regional industry coalitions in Germany (IN4climate. NRW, 2020; Mathews et al., 2018). Such geographically bounded and collaborative economic arrangements represent a departure from global value chains consisting of largely independent actors.

Based on the above transaction cost analysis, the following section discusses why spatial closeness may have a moderating effect on the rising CE transaction costs laid out in the previous sections, and thereby facilitate efficient CE advancement.

Uncertainty in CE

Any new economic system is less well understood than the current LE paradigm, generally increasing uncertainty levels, and most firms, with the possible exception of comprehensive conglomerates, will struggle to implement CE alone. Instead, the CE introduces high and complex dependencies into a network of suppliers, customers, partners, and other stakeholders with complementary goals. Both internal and external collaboration are critical to an efficient CE, which is likely to weaken MNE's control and create uncertainties.

Inbound logistics and procurement activities synchronizing secondary resource supply with new partners are essential to close material loops, but the key concept of industrial symbiosis may provoke planning uncertainties and chain complexity, driving transaction costs. Chertow (2000) explicitly specifies "geographical proximity" (co-location) as an advantageous condition for industrial symbiosis, and subsequent studies support this claim (Boons et al., 2011; Domenech et al., 2019). Primarily high-value, scarce resources, such as rare earth elements and critical metals of standardized quality, thereby curbing uncertainties, are traded on global markets (Domenech et al., 2019).

Slowing CE strategies, such as pivoting to novel PaaS business models, also carry uncertainties in terms of marketing, sales, and service. The transition to new service-based value propositions will create intensified customer relations and reciprocal collaboration. Customers may share products with local peers as part of an access-based approach, and local experts perform repairs. As ownership stays with the producer, they are responsible for managing the resulting complex stakeholder relationships.

The need for intense collaboration is a primary driver of uncertainty in a complex CE, yet collaborations can be more predictable at the local level, so geographically limited economic activity allows for better integration and collaboration along the value loop (Theyel & Hofmann, 2021). Geographic proximity boosts the intensity and variety of inter-firm collaboration, thereby reducing transaction costs (Theyel, 2012). The close links between agglomerated (rather than dispersed) suppliers enable efficient industrial symbioses between secondary resource providers and access-based business model participants.

Information and Knowledge in CE

Information search is a central CE transaction cost driver at the setup and operation stages. To maximize the benefits of industrial symbiosis, procurement, engineers, and strategists must understand resource availabilities, side-products, and practices of previously unfamiliar industries and companies (Boons et al., 2011). Designers then strive to develop alternative materials and durable constructions, sales and services create new business models, and HR creates collaborative working environments with new partners. During CE operations, exchanging information between involved stakeholders is critical to allow efficient and effective closing and slowing activities. Industrial symbiosis depends on timely and accurate information regarding material flows, sharing services, repairs, product condition, and customer relations. An asymmetry in knowledge availability creates additional hidden costs and limits CE efficiency.

Spatial closeness can contribute to effective information building and sharing, particularly as product and process innovation benefits from exchanges, discussions, and initial joint testing, all eased by co-location choices (Theyel, 2012). Learning and adjustments in a CE depend on collaborations between activities and firms and can benefit from local alliances to pilot the new economic approach (Kirchherr et al., 2018). The geographically bounded advantages in knowledge transfer arise from the mechanism of socialization, in which shared values and similar organizational cultures facilitate exchanges (Zeng et al., 2018). These informal conditions are more likely provided in local proximity, reducing transacting costs (North, 1990). In a recent study, Theyel and Hofmann (2021) find that reshoring and the geographic concentration of corporate activities increase agility and innovation potential, both of which are necessary for a CE transition.

Behavioral Challenges in CE

Human misbehavior drives transaction costs, and in a CE, several economic activities are potentially exposed. For example, new resources and value chains are prone to opportunistic supplier behavior at inbound logistics, as these actors potentially sense an overpriced outlet for their by-products. Transaction costs rise significantly if tailored cooperation becomes subject to unreliable behavior and equivalent alternatives are unavailable. In a PaaS business model, products that remain within the ownership and control of the manufacturer are vulnerable to customers behaving opportunistically and damaging products that do not belong to them.

In geographically bounded settings, human misbehavior may be limited for two reasons. First, possible synergies around materials or by-products, cooperation partners, and PaaS offers, are limited to a local level, which may reduce opportunistic and untrustworthy behavior because it may backfire and become common knowledge within the local area. This would then lead to the bad actor being frozen out of the local CE initiatives. Both customers and companies are mutually dependent on the success of access-based business models and would both suffer if such a model collapsed. Second, the literature suggests that trust counters opportunism (Verbeke & Greidanus, 2012). Trustworthy personal relations are easier to build and maintain in geographic and institutional proximity (Theyel, 2012), and a physical and psychological closeness makes it easier to predict behaviors when circumstances become unforeseeable and challenging. However, trust and local restrictiveness may raise the challenges that follow with embeddedness (Granovetter, 1985). Embeddedness is generally valuable as it facilitates interacting among actors, but it also has a downside, as it may restrict attention. CE opportunities may be missed if there is too close a focus on trusted peers.

Contractual Setups in CE

Introducing novel links between CE partners, such as material suppliers, collectors, remanufacturers, and customers, requires multiple new contracts, and contractual partners operating in the same formal and informal institutional context may somewhat simplify negotiations and contract drafting (North, 1990). Fewer requirements for legal adaptations to deviating national laws and a similar cultural understanding of negotiation customs facilitate the contract development process and lower transaction costs. Operating in familiar legal contexts also eases potential contract enforcement efforts.

Transport in CE

Establishing reverse supply chains instead of managing linear, one-way ones naturally increases the number of transportation legs. Inbound and outbound logistics should be coordinated to optimize routes, while businesses deploying successful, access-based business models should strive to offer the best possible levels of availability and product condition, involving back-office logistics. PaaS companies must also factor in the fact that their products are being intensely shared by users, and that will require the products to be regularly returned to the owner for repairs and upgrades.

The more geographically concentrated these processes are the lower the direct transaction costs; they may even fall below those of the LE, where natural resources are often sourced and processed in multiple distant locations.

Transaction Cost Drivers in a Spatially Bounded CE

One way to boost collective efficiency in a CE is to reduce transaction costs. Transaction cost theory fundamentally focuses on efficiency in organizing transactions (including waste minimization) by choosing the adequate organizational form (Ketokivi & Mahoney, 2017; Williamson, 1985). Relying on transaction cost theory to analyze local CE advantages suggests that initiating CE in a spatially bounded manner may be the most efficient approach. Drawing on Williamson (2000), geographically bounded and institutionally homogeneous CE arrangements may moderate the relative increase in transaction costs, reduce uncertainty, limit complexity, ease information gathering, curb opportunistic behavior, scale down bureaucratic contracting efforts, and facilitate expectation management (Williamson, 1975, 1985). Thus, the second proposition is:

Proposition 2. Spatial closeness of CE settings curbs the relatively higher CE transaction costs.

Fig. 15.2 summarizes the propositions.

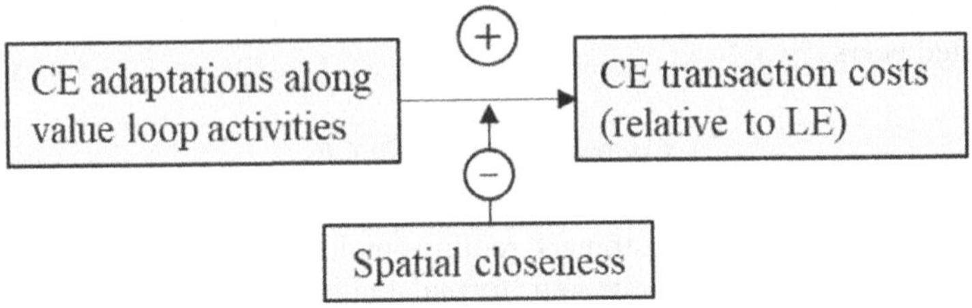

Fig. 15.2. CE Adaptation, Spatial Closeness, and Relative CE Transaction Costs.

CONCLUSION

This study uses transaction cost theory as a lens through which to assess the CE, and the transition to a CE paradigm, and finds that while transactions are costlier in a CE than in the current LE value chains, this can be at least partially mitigated by implementing a geographical closeness for the CE participants. The analysis segregates and analyzes governance costs while assuming that all other cost drivers are equal.

These propositions are relevant to the current sustainability debates within strategic management, particularly in international business reshoring debates. MNEs in the CE transition may factor the mainly advantageous efficiency effects of local CE engagement into their spatial choices, even to the extent of making strategic reshoring decisions. Early CE movers may actively de-internationalize to benefit from spatial proximity in the closed-loop ecosystem and create a competitive advantage, while managers confronted with CE decisions may consider piloting initiatives at geographically bounded locations.

De-internationalizing MNEs may display higher levels of circularity in their value chains than those that are predominantly acting globally; in other words, the de-globalization trend may boost the CE transition. It is generally accepted that de-globalization likely abridges access to the global resource market. With the uneven distribution of natural resources, corporations focusing on local value creation must rely on locally available resources, emphasizing the need for circularity (European Commission, 2020a, 2020b). The EU warns of growing supplier dependencies and potential shortages of raw material and sees the CE as one possible solution (European Commission, 2020a, 2020b). Thus, geographically bounded settings may lead to increased circularity, driven by necessity and the advantages of local CE settings in terms of organizational efficiency.

The propositions regarding the high transaction costs within the CE and the moderating potential of localized value loops demand empirical support. Future studies may analyze the relationship between decreasing multinationality and local economic circularity by comparing companies with varying degrees of multinationality regarding their level of value chain circularity; this might further inform the relationship between international business and circularity.

By de-globalizing economic activity, companies often reduce the scale of their business, which of course reduces value chain costs (Porter, 1985). On the other hand, the limited size of a local CE can negatively affect transaction cost efficiency, since a regional CE limits resource availability when compared to the global market suggesting small-number circumstances that could provoke higher transaction costs (Williamson, 1975). Yet, this circumstance leads to a higher frequency of exchange between the local parties, which justifies the contractual costs (Williamson, 1985). A reliance on new secondary resources, especially if adapted to local synergy partners, can result in the firm being geographically bounded and less flexible in using global alternatives in uncertain circumstances. Thus, limited resource availability in regionally bounded and contractually limiting CEs may also lead to higher transaction costs.

The new economic system must be scaled to significantly impact the global climate. Eventually, an understanding of the mechanisms, cost barriers, and advantages at concise domestic scales could enable the re-internationalization of corporate operations under the new and tested economic system (Bernini et al., 2016; Surdu et al., 2019; Welch & Welch, 2009). Scholars call for the globalization of the CE by, for example, sharing experiences and data and developing integrated global policies (Geng et al., 2019). Multiple local CEs close to the end user (multi-domestic archetype) may be more efficient for scaling CE activities than international, global, or transnational MNE organizations (Benito & Fehlner, 2022).

This study contributes to various streams of literature. First, it addresses CE theory development by bringing efficiency concerns into the phenomenon's analysis to explain how rising transaction costs may pose a barrier to MNEs' transition to the CE. Second, it demonstrates the applicability of transaction cost theory to the novel organizational phenomenon of economic circularity. Despite the CE being characterized by altered transactions across value chain activities, the theory can also help predict MNE strategic behavior. Third, this study underlines the importance of discussing the CE transition in management literature, particularly regarding strategic localization choices.

Increased localization to reduce CE transaction costs is a trade-off to reduce production costs in the globally dispersed value chains that are a core benefit of the global factory (Buckley & Ghauri, 2004). However, this examination segregates and analyzes governance costs and assumes all other cost drivers to be unchanged. Future studies can consider how reduced CE material costs may outweigh rising transaction costs. By replacing costly primary resource exploitation and processing with cycled materials, profitability opportunities and more resilient supply chains are established. Furthermore, operation costs, the most significant cost category according to Porter (1985), may be reduced by reusing components and parts.

Potential areas of future research include alternative managerial strategies to curb rising CE transaction costs. Technological advancements, such as automation and digitalization, could reduce the need for collaborators to be spatially close for efficient CE implementation, while digital exchange can ease the management of complex interrelations between interdependent actors working toward circular value creation (Fjeldstad et al., 2012). Blockchain technology, for example, can create transparency and counter information asymmetries in material and product management, and thereby reduce CE transaction costs (Upadhyay et al., 2021).

ACKNOWLEDGEMENTS

I am particularly grateful for the guidance received by Randi Lunnan and Gabriel R. G. Benito of BI Norwegian Business School. I also appreciate the feedback received at EURAM 2021, EIBA 2022, and the UT Dallas 2023 "Sustainability as a Solution to Global Challenges" conference, and for the many valuable comments and suggestions on earlier versions of this chapter provided by colleagues at SCANCOR (Stanford) and at BI Norwegian Business School.

REFERENCES

Ansari, S., Wijen, F., & Gray, B. (2013). Constructing a climate change logic: An institutional perspective on the "tragedy of the commons". *Organization Science*, *24*(4), 1014–1040. https://doi.org/10.1287/orsc.1120.0799

Argyres, N., Mahoney, J. T., & Nickerson, J. (2019). Strategic responses to shocks: Comparative adjustment costs, transaction costs, and opportunity costs. *Strategic Management Journal*, *40*(3), 357–376. https://doi.org/10.1002/smj.2984

Benito, G. R. G., Cuervo-Cazurra, A., Mudambi, R., Pedersen, T., & Tallman, S. (2022). The future of global strategy. *Global Strategy Journal*, *12*(3), 421–450. https://doi.org/10.1002/gsj.1464

Benito, G. R. G., & Fehlner, C. (2022). Multinational enterprises and the circular economy. In H. Merchant (Ed.), *The new frontiers of international business* (pp. 309–327). Springer. https://doi.org/10.1007/978-3-031-06003-8_15

Benito, G. R. G., & Welch, L. S. (1997). De-internationalization. *Management International Review*, *37*(SI 2), 7–25. http://www.jstor.org/stable/40228430

Bernini, M., Du, J., & Love, J. H. (2016). Explaining intermittent exporting: Exit and conditional re-entry in export markets. *Journal of International Business Studies*, *47*(9), 1058–1076. https://doi.org/10.1057/s41267-016-0015-2

Blackburn, O., Ritala, P., & Keränen, J. (2022). Digital platforms for the circular economy: Exploring meta-organizational orchestration mechanisms. *Organization & Environment*, *36*(2), 253–281. https://doi.org/10.1177/10860266221130717

Bocken, N. M. P., De Pauw, I., Bakker, C., & Van Der Grinten, B. (2016). Product design and business model strategies for a circular economy. *Journal of Industrial and Production Engineering*, *33*(5), 308–320. https://doi.org/10.1080/21681015.2016.1172124

Boons, F., Spekkink, W., & Mouzakitis, Y. (2011). The dynamics of industrial symbiosis: A proposal for a conceptual framework based upon a comprehensive literature review. *Journal of Cleaner Production*, *19*(9–10), 905–911. https://doi.org/10.1016/j.jclepro.2011.01.003

Bressanelli, G., Adrodegari, F., Perona, M., & Saccani, N. (2018). Exploring how usage-focused business models enable circular economy through digital technologies. *Sustainability*, *10*(3), 639. https://doi.org/10.3390/su10030639

Buckley, P. J., & Ghauri, P. N. (2004). Globalisation, economic geography and the strategy of multinational enterprises. *Journal of International Business Studies*, *35*(2), 81–98. https://doi.org/10.1057/palgrave.jibs.8400076

Cannella, S., Bruccoleri, M., & Framinan, J. M. (2016). Closed-loop supply chains: What reverse logistics factors influence performance? *International Journal of Production Economics*, *175*, 35–49. https://doi.org/10.1016/j.ijpe.2016.01.012

Chertow, M. R. (2000). Industrial symbiosis: Literature and taxonomy. *Annual Review of Energy and the Environment*, *25*(1), 313–337. https://doi.org/v10.1146/annurev.energy.25.1.313

Chertow, M. R. (2008). "Uncovering" industrial symbiosis. *Journal of Industrial Ecology*, *11*(1), 11–30. https://doi.org/10.1162/jiec.2007.1110

Circle Economy. (2023). *The circularity gap report 2023*. Circle Economy. https://www.circularity-gap.world/2023

Ciulli, F., & Kolk, A. (2019). Incumbents and business model innovation for the sharing economy: Implications for sustainability. *Journal of Cleaner Production*, *214*, 995–1010. https://doi.org/10.1016/j.jclepro.2018.12.295

Ciulli, F., Kolk, A., & Boe-Lillegraven, S. (2020). Circularity brokers: Digital platform organizations and waste recovery in food supply chains. *Journal of Business Ethics*, *167*(2), 299–331. https://doi.org/10.1007/s10551-019-04160-5

Cleveland, C. J., & Ruth, M. (1997). When, where, and by how much do biophysical limits constrain the economic process? A survey of Nicholas Georgescu-Roegen's contribution to ecological economics. *Ecological Economics*, *22*(3), 203–223. https://doi.org/10.1016/S0921-8009(97)00079-7

Coase, R. H. (1937). The nature of the firm. *Economica N.S.*, *4*(16), 386–405.

Cuypers, I. R. P., Hennart, J.-F., Silverman, B. S., & Ertug, G. (2020). Transaction cost theory: Past progress, current challenges, and suggestions for the future. *Academy of Management Annals*, *15*(1), 111–150. https://doi.org/10.5465/annals.2019.0051

Di Stefano, C., Elia, S., Garrone, P., & Piscitello, L. (2023). The Circular Economy as a new production paradigm to enhance resilience of MNEs and the economic system. *AIB Insights*, *23*(3). https://doi.org/10.46697/001c.74163

Diaz Lopez, F. J., Bastein, T., & Tukker, A. (2019). Business model innovation for resource-efficiency, circularity and cleaner production: What 143 cases tell us. *Ecological Economics*, *155*, 20–35. https://doi.org/10.1016/j.ecolecon.2018.03.009

Domenech, T., Bleischwitz, R., Doranova, A., Panayotopoulos, D., & Roman, L. (2019). Mapping industrial symbiosis development in Europe: Typologies of networks, characteristics, performance and contribution to the circular economy. *Resources, Conservation and Recycling*, *141*, 76–98. https://doi.org/10.1016/j.resconrec.2018.09.016

Ellen MacArthur Foundation. (2013). Towards the circular economy Vol. 1: An economic and business rationale for an accelerated transition. https://www.ellenmacarthurfoundation.org/publications/towards-the-circular-economy-vol-1-an-economic-and-business-rationale-for-an-accelerated-transition

Ellen MacArthur Foundation. (2019). *Completing the picture: How the circular economy tackles climate change*. https://ellenmacarthurfoundation.org/completing-the-picture

Esposito, M., Tse, T., & Soufani, K. (2018). Introducing a circular economy: New thinking with new managerial and policy implications. *California Management Review*, *60*(3), 5–19. https://doi.org/10.1177/0008125618764691

European Commission. (2020a). *A new circular economy action plan. For a cleaner and more competitive Europe*. https://eur-lex.europa.eu/resource.html?uri=cellar:9903b325-6388-11ea-b735-01aa75ed71a1.0017.02/DOC_1&format=PDF

European Commission. (2020b). *Study on the EU's list of critical raw materials – Final Report*. European Union.

Fjeldstad, Ø. D., Snow, C. C., Miles, R. E., & Lettl, C. (2012). The architecture of collaboration. *Strategic Management Journal*, *33*(6), 734–750. https://doi.org/10.1002/smj.1968

Geissdoerfer, M., Pieroni, M. P., Pigosso, D. C., & Soufani, K. (2020). Circular business models: A review. *Journal of Cleaner Production*, *277*, 123741. https://doi.org/10.1016/j.jclepro.2020.123741

Geissdoerfer, M., Savaget, P., Bocken, N. M. P., & Hultink, E. J. (2017). The circular economy – A new sustainability paradigm? *Journal of Cleaner Production*, *143*, 757–768. https://doi.org/10.1016/j.jclepro.2016.12.048

Geng, Y., Sarkis, J., & Bleischwitz, R. (2019). How to globalize the circular economy. *Nature*, *565*, 153–155. https://doi.org/10.1038/d41586-019-00017-z

George, G., Schillebeeckx, S. J., & Liak, T. L. (2015). The management of natural resources: An overview and research agenda. *Academy of Management Journal*, *58*(6), 1595–1613. https://doi.org/10.5465/amj.2015.4006

Gereffi, G., Humphrey, J., & Sturgeon, T. (2005). The governance of global value chains. *Review of International Political Economy*, *12*(1), 78–104. https://doi.org/10.1080/09692290500049805

Ghauri, P., Strange, R., & Cooke, F. L. (2021). Research on international business: The new realities. *International Business Review*, *30*(2), 101794. https://doi.org/10.1016/j.ibusrev.2021.101794

Ghisellini, P., Cialani, C., & Ulgiati, S. (2016). A review on circular economy: The expected transition to a balanced interplay of environmental and economic systems. *Journal of Cleaner Production*, *114*, 11–32. https://doi.org/10.1016/j.jclepro.2015.09.007

Govindan, K., Soleimani, H., & Kannan, D. (2015). Reverse logistics and closed-loop supply chain: A comprehensive review to explore the future. *European Journal of Operational Research*, *240*(3), 603–626. https://doi.org/10.1016/j.ejor.2014.07.012

Granovetter, M. (1985). Economic action and social structure: The problem of embeddedness. *American Journal of Sociology*, *91*(3), 481–510. http://www.jstor.org/stable/2780199

Guide, V. D. R., & Van Wassenhove, L. N. (2009). The evolution of closed-loop supply chain research. *Operations Research*, *57*(1), 10–18. https://doi.org/10.1287/opre.1080.0628

Gupta, S., Wang, Y., & Czinkota, M. (2021). Reshoring and sustainable development goals. *British Journal of Management*, *32*(3), E6–E9. https://doi.org/10.1111/1467-8551.12500

Gygli, S., Haelg, F., Potrafke, N., & Sturm, J.-E. (2019). The KOF globalisation index – Revisited. *The Review of International Organizations*, *14*(3), 543–574. https://doi.org/10.1007/s11558-019-09344-2

Hamari, J., Sjöklint, M., & Ukkonen, A. (2016). The sharing economy: Why people participate in collaborative consumption. *Journal of the Association for Information Science and Technology*, *67*(9), 2047–2059. https://doi.org/10.1002/asi.23552

Hansen, E. G., & Revellio, F. (2020). Circular value creation architectures: Make, ally, buy, or laissez-faire. *Journal of Industrial Ecology*, *24*(6), 1250–1273. https://doi.org/10.1111/jiec.13016

Herczeg, G., Akkerman, R., & Hauschild, M. Z. (2018). Supply chain collaboration in industrial symbiosis networks. *Journal of Cleaner Production*, *171*, 1058–1067. https://doi.org/10.1016/j.jclepro.2017.10.046

Hofstetter, J. S., De Marchi, V., Sarkis, J., Govindan, K., Klassen, R., Ometto, A. R., Spraul, K. S., Bocken, N., Ashton, W. S., Sharma, S., Jaeger-Erben, M., Jensen, C., Dewick, P., Schröder, P., Sinkovics, N., Ibrahim, S. E., Fiske, L., Goerzen, A., & Vazquez-Brust, D. (2021). From sustainable global value chains to circular economy—Different silos, different perspectives, but many opportunities to build bridges. *Circular Economy and Sustainability*, *1*(1), 1–27. https://doi.org/10.1007/s43615-021-00015-2

IN4climate.NRW. (2020). *About us. How we work*. Retrieved June 13 from https://www.in4climate.nrw/en/about-us/how-we-work/

Ketokivi, M., & Mahoney, J. T. (2017). Transaction cost economics as a theory of the firm, management, and governance. In *Oxford research encyclopedia of business and management*. Oxford University Press. https://doi.org/10.1093/acrefore/9780190224851.013.6

Kirchherr, J., Piscicelli, L., Bour, R., Kostense-Smit, E., Muller, J., Huibrechtse-Truijens, A., & Hekkert, M. (2018). Barriers to the circular economy: Evidence from the European Union (EU). *Ecological Economics*, *150*, 264–272. https://doi.org/10.1016/j.ecolecon.2018.04.028

Kirchherr, J., Reike, D., & Hekkert, M. (2017). Conceptualizing the circular economy: An analysis of 114 definitions. *Resources, Conservation and Recycling*, *127*, 221–232. https://doi.org/10.1016/j.resconrec.2017.09.005

Kokoulina, L., Ermolaeva, L., Patala, S., & Ritala, P. (2019). Championing processes and the emergence of industrial symbiosis. *Regional Studies*, *53*(4), 528–539. https://doi.org/10.1080/00343404.2018.1473568

Korhonen, J., Honkasalo, A., & Seppälä, J. (2018). Circular economy: The concept and its limitations. *Ecological Economics*, *143*, 37–46. https://doi.org/10.1016/j.ecolecon.2017.06.041

Lacy, P., Long, J., & Spindler, W. (2020). *The circular economy handbook. Realizing the circular advantage*. Palgrave Macmillan.

Lahti, T., Wincent, J., & Parida, V. (2018). A definition and theoretical review of the circular economy, value creation, and sustainable business models: Where are we now and where should research move in the future? *Sustainability*, *10*(8), 2799. https://doi.org/10.3390/su10082799

Liu, J., Feng, Y., Zhu, Q., & Sarkis, J. (2018). Green supply chain management and the circular economy. *International Journal of Physical Distribution & Logistics Management*, *48*(8), 794–817. https://doi.org/10.1108/ijpdlm-01-2017-0049

Macher, J. T., & Richman, B. D. (2008). Transaction cost economics: An assessment of empirical research in the social sciences. *Business and Politics*, *10*(1), 1–63. https://doi.org/10.2202/1469-3569.1210

Mathews, J. A. (2014). *Greening of capitalism: How Asia is driving the next great transformation*. Stanford University Press.

Mathews, J. A., Tan, H., & Hu, M.-C. (2018). Moving to a circular economy in China: Transforming industrial parks into eco-industrial parks. *California Management Review*, *60*(3), 157–181. https://doi.org/10.1177/0008125617752692

McDonough, W., & Braungart, M. (2010). *Cradle to cradle: Remaking the way we make things*. North Point Press.

Mitra, S., & Datta, P. P. (2014). Adoption of green supply chain management practices and their impact on performance: An exploratory study of Indian manufacturing firms. *International Journal of Production Research*, *52*(7), 2085–2107. https://doi.org/10.1080/00207543.2013.849014

Montiel, I., Cuervo-Cazurra, A., Park, J., Antolin-Lopez, R., & Husted, B. W. (2021). Implementing the United Nations' sustainable development goals in international business. *Journal of International Business Studies*, *52*(5), 999–1030. https://doi.org/10.1057/s41267-021-00445-y

Murray, A., Skene, K., & Haynes, K. (2017). The circular economy: An interdisciplinary exploration of the concept and application in a global context. *Journal of Business Ethics*, *140*(3), 369–380. https://doi.org/10.1007/s10551-015-2693-2

North, D. C. (1990). *Institutions, institutional change and economic performance*. Cambridge University Press.

Nyberg, D., & Wright, C. (2020). Climate-proofing management research. *Academy of Management Perspectives*, *36*, 713–728. https://doi.org/10.5465/amp.2018.0183

Nygaard, A. (2019). From linear to circular organization. The political economy of sustainable strategy. In G. R. G. Benito & R. Lunnan (Eds.), *Voyages of a scholar: Navigating companies, channels, and clusters. A tribute to Torger Reve* (pp. 131–146). Fagbokforlaget.

OECD. (2020). *FDI flows (indicator)*. https://doi.org/10.1787/99f6e393-en

Parmigiani, A., Klassen, R. D., & Russo, M. V. (2011). Efficiency meets accountability: Performance implications of supply chain configuration, control, and capabilities. *Journal of Operations Management*, *29*(3), 212–223. https://doi.org/10.1016/j.jom.2011.01.001

Porter, M. E. (1985). *Competitive advantage: Creating and sustaining superior performance*. Free Press.

Ranta, V., & Saari, U. (2019). Circular economy: Enabling the transition towards sustainable consumption and production. In W. Leal Filho, A. M. Azul, L. Brandli, P. G. Özuyar, & T. Wall (Eds.), *Responsible consumption and production (The encyclopedia of the UN sustainable development goals, pp. 1–12)*. Springer. https://doi.org/10.1007/978-3-319-71062-4_3-1

Reike, D., Vermeulen, W. J., & Witjes, S. (2018). The circular economy: New or refurbished as CE 3.0? — Exploring controversies in the conceptualization of the circular economy through a focus on history and resource value retention options. *Resources, Conservation & Recycling*, *135*, 246–264.

Rindfleisch, A., & Heide, J. B. (1997). Transaction cost analysis: Past, present, and future applications. *Journal of Marketing*, *61*(4), 30–54. https://doi.org/10.1177/002224299706100403

Seuring, S. (2004). Integrated chain management and supply chain management comparative analysis and illustrative cases. *Journal of Cleaner Production*, *12*(8–10), 1059–1071. https://doi.org/10.1016/j.jclepro.2004.02.006

Shelanski, H. A., & Klein, P. G. (1995). Empirical research in transaction cost economics: A review and assessment. *Journal of Law, Economics, and Organization*, *11*, 335–361. https://doi.org/10.2307/765001

Snihur, Y., & Bocken, N. (2022). A call for action: The impact of business model innovation on business ecosystems, society and planet. *Long Range Planning*, *55*(6), 102182. https://doi.org/10.1016/j.lrp.2022.102182

Stabell, C. B., & Fjeldstad, Ø. D. (1998). Configuring value for competitive advantage: on chains, shops, and networks. *Strategic Management Journal*, *19*(5), 413–437. https://doi.org/10.1002/(SICI)1097-0266(199805)19:5<413::AID-SMJ946>3.0.CO;2-C

Stuchtey, M., Enkvist, P.-A., & Zumwinkel, K. (2016). *A good disruption: Redefining growth in the twenty-first century*. Bloomsbury Publishing.

Surdu, I., Mellahi, K., & Glaister, K. W. (2019). Once bitten, not necessarily shy? Determinants of foreign market re-entry commitment strategies. *Journal of International Business Studies*, *50*(3), 393–422. https://doi.org/10.1057/s41267-018-0167-3

Ter Huurne, M., Ronteltap, A., Corten, R., & Buskens, V. (2017). Antecedents of trust in the sharing economy: A systematic review. *Journal of Consumer Behaviour*, *16*(6), 485–498. https://doi.org/10.1002/cb.1667

Theyel, G. (2012). Production location and sustainability. *International Journal of Agile Systems and Management*, *5*(3), 267–275. https://doi.org/10.1504/IJASM.2012.047680

Theyel, G., & Hofmann, K. H. (2021). Manufacturing location decisions and organizational agility. *Multinational Business Review*, *29*(2), 166–188. https://doi.org/10.1108/mbr-02-2020-0035

Tuladhar, S., Yuan, M., & Montgomery, W. D. (2016). *An economic analysis of the circular economy* [Paper presentation]. 19th annual conference on global economic analysis, Washington DC, USA. https://www.gtap.agecon.purdue.edu/resources/res_display.asp?RecordID=4937

UNCTAD. (2020). *International production beyond the pandemic* [World Investment Report, Issue 2020]. https://unctad.org/system/files/official-document/wir2020_overview_en.pdf

United Nations Environment Programme. (2019). *Global resources outlook 2019: Natural resources for the future we want*. https://www.resourcepanel.org/reports/global-resources-outlook

Upadhyay, A., Mukhuty, S., Kumar, V., & Kazancoglu, Y. (2021). Blockchain technology and the circular economy: Implications for sustainability and social responsibility. *Journal of Cleaner Production*, *293*, 126130. https://doi.org/10.1016/j.jclepro.2021.126130

Verbeke, A., & Greidanus, N. (2012). The end of the opportunism versus trust debate: Bounded reliability as a new envelope concept in research on MNE governance. In A. Verbeke & H. Merchant (Eds.), *Handbook of research on international strategic management* (pp. 31–63). Edward Elgar Publishing. https://doi.org/10.4337/9781781009147.00008

Welch, C. L., & Welch, L. S. (2009). Re-internationalisation: Exploration and conceptualisation. *International Business Review*, *18*(6), 567–577. https://doi.org/10.1016/j.ibusrev.2009.07.003

Werning, J. P., & Spinler, S. (2020). Transition to circular economy on firm level: Barrier identification and prioritization along the value chain. *Journal of Cleaner Production*, *245*, 118609. https://doi.org/10.1016/j.jclepro.2019.118609

Williamson, O. E. (1975). *Markets and hierarchies: Analysis and antitrust implications*. Free Press.

Williamson, O. E. (1979). Transaction-cost economics: The governance of contractual relations. *The Journal of Law and Economics*, *22*(2), 233–261. https://doi.org/10.1086/466942

Williamson, O. E. (1985). *The economic institutions of capitalism*. Free Press.

Williamson, O. E. (1991). Comparative economic organization: The analysis of discrete structural alternatives. *Administrative Science Quarterly*, *36*(2), 269–296. https://doi.org/10.2307/2393356

Williamson, O. E. (1996). *The mechanisms of governance*. Oxford University Press.

Williamson, O. E. (2000). The new institutional economics: Taking stock, looking ahead. *Journal of Economic Literature*, *38*(3), 595–613. https://doi.org/10.1257/jel.38.3.595

Wittneben, B. B., Okereke, C., Banerjee, S. B., & Levy, D. L. (2012). Climate change and the emergence of new organizational landscapes. *Organization Studies*, *33*(11), 1431–1450. https://doi.org/10.1177/0170840612464612

Yu, H., Bansal, P., & Arjaliès, D.-L. (2023). International business is contributing to environmental crises. *Journal of International Business Studies*, *54*, 1151–1169. https://doi.org/10.1057/s41267-022-00590-y

Zeng, R., Grøgaard, B., & Steel, P. (2018). Complements or substitutes? A meta-analysis of the role of integration mechanisms for knowledge transfer in the MNE network. *Journal of World Business*, *53*(4), 415–432. https://doi.org/10.1016/j.jwb.2018.02.001

CHAPTER 16

DISENTANGLING THE EFFECTS OF CIRCULAR ECONOMY PRINCIPLES ON GVCs' ACTIVITIES, CONFIGURATION, AND GOVERNANCE

Cristina Di Stefano, Stefano Elia, Paola Garrone and Lucia Piscitello

Politecnico di Milano, Italy

ABSTRACT

Global value chains (GVCs) have been challenged by several emerging macro-trends during the last years. Among them, sustainability of production and consumption patterns is becoming a central theme given the necessity to mitigate the degradation of the environment and the over-exploitation of scarce natural resources. In this respect, scholars and practitioners increasingly propose the circular economy (CE) approach as a systemic solution to overcome the conventional linear "take–make–use–dispose" model underlying the structure of contemporary global economy. However, the international business (IB) community has introduced the topic of CE only marginally in its debate. The aim of the present study is to fill this research gap identifying the opportunities for integrating IB and CE principles. Thus, the main objective is to investigate whether and how the adoption of the CE paradigm by multinational enterprises

Walking the Talk? MNEs Transitioning Towards a Sustainable World
Progress in International Business Research, Volume 18, 319–333

ISSN: 1745-8862/doi:10.1108/S1745-886220240000018020

(MNEs) may affect activities, geographical configuration, and governance of their relevant GVCs.

The authors address the issue from a conceptual point of view, identifying direct and indirect impacts of CE adoption on GVC, relative enablers, and possible broader implications. Lastly, the authors propose some reflections for future investigations.

Keywords: Circular economy; CE transition; CE impacts; 3Rs; circularity; GVCs activities; GVCs geographical configuration; GVCs governance; MNEs

1. INTRODUCTION

In the last years, several emerging issues are imposing a revision of GVCs' configuration (UNCTAD, 2020); one of them is the necessity to mitigate the degradation of the environment and the over-exploitation of scarce natural resources, which is imposing increased attention on the sustainability of production and consumption. This is the reason why a CE approach is gaining momentum among scholars, practitioners, and decision-makers who recognize the urgency to overcome the conventional linear "take–make–use–dispose" model underlying the current structure of the global economy.

However, the IB community has introduced the topic of CE in its debate only recently, even if the possible adoption of a CE paradigm by MNEs may have far-reaching implications for GVCs' (re)configuration (Hofstetter et al., 2021). CE scholars have recognized that the transition to a circular paradigm is a systemic solution that involves "ecosystems of actors" (Pieroni et al., 2019) and networks of different sectors and markets (Kanda et al., 2021). They have also considered roles, relationships, and dynamics connecting the different stakeholders involved in the CE transition (Geissdoerfer et al., 2018; Ho et al., 2022; Lieder & Rashid, 2016). Nonetheless, they have never adopted a GVC perspective in their analysis, and have only marginally investigated the adoption of CE practices by MNEs (Calzolari et al., 2021).

Most MNEs have not yet started the CE transition journey, and even those MNEs that have introduced CE practices in their processes or developed product and process innovations aimed at circularity are still far from a truly circular transformation of their strategies and operations. The aim of the present study is to fill this research gap and to identify the opportunities for integrating IB and CE principles. We believe that IB scholars need to include the CE approach in their analysis of GVCs' reconfiguration. At the same time, however, understanding the disruptive potential of CE asks for a new outlook in which MNEs are considered a central actor in the transition from a linear to a circular paradigm, rather than a pillar of the old linear system and as such a barrier to overcome. Thus, our main objective is to investigate whether and how the adoption of the CE paradigm by MNEs may affect activities, geographical configuration, and governance of the relevant GVCs.

The chapter is organized as follows: in the next section, we provide an overview of the CE paradigm focusing on the main principles of reduce, reuse, and recycle and we investigate the adoption of the CE principles by an MNE and the likely impacts on its GVC, also providing some examples of CE practices in the manufacturing sector. Then, we introduce some broader implications. In the last section, we propose some avenues for future investigations.

2. THE CE PRINCIPLES

CE is a broad term and according to Kirchherr et al. (2017), over 100 definitions of the concept have been already proposed in the literature. In addition, numerous mechanisms have been identified for its implementation, including the "R-imperatives" (Reike et al., 2018). Although different authors have suggested varying numbers of R-imperatives over the years, the most widely recognized and adopted ones are the 3Rs: reduce, reuse, and recycle.

In this work, we focus on the adoption of the 3R imperatives by global lead firms, which are increasingly asked to take a role in the necessary transition toward a sustainable production and consumption paradigm, given the far-reaching repercussions of their strategic choices on the economic system (Kolk et al., 2017; Montiel et al., 2021).

MNEs can introduce circularity in the GVCs by adopting at least one of the 3R (reduce, reuse, and recycling) principles, and depending on the selected R, the impact on the GVCs will differ.

Thus, the possible adoption of a CE approach by the MNE not only affects its business model, but also requires a rethinking of all the flows of its GVC and the roles of the actors involved (Hofstetter et al., 2021).

In the following sections, we present the 3R principles and analyze potential modifications that their adoption by MNEs may cause in GVCs, namely in their activities, localization, and governance (Fernandez-Stark & Gereffi, 2019). In the GVC framework, value chain activities are internationally distributed among several firms and are organized and coordinated by MNEs adopting hierarchical governance or different network-based forms of internationalization (captive, relational, or modular; Gereffi et al., 2005). The relationship between the MNE and its network evolves over time (Buckley & Casson, 2020; Kano et al., 2020; Pananond et al., 2020; Strange & Humphrey, 2019), and MNEs adapt the organization of their value chain activities as well as their ownership and location strategies to cope with the evolving circumstances and maintain competitiveness. We look at the circular redesign of value chains as a possible strategic response of MNEs to far-reaching circumstances such as the scarcity of some essential resources and the increase in internal and external costs caused by waste and/or other current contingencies.

Specifically, we elucidate the effects of the 3Rs' adoption in terms of impacts and enablers. Impacts are well-defined effects of R principles on GVCs. They may be direct when they cause a change in the focal GVC aspect, for example, if a new actor is needed or activities undertaken by the MNE should be changed, or they may be indirect when stemming from side consequences occurring in other GVC aspects, for example, a modification of a GVC activity that causes a change in GVC geography. Since an impact may not occur automatically, but it may require certain complementary conditions to occur, we highlight such conditions and label them enablers. For example, the development of new business models by MNEs may be an enabler of activities' modification.

The results of our conceptual elaboration are summarized in Table 16.1.

2.1. Reduce

The "reduce" principle is relevant to both consumers and producers. For consumers, this means making choices that prevent or delay the creation of waste, such as opting to buy and consume less (also known as the "refuse principle," Ghisellini et al., 2016), taking better care of products, or using them for a longer period of time (Reike et al., 2018).

Producers can reduce their inputs by implementing product redesigns, process innovations, and other strategies that improve cost and resource efficiency without sacrificing output levels. Product dematerialization, or replacing physical components with non-material alternatives, is one such approach that reduces material consumption and limits waste production (Kasulaitis et al., 2019). This trend has been observed in various industries that have replaced physical products with digital alternatives, such as music streaming and online newspapers replacing physical CDs and paper-based newspapers.

In terms of input and raw material selection, a reduction strategy involves using fewer natural resources by replacing non-renewable resources with renewable ones or adopting secondary raw materials in the production process. Finally, the "reduce" principle can also refer to innovations that improve product efficiency (Hollander et al., 2017). However, it is important to note that a rebound effect may occur if increased efficiency results in lower production costs and product prices, which could lead to higher consumer demand and nullify savings (Zink & Geyer, 2017). To avoid this negative feedback and reduce product consumption, sharing or leasing solutions may be supported (Henry et al., 2021).

2.1.1. The MNE's Adoption of the Reduce Principle and the Impacts on Related GVC

There are three main impacts of the adoption of the reduce principle by a MNE on its GVCs.

In terms of GVC activities, the adoption of a reduce approach by the MNE implies, at first, a product redesign aimed at material saving and efficiency (*Impact_Red1*) (Bocken et al., 2016). Different options may be considered, for example, substituting product materials or dematerializing components, adopting solutions that increase materials efficiency, extend the durability of products, and make products attractive to consumers for a longer period (Bakker et al., 2014; Spreafico & Landi, 2022). For instance, the American technology company Cisco has embraced circularity in product design by selecting secondary raw materials and by reducing the use of non-recyclable materials in their products. Also, it

Table 16.1. Impacts of CE Adoption by the MNE on GVC Aspects.

CE Principles	GVC Aspects		
	Activities	Localization	Governance
Reduce	• *Impact_Red1:* Design for material saving and efficiency (Enablers: Reskilling of product designers and developers; aware customers; certification of product compliance) • *Impact_Red2:* Product/process innovation and adoption of resource-efficient technologies (Enablers: Reskilling of product designers and developers)	• *Impact_Red3 (Indirect through Activities):* If scale economies in upstream activities are significant, downsizing may result in regionalization of the GVC	• *Impact_Red4A (Indirect through Activities):* R&D partnership for innovation • *Impact_Red4B (Indirect through Activities):* R&D activities conducted by the MNE
Reuse	• *Impact_Reu1*: Introduction of a new activity, namely repair (Enablers: Design for repair and replacement of single components and viability of components for replacement) • *Impact_Reu2*: Development of the customer–repairer loop (Enabler: Aware customers) • *Impact_Reu3*: Emergence of a repairer - *(A)* independent from the MNE (Enabler: Repair skills and entrepreneurial attitude) - *(B)* coordinated with the MNE (Enabler: Adoption of a new business model by the MNE)	• *Impact_Reu4*: Location of repairers in a country different from the one in which the final product was consumed/used (Enabler: Given the labor-intensive nature of repair activities, countries should have low labor costs) • *Impact_Reu5 (Indirect through Activities):* if scale economies in upstream activities are significant, downsizing may result in regionalization of the GVC	• *Impact_Reu6A*: Coordination between the MNE and the repairers, through vertical integration or non-equity agreements (depending on transaction costs and volumes) (Enabler: Assets specificity) • *Impact_Reu6B:* Independent repairers develop market relationships with customers (Enabler: Certification by producers or other agents if customers suffer from adverse selection toward the quality of the repaired product (eventual, depending on the product price of features, distance, etc.)
Recycle	• *Impact_Rec1*: Design for end of life (Enabler: Reskilling of product designers and developers) • *Impact_Rec2*: Development of recycling loop - *(A)* Outer recycling loop (external to the MNE) (Enabler: Economies of scale of the recyclers) - *(B) (residual)* Inner recycling loop (internal to the MNE) (Enabler: Sufficient quantity of waste for recycling)	• *Impact_Rec3 (Indirect through Activities)*: Possible location of recycling plant in a country different from the one in which the final product was consumed/used (until technologies are enough developed) (Enabler: Given the labor-intensive nature of recycling activities, external countries should have low labor cost	• *Impact_Rec4A (Indirect through Activities)* Partnership among MNEs in the same sector or with a recycler (or waste collector agent) • *Impact_Rec4B (Indirect through Activities)* Vertical integration of the recycling activity in the MNE

co-operates with other actors of the value chain in order to redesign packaging solutions that adopt sustainable materials and protect the products from transport damage while optimizing space efficiency during transport (Cisco, 2021, p. 84). This strategy requires a reskilling of product designers and developers who should adapt their principles and methods to embrace the circularity approach (*Enabler*) (Hollander et al., 2017). Also, consumers should be aware of the need to reduce the negative impacts on the environment of their consumption behaviors (*Enabler*) (Kirchherr et al., 2017). In fact, new product designs should be accepted by consumers since some of the products' features may change or the products' prices may increase given the necessity to adopt different materials or different production processes. Lastly, a new product design may require certifications to demonstrate the compliance of the product with some specific features or performances (*Enabler*).

Alternatively, MNEs can also implement a reduction strategy by introducing product or process innovations that limit the environmental impact of production and consumption processes (*Impact_Red2*). It should be noticed that in a reduce approach, product life extension is not always the preferable solution; negative impacts of a longer usage may be higher than the ones generated by replacing the product with a newer version that incorporates more efficient technologies (Bakker et al., 2014). However, the implementation of sustainable and circular innovations requires dedicated R&D activities. For instance, the sportswear MNE Nike, in order to reduce the consumption of water in the dyeing process, has adopted a "ColourDry" technology that provides the use of recycled carbon dioxide in the process instead of water. The technology has been developed by a Dutch start-up (DyeCoo Textile Systems B.V.) and it has been implemented by one of its contractors in Taiwan (Far Eastern New Century Corp.) (Nike, 2013; The Guardian, 2013).

Lastly, another impact of the reduce strategy is related to the geographical configuration of the GVC. Specifically, the lower need for raw materials or components decreases the importance of economies of scale in upstream activities of the value chain and may induce MNEs to regionalize their supplies, thus reducing environmental impacts and supply costs (Burke et al., 2021) (*Impact_Red3 – Indirect through Activities*).

In terms of governance, the MNEs can conduct research and development (R&D) activities required for redesign or product and process innovations through R&D partnership with external stakeholders such as other companies (as in the case of Cisco or Nike presented above) or research centers (*Impact_Red4A – Indirect through Activities*), or may internally develop innovations in the R&D departments (*Impact_Red4B – Indirect through Activities*).

The main implications of the reduce strategy for GVCs are summarized in the first row of Table 16.1.

2.2. Reuse

The Reuse approach indicates in the first place 'direct reuse' of products, components, and materials; however, adopting an extensive interpretation of the concept, reuse may also follow repairing (refurbishing/remanufacturing/repurposing) activities that allow the extraction of additional value from the items (Galbreth et al., 2013). This principle has been less investigated in the literature than recycling (Ghisellini et al., 2016) while it allows higher value retention (Reike et al., 2018) and should be prioritized and become the norm (Stahel, 2016). In fact, the reduce and reuse strategies may be considered as ex-ante CE applications, as they prevent at the origin the waste creation, while recycling may be seen as an ex-post or end-of-pipe remedy. According to the Ellen MacArthur Foundation (2013), reuse is the CE strategy that presents higher environmental benefits, if potential rebound effects are mitigated (Font Vivanco et al., 2016; Makov & Font Vivanco, 2018).

Reuse of a product or component is "direct" if only small repairs are needed to restore its functions (Agrawal et al., 2015; Loomba & Nakashima, 2012). Refurbishing or remanufacturing activities are performed when product components are replaced or repaired (Lieder & Rashid, 2016). These strategies imply the regeneration of products, or they may even favor their 'upgrading' when components are substituted with a more advanced version (de Brito & Dekker, 2004; Fernández & Kekäle, 2005). However, the adoption of a reuse strategy may not be equally suitable for all the sectors, for example, because a high innovation rate limits the potential adoption of this strategy (Galbreth et al., 2013).

2.2.1. The MNE's Adoption of the Reuse Principle and the Impacts on Related GVC

The adoption of a Reuse paradigm by the MNE may produce three direct impacts on the structure of the GVC. First of all, it determines the introduction of (at least one) new activity (or phase), that is, the repair phase (*Impact_Reu1*). Repair activities are enabled by a product design that allows the repair and replacement of single components (e.g., through modularization of products) (Machado & Morioka, 2021) and easy access to single components for replacement (*Enabler*).

Second, new medium-long circular loops (Reike et al., 2018) emerge: these loops go from the customer (the consumer in the case of final goods, or a business customer in the case of intermediate or capital goods) to the repairer to the customer again (*Impact_Reu2*) and are enabled by the awareness of the consumers and by their willingness to reuse products (*Enabler*). One sector in which the reduce strategy is particularly relevant is electronic and electrical equipment (EEE) since the first use phase is often short (Hischier & Böni, 2021). In Europe, the 'Waste Electrical and Electronic Equipment (WEEE) Directive (European Parliament, 2012) recognizes the relevance of the WEEE in the CE aiming at facilitating collection and processing for reuse and material recovery in the electrical and electronic sector. However, the reuse strategy is also diffused in other industries such as clothing and textile or motor vehicles. The secondhand apparel, footwear, and accessories represent $30 billion to $40 billion in value worldwide and is growing in the last years (BCG, 2020). Focusing on motor vehicles, Eurostat data indicate that in 2019, 95.1% of the parts and materials of motor vehicles were reused and recovered, while 89.6% were reused and recycled in Europe (Eurostat, 2022).

Repairers are new economic actors who emerge in the circular GVC (Ghisellini et al., 2016). They may operate independently from the MNE (*Impact_Reu3A*), if they have the required skills and an entrepreneurial attitude (*Enabler*), or they may engage with the MNE through different types of agreements (*Impact_Reu3B*). In the first case, the MNE is not involved in the new circular loop; in the second one, MNE leads the implementation of the reuse principle in the GVC by adopting a new business model or introducing post-sales repair services (e.g., through a platform) and reverse logistics (*Enabler*). The two situations can coexist; for instance, in the electronic consumer sector, resell is strongly diffused between private individuals that use third-party platforms (e.g., eBay, among the others); additionally, MNEs often sell certified refurbished products on their official websites. For instance, the American company Dell has recently introduced a "Global Dell Outlet" through which it directly resales certified refurbished products backed by a one-year warranty.

The adoption of the reuse paradigm by the MNE may increase the geographic scope of its GVC (*Impact_Reu4*) since repair activities are labor-intensive and are traditionally carried-out in low labor costs countries (Llorente-gonzález & Vence, 2020) (*Enabler*). Also refurbished or remanufactured products are cheaper than new ones and developing countries have large shares and growth rates in the market, as in the case of EEE products (Counterpoint, 2022).

It should be noticed that an opposite reconfiguration of the GVC (i.e., a shortening trend) is also possible: the downsizing of production may induce regionalization of manufacturing activities, particularly if scale economies were significant before the adoption of the reuse paradigm (*Impact_Reu5 – Indirect through Activities*). Moreover, for reusable products that do not require reprocessing steps (repair or refurbishment, repurposing), small decentralized networks may be more appropriate (Agrawal et al., 2015).

Focusing on the changes in the governance structure of the GVC induced by the reuse approach, the new economic actor (the repairer) may be an independent actor *(Impact_Reu6A)* or may operate coordinated with the MNE *(Impact_Reu6B)*, through vertical integration or non-equity or partnership agreements (depending on transaction costs and volumes). Depending on the product's characteristics and on the length of the reuse loop, independent repairers may need to be certified by producers or other agents *(Enabler)* to overcome customers' adverse selection toward the quality of used or repaired products. For instance, the tech company Apple has developed a Repairer Provider Program offering an Apple certification that allows companies to perform out-of-warranty repair services. Certifications require passing exams of the online Authorized Testing Center and should be annually renewed (Apple, 2022). Instead, vertical integration may emerge when the asset specificity of the repair activity is high as in the case of make-to-order products *(Enabler)*.

The second row of Table 16.1 recaps the effects of reuse strategy.

2.3. Recycle

The management of a product's end-of-life involves recycling and recovery activities. Recycling is often considered the primary practice of the CE (Potting et al., 2017) and is the most institutionally supported (Ranta et al., 2018). Moreover, most CE progress indicators focus on recycling activities (Moraga et al., 2019). Nonetheless, recycling practices require more energy, labor, and resources than reuse, and they may result in lower value retention from materials (Reike et al., 2018). Additionally, 100% recycling is impossible, and some waste materials cannot be recycled multiple times or may even be unrecyclable (Ghisellini et al., 2016).

The recycling principle entails the reprocessing of waste materials into products, materials, or substances to be re-employed in the manufacturing of the initial products or directed to other purposes (European Parliament, 2008, p. 10). Primary recycling is a closed loop developed within a single industry where waste is collected and reintroduced in the production process, while secondary recycling is an open loop where recycled materials are used for purposes that differ from their original ones. The latter is less efficient than primary recycling due to the high variability of collected materials (Stahel, 2010).

2.3.1. MNE's Adoption of the Recycle Principle and the Impact on Related GVC

Recycling requires, as a necessary precondition, a product design oriented to this scope; in other words, items should be designed to be disassembled and recycled (Bocken et al., 2016) (*Impact_Rec1*). For this to happen, designers and developers need to acquire all skills that allow them to include evaluations on product end-of-life from the initial phase of product development (De los Rios & Charnley, 2017) (*Enabler*). An example of design for end-of-life is a prototype vacuum cleaner recently developed by the Swedish multinational producer of household appliances Electrolux that is 90% recyclable. To develop such a new design that enables disassembling and recycling of each component, the engineering team worked closely with Stena Recycling, a Swedish MNE that operates in the recycling industry, and explored different material choices and assembling/disassembling solutions (Electrolux, 2022).

A further step to transform waste into secondary raw materials is the activation of a recycling loop. In most cases, recycling loop is external to the MNEs (*Impact_Rec2A*) since the process is more efficiently conducted by recycling companies that can specialize and achieve economies of scale (*Enabler*). Also, collective recycling systems incentivize technology improvements that are essential to increase the recycling rate and quality of secondary raw materials (Rahmani et al., 2021).

However, the recycling loop can also be activated internally by the MNE (Inner loop) (*Impact_Rec2B*) even if the latter can be considered a residual case. In fact, inner loops may be developed if materials are specific to the type of production the company performs, for instance, to recycle unusable defective items or production scrap of the industrial process in the sites of original manufacturing as it often happens in the metal industry (Zhang et al., 2018). Also, the MNE may want to signal its responsibility for sustainable consumption of resources or may have evaluated that internal management of industrial waste is more convenient. Inner recycling loops are efficient if the amount of waste the MNE produces to be recycled is sufficient (*Enabler*) and this minimum dimension for viable recycling business differs, depending on waste types or production techniques adopted in different industries (Lyons, 2007).

Focusing on the GVC geography, outer loops may determine an expansion of its scope (*Impact_Rec3 – Indirect through Activities*); in fact, recycling plants are often located in low labor cost countries since the recycling activity is a labor-intensive one (*Enabler*). However, the waste recycling industry is recently growing in developed countries supported by innovations and new technologies; thus, a regionalization of waste trade may also be possible in the next years.

Inner loops, instead, do not affect GVC geography since, to be efficient, it should be realized in the same site of manufacturing activities.

In terms of governance of the GVC, when outer loops are implemented by the MNE, different types of partnerships among producers and recycling companies may be stipulated (*Impact_Rec4A – Indirect through Activities*) to comply with regulatory requirements, or to answer to the increasing request of environmental attention presented to MNEs by different stakeholders. An example of partnership for circularity is the Circular Fashion Partnership proposed by the Global Fashion Agenda and stipulated by textile manufacturers, recyclers, and fashion brands operating in Bangladesh. The aim of the project is to collect and recycle post-production textile waste and deadstock, and by February 2022, was subscribed by 22 brand and buying agents, 45 factories, and 18 recyclers (Global Fashion Agenda, 2022). Also, the partnership may be stipulated to design products considering the end of life as in the case of Electrolux presented before.

On the other hand, when MNEs activate an inner recycling loop, they vertically integrate the activity and the required assets (*Impact_Rec4B*).

The main implications of the recycle strategy for GVCs are summarized in the third row of Table 16.1.

3. BROAD IMPLICATIONS

The commitment of MNEs to lead the transition to the CE paradigm generates profound transformations of single GVC aspects (Table 16.1). Beyond such specific effects, however, MNE's adoption of CE principles also determines high-level and far-reaching implications for the GVC or the context. The broader consequences of CE adoption deserve further reflections and analyses.

A few examples of the wide implications of CE for GVCs in their entirety and for society at large are related to innovations or redesign of products and processes according to circular principles. By reducing the demand for materials and energy, the circular change mitigates the current over-exploitation of the planet's natural resources and decreases the amount of waste generated by current production and consumption processes. However, rebound effects may occur if innovations or redesign improve product/process efficiency and reduce their price or cost. In turn, this may induce consumers to increase product usage or enables a greater consumption of other goods and services, which in the end diminishes the hypothesized environmental benefits (Font Vivanco et al., 2016; Makov & Font Vivanco, 2018). Secondly, when design or innovation property rights are present, MNEs may implement a strategy of product differentiation and increase product prices, and in this respect, the literature identified a positive relationship between sustainability innovations and a firm's competitiveness (Hermundsdottir & Aspelund, 2021).

MNEs' adoption of the reuse principle also has far-reaching implications. It may determine a production downsizing and a reduction in the total amount of resources consumed in the value chain. However, this effect is not straightforward; for instance, in the EEE sector, Hischier and Böni (2021) indicate that direct reuse should be prioritized for smartphones or laptops since the environmental impact of their production is strong. Instead, for products that present high costs of usage such as washing machines or refrigerators, secondary use may generate rebound effects and other CE strategies should be preferred.

Lastly, the adoption of a recycle principle may determine an increase of the GVC length, and a consequent negative environmental impact resulting from the transport of waste and recycled materials around different countries. Also, the recycling process itself may increase the environmental burden if it is not properly managed. It is well recognized that most global waste trade occurs between developed and developing countries, because of loose environmental regulations and enforcement in these countries; however, they often recycle waste with minimal health and environmental protection. On the other hand, specialization in the recycling industry may become an occasion of growth for developing countries (Qu et al., 2019). However, the recycling industry may also further expand in developed countries in the next years reducing the volume of global waste trade and the connected environmental consequences.

4. CONCLUSIONS

After the investigation conducted in the chapter, we confirm that it is highly necessary to bridge the extant gap between CE and IB scholarships. Despite the incipient stage of the debate, we propose some reflections.

First, to be fruitful for MNEs strategies, the debate should delve in-depth into the CE principles and distinguish clearly the different Rs, which have different implications for GVCs.

Second, CE will not be a paradigm of general scope until MNEs are considered an obstacle to CE inception, or a player alien to CE transition. Indeed, they are expected to play different active roles given their centrality for many economies and for GVCs, such as initiator of the transition and coordinator/ecosystem integrator. Furthermore, new technologies and innovative solutions are necessary in different areas such as product redesign, recycling, and reduce systems, packaging, new repair services, and so on. However, these innovations are only partially available for MNEs and other actors of GVCs. They are underdeveloped or still to be experimented at full scale. Therefore, large firms, and MNEs especially, have an important role to play with their R&D facilities and personnel, financial resources, and superior innovation capabilities, for CE to be adopted.

Third, a matter for further investigation is MNEs' incentives to adopt CE. On the one hand, longer lifespans of products obtained through redesign, and through refurbishment or repair activities related to reuse might reduce the demand for new products, and the revenues for MNEs. However, there are a few reasons why MNEs may see CE as a promising and profitable avenue. MNEs may decide to sponsor CE principles as effective responses to exogenous shocks and crises and a means to move to a more regenerative economy that allows to rebalance the most scarce and uncertain supplies and to enhance own and society's resilience. Diversification of MNEs toward services such as repair or circular product design, or toward refurbishment, remanufacturing, and repurposing opens new sources of revenue. At the same time, the transition to CE may imply reducing the demand for goods coming from higher-income markets, but emerging and developing countries are at the origin of new demand flows that could outweigh the steadier or declining demand from higher-income countries. Overall, MNEs' sales are not destined to decrease because of CE adoption.

Finally, market functioning and the mentioned incentives to CE transition for MNEs cannot be taken for granted. Governmental policies, such as for example the Circular Economy Action Plan of the European Commission with its mix of compulsory targets and economic measures, are necessary to trigger the transition. Additionally, bottom-up policies such as voluntary standards of more advanced producers can complement regulations and fiscal measures, and create snowball dynamics on both the consumption and production sides.

A last reflection should concern the measurement and monitoring of CE adoption at the firm level. Some CE indicators are available at plant, product, or country levels, and are rarely based on objective observations. Unfortunately, today we are not yet able to compare different MNEs in terms of their degree of adoption of CE principles and their position in the circular transition. Development and utilization of firm level, objective CE indicators is a promising area for allowing IB scholars to produce much needed empirical evidence on MNEs adoption of CE and the consequent transformation of GVCs.

REFERENCES

Agrawal, S., Singh, R. K., & Murtaza, Q. (2015). Resources, conservation and recycling: A literature review and perspectives in reverse logistics. *Resources, Conservation & Recycling*, *97*, 76–92. https://doi.org/10.1016/j.resconrec.2015.02.009

Apple. (2022). *Independent repair provider program*. https://support.apple.com/en-asia/irp-program

Bakker, C., Wang, F., Huisman, J., & den Hollander, M. (2014). Products that go round: Exploring product life extension through design. *Journal of Cleaner Production*, *69*, 10–16. https://doi.org/10.1016/j.jclepro.2014.01.028

BCG. (2020). *The consumers behind fashion's growing secondhand market*. https://www.bcg.com/it-it/publications/2020/consumer-segments-behind-growing-secondhand-fashion-market

Bocken, N. M. P., de Pauw, I., Bakker, C., & van der Grinten, B. (2016). Product design and business model strategies for a circular economy. *Journal of Industrial and Production Engineering*, *33*(5), 308–320. https://doi.org/10.1080/21681015.2016.1172124

Buckley, P. J., & Casson, M. (2020). The internalization theory of the multinational enterprise: Past, present and future. *British Journal of Management*, *31*(2), 239–252. https://doi.org/10.1111/1467-8551.12344

Burke, H., Zhang, A., & Wang, J. X. (2021). Integrating product design and supply chain management for a circular economy. *Production Planning & Control*, *34*(11), 1097–1113. https://doi.org/10.1080/09537287.2021.1983063

Calzolari, T., Genovese, A., & Brint, A. (2021). The adoption of circular economy practices in supply chains – An assessment of European multi-national enterprises. *Journal of Cleaner Production, 312*, 127616. https://doi.org/10.1016/j.jclepro.2021.127616

Cisco. (2021). *Purpose report. Our purpose, Our progress*. https://www.cisco.com/c/en/us/about/csr.html

Counterpoint. (2022). *Global refurbished smartphone market beats expectations, grows 15%*. Springer, Berlin, Heidelberg. https://www.counterpointresearch.com/global-refurbished-smartphone-market-2021/

de Brito, M. P., & Dekker, R. (2004). A framework for reverse logistics. In *Reverse logistics* (pp. 3–27). Springer. https://doi.org/10.1007/978-3-540-24803-3_1

De los Rios, I. C., & Charnley, F. J. S. (2017). Skills and capabilities for a sustainable and circular economy: The changing role of design. *Journal of Cleaner Production, 160*, 109–122. https://doi.org/10.1016/j.jclepro.2016.10.130

Electrolux. (2022). *Electrolux introduces new product recyclability tool*. https://www.electroluxgroup.com/en/electrolux-introduces-new-product-recyclability-tool-34279/

Ellen MacArthur Foundation. (2013). *Towards the circular economy 2: Opportunities for the consumer goods sector*. https://emf.thirdlight.com/file/24/qzvD2i1qVnZjTMqzpI2qI32rGA/Towards%20the%20circular%20economy%20Vol.%202%3A%20opportunities%20for%20the%20consumer%20goods%20sector.pdf

European Parliament. (2008). Directive 2008/98/EC of the European Parliament and of the Council of 19 November 2008 on waste and repealing certain directives. *Official Journal of the European Union*. https://eur-lex.europa.eu/legal-content/EN/TXT/?uri=celex%3A32008L0098

European Parliament. (2012). *Directive 2012/19/EU of the European Parliament and of the Council on waste electrical and electronic equipment (WEEE). Official Journal of the European Union.*

Eurostat. (2022). *End-of-life vehicle statistics*. https://eur-lex.europa.eu/legal-content/EN/TXT/?uri=celex%3A32012L0019

Fernandez-Stark, K., & Gereffi, G. (2019). Global value chain analysis: A primer. In S. Ponte, G. Gereffi, & G. Raj- Reichert (Eds.), *Handbook on global value chains* (2nd ed., pp. 54–76). Edward Elgar Publishing. https://doi.org/10.4337/9781788113779.00008

Fernández, I., & Kekäle, T. (2005). The influence of modularity and industry clockspeed on reverse logistics strategy: Implications for the purchasing function. *Journal of Purchasing and Supply Management, 11*(4), 193–205. https://doi.org/10.1016/j.pursup.2006.01.005

Font Vivanco, D., Kemp, R., & van der Voet, E. (2016). How to deal with the rebound effect? A policy-oriented approach. *Energy Policy, 94*, 114–125. https://doi.org/10.1016/j.enpol.2016.03.054

Galbreth, M. R., Boyac, T., & Verter, V. (2013). Product reuse in innovative industries. *Production and Operations Management, 22*(4), 1011–1033. https://doi.org/10.1111/j.1937-5956.2012.01330.x

Geissdoerfer, M., Vladimirova, D., & Evans, S. (2018). Sustainable business model innovation: A review. *Journal of Cleaner Production, 198*, 401–416. https://doi.org/10.1016/j.jclepro.2018.06.240

Gereffi, G., Humphrey, J., & Sturgeon, T. (2005). The governance of global value chains. *Review of International Political Economy, 12*(1), 78–104. https://doi.org/10.1080/09692290500049805

Ghisellini, P., Cialani, C., & Ulgiati, S. (2016). A review on circular economy: The expected transition to a balanced interplay of environmental and economic systems. *Journal of Cleaner Production, 114*, 11–32. https://doi.org/10.1016/j.jclepro.2015.09.007

Global Fashion Agenda. (2022). *Circular fashion partnership*. https://globalfashionagenda.org/circular-fashion-partnership/

Henry, M., Schraven, D., Bocken, N., Frenken, K., Hekkert, M., & Kirchherr, J. (2021). The battle of the buzzwords: A comparative review of the circular economy and the sharing economy concepts. *Environmental Innovation and Societal Transitions, 38*, 1–21. https://doi.org/10.1016/j.eist.2020.10.008

Hermundsdottir, F., & Aspelund, A. (2021). Sustainability innovations and firm competitiveness: A review. *Journal of Cleaner Production, 280*, 124715. https://doi.org/10.1016/j.jclepro.2020.124715

Hischier, R., & Böni, H. W. (2021). Combining environmental and economic factors to evaluate the reuse of electrical and electronic equipment – A Swiss case study. *Resources, Conservation and Recycling, 166*, 105307. https://doi.org/10.1016/j.resconrec.2020.105307

Ho, C. H., Böhm, S., & Monciardini, D. (2022). The collaborative and contested interplay between business and civil society in circular economy transitions. *Business Strategy and the Environment, 31*(6), 2714–2727. https://doi.org/10.1002/bse.3001

Hofstetter, J. S., De Marchi, V., Sarkis, J., Govindan, K., Klassen, R., Ometto, A. R., Spraul, K. S., Bocken, N., Ashton, W. S., Sharma, S., Jaeger-Erben, M., Jensen, C., Dewick, P., Schröder, P., Sinkovics, N., Ibrahim, S. E., Fiske, L., Goerzen, A., & Vazquez-Brust, D. (2021). From sustainable global value chains to circular economy—Different silos, different perspectives, but many opportunities to build bridges. *Circular Economy and Sustainability*, *1*(1), 21–47. https://doi.org/10.1007/s43615-021-00015-2

Hollander, M. C., Den, Bakker, C. A., & Hultink, E. J. (2017). Product design in a circular economy: Development of a typology of key concepts and terms. *Journal of Industrial Ecology*, *21*(3), 517–525. https://doi.org/10.1111/jiec.12610

Kanda, W., Geissdoerfer, M., & Hjelm, O. (2021). From circular business models to circular business ecosystems. *Business Strategy and the Environment*, *30*(6), 2814–2829. https://doi.org/10.1002/bse.2895

Kano, L., Tsang, E. W. K., & Yeung, H. W. C. (2020). Global value chains: A review of the multi-disciplinary literature. *Journal of International Business Studies*, *51*(4), 577–622. https://doi.org/10.1057/s41267-020-00304-2

Kasulaitis, B. V., Babbitt, C. W., & Krock, A. K. (2019). Dematerialization and the circular economy: Comparing strategies to reduce material impacts of the consumer electronic product ecosystem. *Journal of Industrial Ecology*, *23*(1), 119–132. https://doi.org/10.1111/jiec.12756

Kirchherr, J., Reike, D., & Hekkert, M. (2017). Conceptualizing the circular economy: An analysis of 114 definitions. *Resources, Conservation and Recycling*, *127*, 221–232. https://doi.org/10.1016/j.resconrec.2017.09.005

Kolk, A., Kourula, A., & Pisani, N. (2017). Multinational enterprises and the sustainable development goals: What do we know and how to proceed? *Transnational Corporations*, *24*(3), 9–32.

Lieder, M., & Rashid, A. (2016). Towards circular economy implementation: A comprehensive review in context of manufacturing industry. *Journal of Cleaner Production*, *115*, 36–51. https://doi.org/10.1016/j.jclepro.2015.12.042

Llorente-gonzález, L. J., & Vence, X. (2020). How labour-intensive is the circular economy? A policy-orientated structural analysis of the repair, reuse and recycling activities in the European Union. *Resources, Conservation & Recycling*, *162*, 105033. https://doi.org/10.1016/j.resconrec.2020.105033

Loomba, A. P. S., & Nakashima, K. (2012). Enhancing value in reverse supply chains by sorting before product recovery. *Production Planning & Control*, *23*(2–3), 205–215. https://doi.org/10.1080/09537287.2011.591652

Lyons, D. I. (2007). A spatial analysis of loop closing among recycling, remanufacturing, and waste treatment firms in Texas. *Journal of Industrial Ecology*, *11*(1), 43–54. https://doi.org/10.1162/jiec.2007.1029

Machado, N., & Morioka, S. N. (2021). Contributions of modularity to the circular economy: A systematic review of literature. *Journal of Building Engineering*, *44*, 103322. https://doi.org/10.1016/j.jobe.2021.103322

Makov, T., & Font Vivanco, D. (2018). Does the circular economy grow the pie? The case of rebound effects from smartphone reuse. *Frontiers in Energy Research*, *6*(39), 1–11. https://www.frontiersin.org/articles/10.3389/fenrg.2018.00039

Montiel, I., Cuervo-Cazurra, A., Park, J., Antolín-López, R., & Husted, B. W. (2021). Implementing the United Nations' sustainable development goals in international business. *Journal of International Business Studies*, *52*(5), 999–1030. https://doi.org/10.1057/s41267-021-00445-y

Moraga, G., Huysveld, S., Mathieux, F., Blengini, G. A., Alaerts, L., Van Acker, K., de Meester, S., & Dewulf, J. (2019). Circular economy indicators: What do they measure? *Resources, Conservation and Recycling*, *146*, 452–461. https://doi.org/10.1016/j.resconrec.2019.03.045

Nike. (2013). *NIKE, Inc. unveils ColorDry technology and high-tech facility to eliminate water and chemicals in dyeing*. https://about.nike.com/en/newsroom

Pananond, P., Gereffi, G., & Pedersen, T. (2020). An integrative typology of global strategy and global value chains: The management and organization of cross-border activities. *Global Strategy Journal*, *10*(3), 421–443. https://doi.org/10.1002/gsj.1388

Pieroni, M. P. P., McAloone, T. C., & Pigosso, D. C. A. (2019). Business model innovation for circular economy and sustainability: A review of approaches. *Journal of Cleaner Production, 215*, 198–216. https://doi.org/10.1016/j.jclepro.2019.01.036

Potting, J., Hekkert, M., Worrell, E., & Hanemaaijer, A. (2017). *Circular economy: Measuring innovation in the product chain* (Issue 2544). PBL Netherlands Environmental Assessment Agency.

Qu, S., Guo, Y., Ma, Z., Chen, W.-Q., Liu, J., Liu, G., Wang, Y., & Xu, M. (2019). Implications of China's foreign waste ban on the global circular economy. *Resources, Conservation and Recycling, 144*, 252–255. https://doi.org/10.1016/j.resconrec.2019.01.004

Rahmani, M., Gui, L., & Atasu, A. (2021). The implications of recycling technology choice on extended producer responsibility. *Production and Operations Management, 30*(2), 522–542. https://doi.org/10.1111/poms.13279

Ranta, V., Aarikka-Stenroos, L., Ritala, P., & Mäkinen, S. J. (2018). Exploring institutional drivers and barriers of the circular economy: A cross-regional comparison of China, the US, and Europe. *Resources, Conservation and Recycling, 135*, 70–82. https://doi.org/10.1016/j.resconrec.2017.08.017

Reike, D., Vermeulen, W. J. V., & Witjes, S. (2018). The circular economy: New or refurbished as CE 3.0? — Exploring controversies in the conceptualization of the circular economy through a focus on history and resource value retention options. *Resources, Conservation and Recycling, 135*, 246–264. https://doi.org/10.1016/j.resconrec.2017.08.027

Spreafico, C., & Landi, D. (2022). Using product design strategies to implement circular economy: Differences between students and professional designers. *Sustainability, 14*(3), 1122. https://doi.org/10.3390/su14031122

Stahel, W. R. (2010). *The performance economy*. Palgrave Macmillan.

Stahel, W. R. (2016). The circular economy. *Nature, 531*(7595), 435–438. https://doi.org/10.1038/531435a

Strange, R., & Humphrey, J. (2019). What lies between market and hierarchy? Insights from internalization theory and global value chain theory. *Journal of International Business Studies, 50*(8), 1401–1413. https://doi.org/10.1057/s41267-018-0186-0

The Guardian. (2013). *Nike ColorDry adds water-free dyed fabric to sustainable materials menu.* https://www.theguardian.com/sustainable-business/nike-colordry-water-free-sustainable-materials

UNCTAD. 2020. *World investment report 2020*. United Nation Pubblications. https://doi.org/10.18356/920f7642-en

Zhang, Q., Xu, J., Wang, Y., Hasanbeigi, A., Zhang, W., Lu, H., & Arens, M. (2018). Comprehensive assessment of energy conservation and CO_2 emissions mitigation in China's iron and steel industry based on dynamic material flows. *Applied Energy, 209*, 251–265. https://doi.org/10.1016/j.apenergy.2017.10.084

Zink, T., & Geyer, R. (2017). Circular economy rebound. *Journal of Industrial Ecology, 21*(3), 593–602. https://doi.org/10.1111/jiec.12545

CHAPTER 17

THE ROLE OF SUSTAINABILITY IN BACKSHORING DECISIONS – A CONCEPTUAL FRAMEWORK

Lise Lillebrygfjeld Halse

Molde University College, Norway

ABSTRACT

Even though sustainability appears to be a relevant driver for the relocation of production, this has only to a limited extent been studied as an independent motive or a result of backshoring. This study explores the literature on backshoring and sustainability and discusses some empirical cases to shed light on the connection between sustainability and backshoring. This study argues that sustainability issues may require a broader perspective than the dominant economic logic framing previous studies on backshoring. Institutional theory is suggested as a theoretical framework for analyzing how sustainability can play a role in companies' backshoring decisions.

Keywords: Offshoring; backshoring; sustainability; institutional isomorphism; motives

Walking the Talk? MNEs Transitioning Towards a Sustainable World
Progress in International Business Research, Volume 18, 335–350

Published under exclusive licence by Emerald Publishing Limited
ISSN: 1745-8862/doi:10.1108/S1745-886220240000018021

INTRODUCTION

In recent years, environmental and social sustainability has received growing attention in all aspects of social life, as well as in practice and research. The international business (IB) literature paid attention to sustainability issues at a relatively early stage (Rugman & Verbeke, 1998, 2001), especially associated with sustainability policy and regulations that MNEs must deal with. Furthermore, sustainability has mainly been addressed under the corporate social responsibility heading (Kolk, 2016; Paul & Parra, 2021). However, despite the growing interest in the IB field, and that decisions regarding the localization of production affect companies' sustainability performance (Kolk, 2016), the literature that addresses the role of sustainability in backshoring decisions is scarce (Ashby, 2016; Orzes & Sarkis, 2019). According to Orzes and Sarkis (2019), the relationship between backshoring and sustainability is a "fundamentally unexplored relationship" (p. 482). Considering the increased and timely focus on sustainability, more research has therefore been called for to shed light on the relationship between sustainability and backshoring (Ashby, 2016; Di Stefano & Fratocchi, 2019; Orzes & Sarkis, 2019). Hence, the aim of this study is to explore the role of sustainability as a motive and a result of the backshoring decision.

THE PHENOMENON OF BACKSHORING

The backshoring or reshoring decision can be considered as the opposite of the decision to outsource production to another country. Backshoring thus involves moving activities "home" to where the parent company is located. Ellram (2013) defines reshoring as follows: "Moving manufacturing back to the country to [the firm's] parent company" (p. 3). Gray et al. (2013) have expanded this definition by including the owner dimension and propose a typology with four situations: internal reshoring, outsourced reshoring, reshoring for outsourcing, and reshoring for insourcing. In this chapter, the term "backshoring" is used, which includes all cases of bringing back production to the home country, regardless of ownership. This includes situations where production is moved from abroad to the parent companies' own organization or another organization in their home country, either from foreign subsidiaries or independent foreign suppliers.

Benito and Welch (1997) published one of the first studies looking at the withdrawal or reduction of international operations which they called "de-internationalization." In their theoretical study, they focus especially on leadership and reflect on why this strategy is so unusual. They believe that this may be due to negative and undesirable attitudes toward foreign market exit, since there is some form of admission of previously failed decisions related to this strategy that causes managers to avoid correcting these decisions. Furthermore, managers and organizations build commitment to their international operations over time, which can be difficult to reverse. This study probably reflects the general attitude at the time when building IB operations was a norm (Musteen, 2016). In recent years, however, the general attitude to this issue seems to have changed, and thus probably also the perceptions of managers. Several countries have introduced certain initiatives to bring production back, based on the assumption that production is essential to a country's prosperity (Roos & Kennedy, 2014) and innovation (Pisano & Shih, 2012). During the previous US presidential term, this was particularly prominent on the political agenda, with promises to revive manufacturing and stop the offshoring of American jobs. Furthermore, with the recent experience with the vulnerability of global value chains during the COVID-19 pandemic, it is claimed that we will see a movement toward a radically different global economy with shorter supply chains and fewer production steps (Liversey, 2012). From this perspective, it is beneficial to concentrate business activities within smaller geographical areas.

While there is extensive literature studying global sourcing, the literature on backshoring is still relatively limited (Stentoft et al., 2016). Literature reviews show that most studies have mainly focused on the scope and drivers of this phenomenon (Kinkel, 2012; Kinkel & Maloca, 2009), the decision-making processes (Tate & Bals, 2014), and the consequences of the decisions (Kinkel, 2014). The studies mostly apply the same theoretical frameworks used to explain and analyze global offshoring (Foerstl et al., 2016), mainly resting on transaction cost theory (Williamson, 1975) and the resource-based view (Prahalad & Hamel, 1990). Based on a systematic literature review, Wiesmann et al. (2017) conclude that no consensus has been reached on theory in the field. This is supported by a recently published conceptual study conducted by McIvor and Bals (2021), claiming that there is a lack of theoretical frameworks that explain the various and coherent factors in decisions related to backshoring.

According to Canham and Hamilton (2013), decisions to move production back to high-cost locations may indicate that companies are considering a wider range of resource-based factors beyond the initial cost focus that was prominent when the organizations initially outsourced production. Based on findings from a survey in three Scandinavian countries (Sweden, Denmark, and Finland), Johansson et al. (2019) claim that for offshoring, costs, market, development, trade policy, and external influence are relevant factors. For backshoring, they find that quality stands out as the most important factor. Other factors include development, proximity to the market, costs and external influence, and trade policy, as well as experience in backshoring such as quality, delivery, and benefits related to flexibility.

The scarcity of studies focusing on sustainability in the backshoring literature may indicate that sustainability aspects so far have not been important in companies' sourcing decisions, or that this phenomenon has been overlooked by IB scholars. Consequently, there is a need for studies exploring the role of sustainability in backshoring decisions.

SUSTAINABILITY AND BACKSHORING

The concept of sustainability is increasingly powerfully present but is also an unclear and complex concept. According to the Brundtland report, sustainable development is defined as development that "meets the needs of the present without compromising the ability of future generations to meet their own needs"

(United Nations General Assembly, 1987). The division of the concept of sustainability into three pillars has been widely applied (Elkington, 1999), namely the economic, social, and environmental pillars which are explicitly embedded in the United Nations Sustainable Development Goals. According to these, companies should operate and report within three dimensions of sustainability, the economic, social, and environmental. The triple bottom line assumes that it is possible to contribute along all three dimensions at the same time. However, it has been shown that environmental and social sustainability may conflict with the goal of achieving economic profit (Glover et al., 2014).

Despite the increasing interest in sustainable supply chains (Koberg & Longoni, 2019) and sustainable global sourcing (Jia & Jiang, 2018), the literature that addresses the role of sustainability in backshoring decisions is scarce (Ashby, 2016; Orzes & Sarkis, 2019). This is documented in a literature review carried out by Fratocchi and Di Stefano (2019) which concludes that "scarce attention has been paid, until now, to the interdependences (if any) among sustainability issues and the firm's decision to backshore its manufacturing activities" (p. 451). According to Fratocchi and Di Stefano (2019), only a few publications focus specifically on sustainability as a specific and independent result of backshoring, but increasingly assume relevance. However, they also stress that the decision to move production activities to their home country may have an impact on sustainability, even if it was not the main motivation in the first place. Di Stefano and Fratocchi (2019) claim that little attention has been paid to more specific elements of sustainability, such as the role and environmental certifications of consumers and suppliers (footprints), and argue that sustainability is not recognized as an individual result of moving production to their home country. They propose a model for further research where social and environmental sustainability can represent a motivation/driver, barrier/facilitator, and outcome for backshoring decisions. While the model shows that sustainability may play a role in different stages of the backshoring process, its theoretical foundation and how sustainability can motivate the backshoring decision is however not clear.

Backshoring is often explained as a localization and/or cost-related choice in line with transactional cost theory and resource-based theory, where one sees reduced cost benefits from operating abroad, varying demand, or a question related to managing networks and ownership issues (Wiesmann et al., 2017). From an economic logics perspective, companies make strategic choices that ensure short- and/or long-term economic sustainability. The trend of global sourcing from Western countries has its foundation in economic considerations, resting on lower cost levels in some regions, for example, in Asia and East Europe. Increased attention toward sustainability aspects of transportation and production may alter these considerations. For example, cost of global gas emissions associated with transport and manufacturing involves increased costs for the company (Akyelken & Keller, 2014; Sawhney & Rastogi, 2015). The cost of emissions may vary across locations, altering country specific advantages (Cole, 2004; Kolk & Pinkse, 2008; Li & Zhou, 2017; Sawhney & Rastogi, 2015). The cost of emissions, as carbon pricing, can be related to taxes that the polluter must pay, costs of changing production processes to be more environmentally friendly, and costs associated with complying with regulations. Pricing on emissions and certificates in the countries where companies operate may thus influence decisions on backshoring. Regarding social sustainability, poor working conditions and human rights in several countries have contributed to the attention concerning the problems of production in low-cost locations (Clarke & Boersma, 2017; Lim & Phillips, 2008). More attention attached to social and economic sustainability can lead to stricter requirements associated with salary and working conditions in low-cost locations, which could increase the cost of producing abroad and hence affect backshoring decisions.

In addition to the direct costs associated with environmental and social sustainability, sustainability issues may involve increased transaction costs related to managing, controlling, and monitoring agreements with suppliers or foreign subsidiaries. With a more extensive set of regulatory requirements for sustainability, we could expect that such agreements will include increasingly detailed sustainability requirements. Consequently, it will involve increased costs to handle such agreements. Distance between the parent company and foreign production facilities may also limit the possibilities for managing aspects related to sustainability (Koberg & Longoni, 2019), since increased geographical distance makes it difficult to assess suppliers' commitments and performance in terms of sustainability (Gualandris et al., 2014). In line with this, Ashby (2016) claims that proximity to the home country provides better control over the environmental impact of production processes, in addition to increased visibility related to ethical behavior (Gray et al., 2013). Sustainability issues may also involve investing in specific assets, which may affect transaction costs and hence backshoring decisions.

Sustainability can represent opportunities for MNEs to create competitive advantages. Kolk and Pinkse (2008) discuss how climate change may induce MNEs to transform existing firm-specific advantages (FSA) or build new ones. Examples of this are product differentiation based on improving environmental quality, and in-house development of pollution prevention technologies. The question in this context is how the development of FSA is related to the location of operations, and may indirectly motivate backshoring.

According to backshoring literature, the motives for backshoring are more complex than decisions to outsourcing of production, which has traditionally been driven by cost motives. The competence and level of education of human resources have been considered as a possible motive in this literature. Lund and Steen (2020) highlight regional values ("assets") and embeddedness as important aspects of localization. They point to the education system in Norway as a resource that is important for companies' ability to implement production technologies, which can play a role in backshoring. Furthermore, the Norwegian culture and the way of organizing work ("The Norwegian Model"), and the historically accumulated competence can be regarded as regional values that can enable backshoring. On the other hand, the depletion of knowledge in the home country due to global sourcing may represent a barrier to backshoring (Nujen et al., 2018). Although knowledge related to sustainability has not been explicitly addressed in this context, it can be envisaged that knowledge related to sustainable operations may become increasingly important. We can expect that this knowledge will be important to have "in-house" to ensure that the company operates according to environmental and social sustainability. This knowledge can represent a competitive advantage in a market that is becoming increasingly concerned with sustainability. In the extension of the resource-based perspective, collaboration between the parent company and suppliers to develop knowledge about sustainability will become increasingly important (Gold et al., 2010). Moreover, geographical proximity may facilitate such competence development (Porter, 1998). Resources in the form of knowledge related to sustainability may thus represent an important motive for backshoring.

Above, we have presented some reflections about how sustainability may motivate backshoring, anchored in the theories that have mostly applied for shedding light on backshoring decisions, namely transaction cost theory and the resource-based view. The main limitation of both transactional cost theory and the resource-based view is that one considers the sourcing and localization decision as mainly a rational decision where actors independently are able to gather information and knowledge as the basis for the decision (Canello, 2021; Johanson & Vahlne, 1977), and where there is some certainty regarding the outcome of the backshoring decision. These perspectives further address the individual company and its core competence when putting limits on what the company must keep internally and what can be outsized to suppliers. There is a fundamental atomistic and competitive view associated with these considerations, which contrasts with the idea that organizations are part of a context where organizations together will contribute to the development toward sustainability. An alternative view is that companies' strategic choices will be influenced by institutional factors and elements of the company contexts (Halse et al., 2019). Institutional theory provides a theoretical lens that opens new avenues for considering aspects of backshoring decisions that are influenced by other factors than economic efficiency.

INSTITUTIONAL THEORY AS AN ALTERNATIVE THEORETICAL LENS

Decision makers' motives and perceptions are shaped by the context in which companies and managers are a part (Benito & Welch, 1997; March, 1994). Institutional theory provides a theoretical framework where we can identify and investigate how factors such as culture, social environment, regulation, traditions, and history can influence actors' decisions (DiMaggio & Powell, 1983, 1991; Scott, 2008). Political and economic pressures affect companies' strategies and organizational decision-making processes as companies are seeking to legitimize their practices in the views of other stakeholders. This represents an alternative view to the more risk-oriented assumptions on which the traditional theories in the backshoring literature are based.

Institutional theory highlights legitimization processes where, for example, organizational structures and management models are taken for granted, regardless of whether they are effective or not for the individual company. DiMaggio and Powell (1983) formulate the term institutional isomorphism that describes how organizations are influenced by pressure from the institutional context, and where organizations' strategic choices are part of their cultural and historical context. Isomorphism can be a "constraining process that forces one unit in a population to resemble other units that face the same set of environmental conditions" (p. 149). This contrasts with the supposed rational actor at the center of transaction cost theory and resource-based theory. DiMaggio and Powel describe three types of isomorphism, coercive, normative, and mimetic. Coercive pressure can arise through the influence of actors who have a powerful position in the industry and who are important for driving sustainability forward. This kind of pressure can also come from governments that adopt laws and regulations related to the different dimensions of sustainability. Such laws will not only depend on the localization in which the operations take place but also on laws in the home country and requirements from consumers that may impact the assessment of moving production home. This applies, for example, to requirements for sustainability reporting that may include companies' operations in other countries. Requirements for guarantees of origin and various environmental approvals and certificate systems are forms of coercive isomorphism that affect companies' operations in countries other than their own.

Normative drivers ensure that organizations' actions are perceived as legitimate by the environment. Normative pressures related to sustainability may emerge from customers, other interest groups or from currents in society that influence what is perceived as important and correct. For example, customers may question the companies' activities at home and abroad in terms of sustainability and perceive local suppliers as being more sustainable (Veit et al., 2018). The norms for what are considered acceptable can change over time, as we have seen in several industries where expectations regarding sustainability have changed considerably recently. Legitimization processes related to increased sustainability in value chains have previously been investigated in different industries and contexts (Glover et al., 2014). Companies and their leaders will seek to gain legitimacy among their stakeholders through decisions that are in line with sustainability principles which affect sourcing decisions.

Mimetic isomorphism refers to the tendency of an organization to copy other organizations due to uncertainty, for example, due to ambiguous and uncertain goals and solutions. Sustainability may represent uncertainty to many companies, and they may therefore seek solutions to show commitment and gain legitimacy related to sustainability. This can play an important role in companies implementing sustainable practices (Sarkis et al., 2011). It is well known that management models "travel" in time and space and are spread when companies copy each other (Røvik, 2011), locally, nationally, and globally. Sourcing strategies have been considered management models that have spread through mimetic isomorphism (Canello, 2021). Likewise, sustainability as an issue may be considered as something that companies want to comply with to gain legitimacy.

In the following, Norwegian cases of backshoring are presented to shed light on sustainability and backshoring, which provide a basis for developing a theoretical framework that can be used to understand the link between these two phenomena.

NORWEGIAN CASES OF BACKSHORING

Norway has an open economy with a high level of cost. From being an industry society, a considerable part of manufacturing has been sourced out to localizations with lower costs of labor in the last decades. As in other Western countries, we have recently seen an increased interest in reviving manufacturing, which has materialized through the establishment of a center for competitive high value manufacturing in Norway (Karlsen, 2019), clusters and catapult centers, among other initiatives. Despite this, backshoring is still not very common in the Norwegian context. According to European Reshoring Monitor (Eurofound), only 15 different companies in Norway have carried out backshoring (until 2019 when the Eurofound project ended). Several of these cases have been subject to previous studies, and here we will comment upon some of the Norwegian cases relevant to the discussion on sustainability. In the following, four companies that have brought back manufacturing from low-cost locations are presented. Companies 1 and 2 are located in a Norwegian maritime cluster, encompassing supply chains that until recently has primarily delivered advanced vessels to the oil and gas industry. After the oil crisis in 2014, cluster companies have oriented themselves into several other market segments. Table 17.1 gives a brief presentation of the companies.

Company 1 is a shipbuilding company that until recently mainly delivered advanced and complex vessels to the petroleum offshore industry. The company started its backshoring journey in 2012, which was motivated by reduced lead time, knowledge development, and norms in the institutional context (Halse et al., 2019). The company was later restructured with a new strategy involving recycling and repair of vessels. In a recent study, a respondent from this company

Table 17.1. Case Companies.

Company	Industry	Year, Backshoring Decision	Main Motives for Backshoring Decision
Company 1	Maritime, shipbuilding	2012	Lead time, manufacturing competence, norms associated with regional manufacturing
Company 2	Maritime, equipment supplier	2016	Free capacity at home, product development, norms associated with local employment
Company 3	Construction, building systems	2016	Trustworthiness and legitimacy regarding sustainability issues
Company 4	Marine, plastic components/modules	2011	Innovation and product development with customer

claimed that the company's backshoring strategy was important to build the knowledge required by their recent recycling strategy (Halse & Klymenko, 2022). The backshoring process involved introducing automatization technology in their manufacturing, which has been key to their recent strategy of recycling and repair:

> We were able to use the robotization we had here and our great advantage of having a much greater flexibility at the shipyard. We have everyday control instead of being tied to those contracts abroad. [...] At the same time we are building expertise and is future-oriented, and will be competitive in the future.

This illustrates that even though sustainability was not the main issue at the outset, backshoring laid the foundation for pursuing a new strategy in accordance with a growing market associated with sustainability. Manufacturing knowledge was an important element of the original backshoring decision, which turned out to be of great importance when the company several years later oriented itself to be "a full lifecycle shipyard" (Halse & Klymenko, 2022).

Company 2 delivers ship equipment to ship owners and the oil and gas industry. After a period when the company offshored parts of its production, the company initiated a strategy for bringing back manufacturing around 2016 (Halse et al., 2019). Initially, the decision to backshore was based on available capacity at home, and the need for product development: the underlying norm for the decision was to sustain activities at home when the company experienced reduced market demand in the aftermath of the oil crisis. The respondent points out that environmental sustainability was not an important issue when they decided to bring back manufacturing, as it was not "fashionable" at that time (Halse & Klymenko, 2022). The company still has its main customers within the oil and gas industry where the focus is primarily on cost:

> Price comes first, in addition they expect you to deliver on sustainability. However, we see that it is mostly about being good at selling ideas.

The quote may illustrate that sustainability issues have not yet penetrated this market.

Company 3 is a manufacturer of wooden building systems that claims that its reputation regarding environmental sustainability was important when it decided to bring back manufacturing from Eastern Europe in 2016 (Halse & Klymenko, 2022). In this case, climate gas emissions associated with transport and manufacturing abroad conflicted with the green marketing of the company's products. This was the main reason why the company decided to explore the opportunities to build manufacturing capacity at home:

> When we sourced from Poland the freight and energy mix ruined our environmental account. We found out that either we had to stop with it, or we had to produce it ourselves.

The company decided to build a new factory locally where the product received their own quality label "Norwegian made," confirming that the products are made from renewable, locally sourced raw materials.

Company 4 is a manufacturer of customized plastic components for different applications. After a process of persuasion with a large customer in the Norwegian fish farming industry, the company started to produce a small component to a larger construction that was produced in China. This was the first stage in the process leading to the development of recycled products on a larger scale. From only delivering smaller components, further product development has led to local manufacturing of the complete construction that was produced in China before. A company manager emphasizes that the long-term collaboration and shared culture with the Norwegian customer was essential for developing a product with recycled materials. Backshoring of production from China to Norway was the first stage of the process that led to the development of sustainable products and a new manufacturing process.

> Sustainability was not the initial issue in the process with the customer, but we have worked hard to get it on the agenda. [...] We will soon deliver the first complete product which is produced in 100 per cent of recycled plastic, plastic that comes from previously discarded products, so it is completely circular.

This case illustrates that even though sustainability was not an initial motive for backshoring but became an important outcome of the decision. The original backshoring decision was based on a need for proximity between the manufacturer and the customer to be able to develop new innovative solutions, which later developed to be oriented toward sustainable products.

REFLECTIONS

There are some reflections to be made from the above four cases. First, sustainability appears to be an emerging issue, which also is reflected in the scarcity of empirical studies in the field of backshoring. The companies have reported several motives for backshoring, however only one of them reported sustainability as the first and direct motive for backshoring. In this case, reputation and market considerations were the underlying reasons for backshoring. The company decided to backshore because it was not trustworthy in promoting green products while at the same time having a large carbon footprint related to manufacturing and transportation outside of Norway. An important observation here is that this decision did not appear to be based on detailed calculations of climate gas emission but was mainly concerned with norms concerning sustainability and how the company was seen by customers and other actors. This illustrates how legitimacy plays a role in motivating backshoring (DiMaggio & Powell, 1991). This aspect is also reflected in the case (Company 2) where sustainability appears not to be an essential topic yet, which may reflect the general attitudes in the industry.

Two of the companies that did not report sustainability as an initial backshoring motive reported that this earlier backshoring decision has been important when they later changed their strategy in the sustainable direction, and to innovate products that are more sustainable. These findings could be considered to be in line with the resource-based view, where resources are linked to people, knowledge, processes, and routines that the company develops over time, and the sharing of these between employees (Kogut & Zander, 1992). However, the network and institutional aspects of knowledge related to backshoring are not fully explained by the main theories within the backshoring literature, which calls for further theoretical development in the field.

CONCEPTUAL MODEL

Based on the above reflections and case descriptions, a conceptual model is proposed (Fig. 17.1).

The model is inspired by institutional theory and the model published in the study by Di Stefano and Fratocchi (2019, p. 134), where environmental and social sustainability are shown as motivations/drivers before the backshoring decision, a barrier or enabling factor when implementing the decision, and as an outcome of the decision. The model is further developed by including institutional theory in the model to explain how environmental and social sustainability may affect the motivation for the backshoring decision. Furthermore, the outcome of backshoring is refined by dividing between indirect and direct sustainability outcomes of backshoring in line with empirical findings from previous case studies in a Norwegian context described above.

In the further development of the original model, three main processes underlying sustainability motives for backshoring decisions have been introduced. First, formal (explicit) requirements posed by customers, regulative bodies, or other stakeholders, may lead to a direct motive for backshoring. Requirement associated with climate gas emission is an example of this, where backshoring may reduce emissions associated with transport, or cleaner energy sources in the home country. Requirements associated with social sustainability issues such as workers' rights at the offshoring location may also motivate backshoring.

Secondly, norms in the industry, among managers or the companies' context (e.g., a cluster), may shape backshoring motives in a sustainable direction. This is the situation when sustainability issues are not direct requirements from stakeholders but represents more general expectations directed toward the company

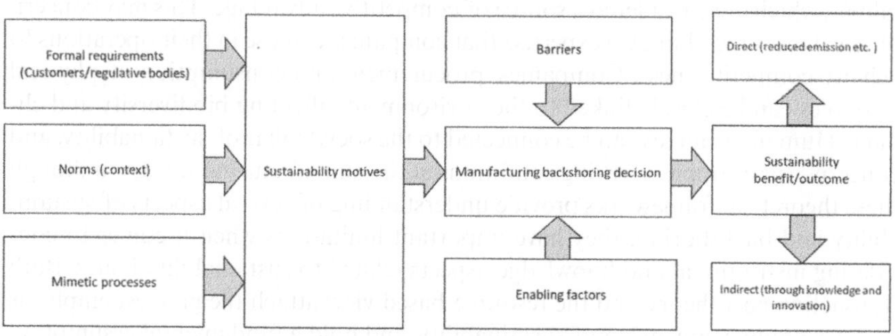

Fig. 17.1. How Sustainability Can Affect Backshoring Decision and Be an Outcome of Backshoring (Based on DiMaggio & Powell, 1991; Di Stefano & Fratocchi, 2019).

(Halse et al., 2019). For example, companies that are part of a cluster that pay particular attention to sustainability issues, may experience that they are expected to make strategic decisions in line with this. The backshoring decision is an example of such strategic decisions. The findings in this study show that there has been a change in norms regarding sustainability the recent years. However, it could take some time before these norms materialize into backshoring decisions, as there could be barriers or inertia in the process, as illustrated by Benito and Welch (1997). Thirdly, managers tend to mimic other organizations (DiMaggio & Powell, 1983). When other companies become more concerned about sustainable issues and choose to relocate their operations, other companies may follow to be "fashionable," to gain legitimacy and to reduce uncertainty.

I the model we have included barriers and enablers that companies may experience in the backshoring process, in accordance with the model presented by Di Stefano and Fratocchi (2019). Moreover, and according to their model, sustainability may be an outcome of backshoring. Based on the findings in this study, this is further elaborated in the conceptual model in Fig. 17.1. Sustainability may be a direct outcome of backshoring, for example, by reducing climate gas emission but can also be an indirect outcome of backshoring as companies build knowledge and increase their innovativeness, enabling them in developing new sustainable products and services.

CONCLUDING REMARKS

The main theories in the backshoring literature have been transaction cost theory and the resource-based view. Transaction cost theory is well suited to addressing the economic dimension of sustainability and can shed light on the role of the other two sustainability dimensions in backshoring decisions when these are associated with cost. Example of this is the pricing of greenhouse gas emissions and coordination costs associated with implementing and managing increased sustainability requirements. The resource-based view is primarily concerned with the composition and development of an organization's internal resources and capabilities, which are considered a source of competitive advantage. This may concern physical resources, but also expertise that companies can use in their operations to achieve competitiveness. Companies' procurement and consumption of physical resources can be closely linked to the environment, affecting biodiversity and climate. Human resources can be connected to the social pillar of sustainability, and concerns, for example, working conditions at the manufacturing site. Even though these theoretical frameworks provide understanding of several aspects of sustainability and backshoring, they have important limitations when it comes to considering institutional and knowledge aspects related to sustainability issues. Both transaction cost theory and the resource-based view attach the greatest emphasis to the economic dimension of sustainability, and hold a fundamental assumption that actors make rational choices based on complete information. The findings from Norwegian backshoring cases indicate that there are also factors and mechanisms that are not well explained through these theoretical frameworks.

In this study, institutional theory is introduced to provide a more nuanced explanation of the role of sustainability in backshoring decisions. In addition to formal requirements from customers and regulative bodies, the framework includes norms and mimetic processes underlying sustainability motives leading to backshoring. As we have seen from the cases presented in this study, these aspects may play an important role when considering the role of sustainability in backshoring decisions. Moreover, previous backshoring decisions may prepare companies for their later orientation toward environmental sustainability, showing that sustainability may be an indirect outcome of backshoring.

The study demonstrates that by introducing institutional theory we can account for changes in the underlying motives over time. This is especially prominent regarding sustainability issues, which have become increasingly important in society. What is considered important changes over time, and thus also motives underlying backshoring decisions. Moreover, as an increasing number of companies backshore and associate this with increased sustainability, we may expect an amplified effect when these firms create norms and mimetic processes that motivate other firms to follow. In that sense, we may expect that sustainability will play an increasingly important role in backshoring decisions.

This study contributes to the backshoring literature by elucidating how institutional theory can capture key aspects of sustainability in backshoring decisions. Institutional theory and isomorphism represent approaches that are well suited to address other factors than the rational ones, and the context in which organizations' choices are made. In this way, the proposed framework can explain how the three dimensions of sustainability can influence organizations' decisions concerning backshoring. This represents a first step in the development of theoretical frameworks that can provide a more nuanced understanding of the backshoring phenomenon. The study presented in this chapter contributes to the emerging literature on backshoring and sustainability by advancing current theoretical frameworks on analyzing the role of sustainability of firm's backshoring decisions.

The study adopts only four illustrative cases from the Norwegian context. Further studies should provide rich empirical studies demonstrating the framework and how sustainability plays into backshoring decisions, and how these decisions can result in increased sustainability.

REFERENCES

Akyelken, N., & Keller, H. (2014). Framing the nexus of globalisation, logistics and manufacturing in Europe. *Transport Reviews*, *34*(6), 674–690.

Ashby, A. (2016). From global to local: Reshoring for sustainability. *Operations Management Research*, *9*(3–4), 75–88.

Benito, G. R. G., & Welch, L. S. (1997). De-internationalization. *MIR: Management International Review*, *37*, 7–25.

Canello, J. (2021). Mimetic isomorphism, offshore outsourcing and backshoring decisions among micro and small enterprises. *Regional Studies*, *56*(5), 719–736.

Canham, S., & Hamilton, R. T. (2013). SME internationalisation: Offshoring, "backshoring", or staying at home in New Zealand. *Strategic Outsourcing: An International Journal*, *6*(3), 277–291.

Clarke, T., & Boersma, M. (2017). The governance of global value chains: Unresolved human rights, environmental and ethical dilemmas in the apple supply chain. *Journal of Business Ethics*, *143*(1), 111–131.

Cole, M. A. (2004). Trade, the pollution haven hypothesis and the environmental Kuznets curve: Examining the linkages. *Ecological Economics*, *48*(1), 71–81.

Di Stefano, C., & Fratocchi, L. (2019). Manufacturing back-shoring and sustainability: A literature review. *Sinergie Italian Journal of Management*, *37*, 119–143.

DiMaggio, P., & Powell, W. (1983). The iron cage revisited: Institutional isomorphism and collective rationality in organizational fields. *American Sociological Review*, *48*(2), 147–160.

DiMaggio, P. J., & Powell, W. W. (1991). The iron cage revisited: Institutional isomorphism and collective rationality in organizational fields. In P. J. DiMaggio & W. W. Powell (Eds.), *The new institutionalism in organizational analysis* (pp. 62–82). University of Chicago Press.

Elkington, J. (1999). *Cannibals with forks: The triple bottom line of 21st century business*. Capstone.

Ellram, L. M. (2013). Offshoring, reshoring and the manufacturing location decision. *Journal of Supply Chain Management*, *49*(2), 3–5.

Eurofound. (n.d.). *European reshoring monitoring: Reshoring cases*. https://reshoring.eurofound.europa.eu/reshoring-cases

Foerstl, K., Kirchoff, J. F., & Bals, L. (2016). Reshoring and insourcing: Drivers and future research directions. *International Journal of Physical Distribution & Logistics Management*, *46*(5), 492–515.

Fratocchi, L., & Di Stefano, C. (2019). Does sustainability matter for reshoring strategies? A literature review. *Journal of Global Operations and Strategic Sourcing*, *12*, 449-476.

Glover, J. L., Champion, D., Daniels, K. J., & Dainty, A. J. D. (2014). An institutional theory perspective on sustainable practices across the dairy supply chain. *International Journal of Production Economics*, *152*, 102–111.

Gold, S., Seuring, S., & Beske, P. (2010). Sustainable supply chain management and inter-organizational resources: A literature review. *Corporate Social Responsibility and Environmental Management*, *17*(4), 230–245.

Gray, J. V., Skowronski, K., Esenduran, G., & Johnny Rungtusanatham, M. (2013). The reshoring phenomenon: What supply chain academics ought to know and should do. *Journal of Supply Chain Management*, *49*(2), 27–33. https://doi.org/10.1111/jscm.12012

Gualandris, J., Golini, R., & Kalchschmidt, M. (2014). Do supply management and global sourcing matter for firm sustainability performance? An international study. *Supply Chain Management*, *19*(3), 258–274.

Halse, L. L., & Klymenko, O. (2022). Manufacturing backshoring and sustainability: Insights from Norway. In J. Larimo, P. Arte, P. Ghauri, J. Mata, & C. M. P. Sousa (Eds.), *Research handbook on foreign exits, relocation and re-entry: Theoretical perspectives and empirical evidence* (pp. 292–309). Edward Elgar.

Halse, L. L., Nujen, B. B., & Solli-Sæther, H. (2019). The role of institutional context in backshoring decisions. In R. V. Tulder, A. Verbeke, & B. Jankowska (Eds.), *International business in a VUCA world: The changing role of states and firms* (Vol. 14, pp. 251–268). Emerald Publishing Limited.

Jia, F., & Jiang, Y. (2018). Sustainable global sourcing: A systematic literature review and bibliometric analysis. *Sustainability*, *10*, 595.

Johanson, J., & Vahlne, J.-E. (1977). The internationalization process of the firm—A model of knowledge development and increasing foreign market commitments. *Journal of International Business Studies*, *8*(1), 23. https://doi.org/10.1057/palgrave.jibs.8490676

Johansson, M., Olhager, J., Heikkilä, J., & Stentoft, J. (2019). Offshoring versus backshoring: Empirically derived bundles of relocation drivers, and their relationship with benefits. *Journal of Purchasing and Supply Management*, *25*(3), 100509.

Karlsen, A. (2019). Historical examples of entrepreneurial discovery: Revisiting the manufacturing history of Raufoss evolving between exploration and exploitation. In Å. Mariussen, S. Virkkala, H. Finne, & T. M. Aasen (Eds.), *The entrepreneurial discovery process and regional development. New knowledge emergence, conversion and exploitation* (pp. 288–303). Routledge.

Kinkel, S. (2012). Trends in production relocation and backshoring activities. *International Journal of Operations & Production Management*, *32*(6), 696–720.

Kinkel, S. (2014). Future and impact of backshoring—Some conclusions from 15 years of research on German practices. *Journal of Purchasing and Supply Management*, *20*(1), 63–65.
Kinkel, S., & Maloca, S. (2009). Drivers and antecedents of manufacturing offshoring and backshoring—A German perspective. *Journal of Purchasing and Supply Management*, *15*(3), 154–165.
Koberg, E., & Longoni, A. (2019). A systematic review of sustainable supply chain management in global supply chains. *Journal of Cleaner Production*, *207*, 1084–1098.
Kogut, B., & Zander, U. (1992). Knowledge of the firm, combinative capabilities, and the replication of technology. *Organizational Science*, *3*(3), 383–397.
Kolk, A. (2016). The social responsibility of international business: From ethics and the environment to CSR and sustainable development. *Journal of World Business*, *51*(1), 23–34.
Kolk, A., & Pinkse, J. (2008). A perspective on multinational enterprises and climate change: Learning from "an inconvenient truth"? *Journal of International Business Studies*, *39*(8), 1359–1378.
Li, X., & Zhou, Y. (2017). Offshoring pollution while offshoring production? *Strategic Management Journal*, *38*(11), 2310–2329.
Lim, S.-J., & Phillips, J. (2008). Embedding CSR values: The global footwear industry's evolving governance structure. *Journal of Business Ethics*, *81*(1), 143–156.
Liversey, F. (2012). The need for a new understanding of manufacturing and industrial policy in leading economies. *Innovations: Technology, Governance, Globalization*, *7*(3), 193–202.
Lund, H. B., & Steen, M. (2020). Make at home or abroad? Manufacturing reshoring through a GPN lens: A Norwegian case study. *Geoforum*, *113*, 154–164.
March, J. G. (1994). *A primer on decision making: How decisions happen*. Free Press.
McIvor, R., & Bals, L. (2021). A multi-theory framework for understanding the reshoring decision. *International Business Review*, *30*(6), 101827.
Musteen, M. (2016). Behavioral factors in offshoring decisions: A qualitative analysis. *Journal of Business Research*, *69*(9), 3439–3446.
Nujen, B. B., Halse, L. L., Damm, R., & Gammelsæter, H. (2018). Managing reversed (global) outsourcing – The role of knowledge, technology and time. *Journal of Manufacturing Technology Management*, *29*(4), 676–698.
Orzes, G., & Sarkis, J. (2019). Reshoring and environmental sustainability: An unexplored relationship? *Resources, Conservation and Recycling*, *141*, 481–482.
Paul, K., & Parra, C. M. (2021). Corporate social responsibility in international business literature: Results from text data mining of the *Journal of International Business Studies*. *International Journal of Corporate Social Responsibility*, *6*(1), 12.
Pisano, G., & Shih, W. (2012). *Producing prosperity: Why America needs a manufacturing renaissance*. Harvard Business Review Press.
Porter, M. E. (1998). Clusters and the new economics of competition. *Harvard Business Review*, *76*(6), 77–90.
Prahalad, C. K., & Hamel, G. (1990). The core competence of the corporation. *Harvard Business Review*, *68*(3), 79.
Roos, G., & Kennedy, N. (2014). *Global perspectives on achieving success in high and low cost operating environments*. Business Science References.
Rugman, A. M., & Verbeke, A. (1998). Corporate strategies and environmental regulations: An organizing framework. *Strategic Management Journal*, *19*(4), 363–375.
Rugman, A. M., & Verbeke, A. (2001). Environmental policy and international business. In A. M. Rugman & T. L. Brewer (Eds.), *The Oxford handbook of international business* (Vol. 33, pp. 537–557). Oxford University Press.
Røvik, K. A. (2011). From fashion to virus: An alternative theory of organizations' handling of management ideas. *Organization Studies*, *32*(5), 631–653.
Sarkis, J., Zhu, Q., & Lai, K.-h. (2011). An organizational theoretic review of green supply chain management literature. *International Journal of Production Economics*, *130*(1), 1–15.
Sawhney, A., & Rastogi, R. (2015). Is India specialising in polluting industries? Evidence from US–India bilateral trade. *The World Economy*, *38*(2), 360–378.
Scott, W. R. (2008). *Institutions and organizations: Ideas and interests*. Sage Publications.
Stentoft, J., Olhager, J., Heikkilä, J., & Thoms, L. (2016). Manufacturing backshoring: A systematic literature review. *Operations Management Research*, *9*(3), 53–61.

Tate, W. L., & Bals, L. (2014). Outsourcing/offshoring Insights: Going beyond reshoring to rightshoring – Call for papers. *International Journal of Physical Distribution & Logistics Management (IJDPLM)*, *47*(2–3), 106–113. http://www.emeraldgrouppublishing.com/authors/writing/calls.htm?id=5868

United Nations General Assembly. (1987). *Our Common Future report of the world commission on environment and development: Our Common Future*. http://www.un-documents.net/our-common-future.pdf

Veit, C., Lambrechts, W., Quintens, L., & Semeijn, J. (2018). The impact of sustainable sourcing on customer perceptions: Association by guilt from scandals in local vs. offshore sourcing countries. *Sustainability*, *10*, 2519.

Wiesmann, B., Snoei, J. R., Hilletofth, P., & Eriksson, D. (2017). Drivers and barriers to reshoring: A literature review on offshoring in reverse. *European Business Review*, *29*(1), 15–42. https://doi.org/doi:10.1108/EBR-03-2016-0050

Williamson, O. E. (1975). *Markets and hierarchies: Analysis and antitrust implications*. Free Press.

CHAPTER 18

CORPORATE SOCIAL RESPONSIBILITY, IRRESPONSIBILITY, AND FIRMS' FINANCIAL PERFORMANCE: EVIDENCE FROM EMERGING VERSUS DEVELOPED COUNTRIES

Stefano Elia, Gezim Hoxha and Lucia Piscitello

Department of Management, Economics and Industrial Engineering – Politecnico di Milano Via Lambruschini, Milan, Italy

ABSTRACT

This study aims at investigating the effect of corporate social responsibility (CSR) and corporate social irresponsibility (CSI) on corporate financial performance (CFP) in firms headquartered in developed versus emerging countries. Drawing upon stakeholder and legitimacy perspectives, the authors argue that the CSR/CSI–CFP relationship differs depending on the home-countries' level of economic development as this reflects their different sensitivity to sustainability. Indeed, as emerging economies are normally characterized by weaker regulations, they are likely to place lower pressures on companies for superior CSR practices. Therefore, the authors expect the effect of CSR on CFP to be more positive for firms headquartered in advanced than in emerging countries. At the same time, the authors propose a more negative relationship between CSI and CFP for firms headquartered in developed countries due

Walking the Talk? MNEs Transitioning Towards a Sustainable World
Progress in International Business Research, Volume 18, 351–371

ISSN: 1745-8862/doi:10.1108/S1745-886220240000018022

to the higher overall sustainability expectations required to gain legitimacy. The empirical analyses, run on a sample of 1,971 publicly listed firms between 2010 and 2020 from developed and emerging economies, support the expectations, thus confirming that country-specific contextual factors do play a role in shaping both the positive and the negative impact of CSR and CSI on CFP, and that the reactions of stakeholders to responsible and irresponsible behavior are stronger when their sensitivity to sustainability is higher.

Keywords: Corporate social responsibility; corporate social irresponsibility; financial performance; developed markets; emerging markets; sustainability

INTRODUCTION

The concept of CSR has evolved to encompass a wide range of environmental, social, and governance (ESG) issues. Companies are increasingly expected to engage in CSR-related actions and expand their accountability to all stakeholders, including employees, customers, suppliers, communities, and the environment (Campbell, 2007; Nirino et al., 2021). This trend is driven not only by ethical considerations but also by business imperatives such as reputation management, risk mitigation, and stakeholder expectations (Carroll & Shabana, 2010; Orlitzky et al., 2003). Furthermore, companies may face various pressures to engage in CSR, including legal regulations, market forces, non-governmental organization campaigns, and public opinion (Nirino et al., 2021). For instance, the rise of social media and online activism has made it easier for stakeholders to voice their concerns and hold companies accountable for their actions. Moreover, the increasing complexity and interdependence of global markets and societies pose new challenges and opportunities for companies to engage in CSR. For instance, companies may need to navigate different legal, cultural, and ethical norms across countries and regions, as well as respond to global challenges such as climate change, inequality, and human rights abuses (Kolk & van Tulder, 2010; Christensen et al., 2013). Therefore, companies need to develop effective CSR strategies that align with their business goals, values, and stakeholder expectations, while also addressing broader societal and environmental concerns.

Despite the growing importance attributed to CSR, there is little consensus about its influence on CFP. Some studies have found a positive relationship between the two (e.g., Reverte et al., 2016; Wang & Sarkis, 2017), as investments in socially responsible activities may have internal and external benefits for companies such as developing intangible resources associated with employees or enhancing corporate reputation (Reverte et al., 2016). Other studies report a negative relationship, mainly due to the costs incurred to implement CSR activities (e.g., Duque-Grisales & Aguilera-Caracuel, 2021).

At the same time, not all companies are equally committed to CSR, and some even engage in CSI by neglecting their social and environmental responsibilities. CSI has been traditionally measured through ESG controversies

(Nirino et al., 2021), which refer to negative news about corporate ESG issues, such as scandals involving social behavior or product harm, which attract media attention and capture the interest of investors. Such news can lead to doubts about the company's future prospects, pose a risk to its reputation, and potentially affect firm's performance (Nardella et al., 2023; Nirino et al., 2021). Hence, CSI can have significant negative consequences for companies, including damage to their reputation and stakeholder relationships, as well as legal and financial liabilities (Gao et al., 2018). However, some papers find that CSI does not have a significant influence on CFP or that the relationship is even positive (Aouadi & Marsat, 2016; Melinda & Wardhani, 2020; Sachin & Rajesh, 2022).

Hence, the evidence about the relationship between firms' financial performance and both CSR and CSI is, at best, inconclusive. Although previous literature has identified some contingencies and moderators, such as firm-level characteristics, the industry context, and institutional factors (Gras & Krause, 2020; Hull & Rothenberg, 2008; Klassen & McLaughlin, 1996; Russo & Fouts, 1997; Shi & Veenstra, 2021), there are still inconsistencies in the findings across different national contexts. A possible explanation is that the moderating role of country-level contingency factors is particularly critical for at least two reasons. On the one hand, the institutional context in which companies operate, including the legal system and regulatory environment, can influence the relationship between CFP and CSR/CSI by defining different sets of rules (Flammer et al., 2018). On the other hand, cultural values and social norms, such as the importance placed on environmental stewardship and social responsibility, can affect the perceived importance of CSR practices and CSI controversies and their impact on financial performance (Gartenberg et al., 2018; Grewatsch & Kleindienst, 2017; Napier et al., 2023).

Given the ambiguous results of the literature and the pivotal role of the national contexts in the CFP–CSR/CSI relationship, and in response to recent calls for research on the impact of firms' (mis)behavior on financial performance in developed and emerging markets (Cuervo-Cazurra et al., 2021; Linnenluecke, 2022), we relate the impact of CSR and CSI on CFP with the company's home-country. Namely, we claim that companies headquartered in developed countries may face greater scrutiny and pressure to adhere to CSR standards and to avoid CSI than those in emerging economies, where legal and regulatory frameworks may be weaker and where social and environmental issues may be less of a priority for stakeholders (Brammer & Pavelin, 2006). Companies in advanced countries are, indeed, subject to stronger pressure from domestic stakeholders (Napier et al., 2023) to take a proactive approach to CSR and implement effective policies and practices to avoid and mitigate CSI. This implies not only complying with legal and regulatory frameworks, but also engaging with stakeholders, monitoring supply chains, and being transparent about CSR performance (Epstein & Buhovac, 2014). By doing so, companies can build legitimacy, trust, and credibility with stakeholders and potentially gain competitive advantages in the marketplace. Therefore, by integrating the stakeholder theory (Freeman, 1984) with the legitimacy theory (Suchman, 1995), we propose that the effect of CSR on firms' financial performance is more positive in advanced than in emerging countries, while the effect of CSI on firms' financial performance is more negative in advanced than in emerging countries.

We use ESG performance as a proxy for CSR, and ESG controversies as a proxy for CSI. By doing so, we aim to investigate the relationship between the two constructs and firms' financial performance, as measured by Tobin's *Q*. Our empirical analyses rely on a sample of firms observed for a 10-year period, from 2011 to 2020, obtained by Thomson Reuters – Refinitiv Eikon. Namely, the sample is composed of 17,703 firm-year observations. Our findings confirm the more positive and more negative impact of CSR and CSI, respectively, on the CFP for firms headquartered in advanced countries compared to emerging markets firms.

The remainder of the chapter is organized as follows. The next section presents the literature review and the development of the hypotheses. The third section shows the methodology and illustrates the sample and the variables adopted in the econometric analysis. In the fourth section, we perform the analysis and present the econometric findings. The last section presents the discussion and the conclusions together with the implications of the study and its relevance from a theoretical and practical point of view. Limitations and future research avenues are provided at the end of this concluding section.

LITERATURE REVIEW AND HYPOTHESES DEVELOPMENT

CSR and CFP in Advanced Versus Emerging Countries

According to stakeholder theory (Freeman, 1984), businesses create externalities that influence a wide range of people, both inside and outside the company. Stakeholders frequently put pressure on corporations to lessen negative impacts and raise favorable ones as a result of externalities (Delmas, 2001; Delmas & Toffel, 2004). Stakeholder pressure occurs when a company is held accountable for its actions and decisions about product design, sourcing, production, or distribution to stakeholders (Parmigiani et al., 2011). Corporations are not solely responsible for maximizing shareholder value, but they also have obligations to various stakeholders, such as employees, customers, suppliers, and the broader community (Freeman, 1984). Thus, they engage in CSR to maintain and enhance their relationships with stakeholders, rather than solely for financial gains.

To shed light on this mechanism, we combine the stakeholder theory (Freeman, 1984) with the legitimacy theory proposed by Suchman (1995). The legitimacy theory provides, indeed, a complementary framework to explain why firms engage in CSR practices. According to Suchman (1995), legitimacy is defined as "a generalized perception or assumption that the actions of an entity are desirable, proper, or appropriate within some socially constructed system of norms, values, beliefs, and definitions" (p. 574). As noted by Shi and Veenstra (2021), the level of legitimacy is influenced by stakeholders' dominant social values and beliefs, and firms can gain legitimacy by meeting society's expectations. The attainment of legitimacy in a society where a firm operates is a crucial factor that affects a firm's reputation and survival in the market. Compliance with stakeholder expectations can lead to greater legitimacy, support, and appreciation for firms, thus improving their performance. Conversely, non-compliance can cause firms to lose legitimacy and be punished through various means, including boycotts by stakeholders, thus lowering their performance (Napier et al., 2023).

Although the relationship between CSR and financial performance has been a subject of intense debate among scholars and practitioners for decades, previous empirical findings often yield contrasting results (Awaysheh et al., 2020; Lu et al., 2014; Margolis & Walsh, 2003; Zhao & Murrell, 2016). In fact, most studies report a positive and significant correlation between CSR activities and CFP (e.g., Alshehhi et al., 2018; Bhaskaran et al., 2020; De Lucia et al., 2020; Friede et al., 2015). However, other studies have found a negative relationship (e.g., Duque Grisales & Aguilera Caracuel, 2021; Landi & Sciarelli, 2019; Marsat & Williams, 2011) or an insignificant or non-existent relationship (e.g., Atan et al., 2018; Lopez-de-Silanes et al., 2020).

Building on the recent literature that calls for research on the impact of firms' behavior on financial performance in developed and emerging markets firms (Cuervo-Cazurra et al., 2021; Linnenluecke, 2022), we claim that the impact of CSR can vary depending on the country in which a company is headquartered. While several studies have employed institutional and neo-institutional frameworks to examine cross-country differences in CSR disclosure, literature investigating the relationship between CSR and performance across different national contexts remains scarce. A lack of consideration for moderating factors has been suggested as a possible explanation for these mixed results (Endrikat et al., 2014; Javed et al., 2016; Reverte et al., 2016). This is also partly due to the scant availability of CSR data for companies in regions outside of North America and Europe, resulting in analytical challenges (Arun et al., 2022; Orsato et al., 2015; Rajesh & Rajendran, 2020). Thus, it remains unclear if any significant difference can be found in the CSR–financial performance relationship between corporations headquartered in developed markets and corporations headquartered in emerging markets.

The popularity of CSR investments has grown significantly in developed countries with mature markets, leading to its characterization as a "rich world phenomenon" (Chung, 2021). In fact, in developed countries, CSR practices have been pushed by the increased pressure and regulation from governments and other stakeholders to incorporate CSR schemes in business strategies (Manrique & Martí-Ballester, 2017). In contrast, corporations headquartered in emerging economies can spend relatively less on CSR operations as CSR practices are still in their early stages in these countries. Emerging countries lack adequate legislation, penalties, and incentives for CSR practices and efforts.

Within this context, we claim that the implications and impact of CSR practices on financial performance vary between developed and emerging countries (Naeem et al., 2022). Developed countries are more focused on economic and environmental sustainability and sociocultural rights. The business environment, risks, and opportunities significantly differ between developed and emerging economies, leading to different levels of legislation and social pressure and different advantages in planning, executing, and reporting CSR activities and operations (Naeem et al., 2022). Environmental standards in emerging economies are generally weaker than those in advanced economies, and the enforcement of environmental regulations is less strict (Earnhart & Lizal, 2010; Wei et al., 2017). Local regulatory agencies are less developed, and courts tend to be inefficient in handling cases, especially those related to environmental harm, where it can be difficult to assess liability and pollution costs (Khanna & Palepu, 2010). Consequently, emerging market firms may not have the same incentive as their developed country counterparts to engage in CSR initiatives (Esfahbodi et al., 2016). In addition, emerging market companies are not favored by CSR-focused investors due to several factors. First, investments in emerging markets are often perceived as riskier due to weaker formal institutions, less strict regulatory environments, limited protection of shareholder rights, low transparency, and widespread corruption (Bahadori et al., 2021; Hoang, 2018). Moreover, firms operating in emerging markets face additional challenges resulting from volatile government policies that downsize the adoption of CSR practices (Bahadori et al., 2021). Additionally, the limited and opaque information environment in emerging markets make it difficult for stakeholders to obtain reliable information on CSR practices (Mobius & Ali, 2021). Thus, we propose the following hypothesis:

> *H1*. The impact of CSR on CFP is more positive for firms headquartered in developed countries than for firms headquartered in emerging countries.

CSI and CFP in Advanced Versus Emerging Countries

An emerging strand of literature suggests there is a negative relationship between bad CSR-related news and firms' market value. Given that a firm's reputation is a key element to enhance financial performance (Aguilera et al., 2007; Li et al., 2019), if a company reports ongoing controversies, it is likely to experience a decrease in its financial performance. This is particularly true for listed companies as the market can overreact to such events and make the impact even greater. Frooman (1999) tested empirically that the stock market negatively reacts to a firm's involvement in social and environmental irresponsible behavior. This explains why CSI is receiving increasing attention from investors and policy makers. Indeed, CSI scores assess a corporation's exposure to ESG-related negative incidents in worldwide media. These controversies include harmful waste disposal, human rights violations, environmental pollution damage, unethical practices of managers or CEOs, legal disputes, or company lawsuits, etc. Controversial news or scandals related to environmental and social issues could impact a corporation's value as it would reflect suspicion of the corporation's long-term success (Nirino et al., 2021).

As already done for the relationship between CSR and firm's performance discussed in the previous section, we claim that the (negative) impact of CSI on firm's performance is contingent upon the development of the country in which the firm is headquartered. Harjoto et al. (2021) propose that CSI events can significantly impact a company's financial performance, especially in countries where stakeholders have a greater level of trust in major corporations. This is because when a company engages in socially irresponsible behavior, it damages the trust that stakeholders had in the company and, hence, its legitimacy to operate in that country. In countries where stakeholders hold a high level of trust in major corporations, the negative impact of CSI events can be more significant as such events are seen as actions that violate the public's trust in corporations. Developed countries tend to have more rigid informal institutional constraints and lower tolerance toward CSI events. Stakeholders in developed countries hold companies accountable for their actions and demand greater CSR. Thus, stakeholders in developed countries punish hard the companies for irresponsible actions, and this will be reflected in the company's market value. In contrast, emerging economies have normally weaker informal institutional constraints and less intransigent attitude toward CSI events (e.g., Tamayo-Torres et al., 2019; Wang & Sarkis, 2013). This could be due to factors such as weaker regulatory frameworks and different cultural norms. As a result, companies in emerging economies may face fewer reactions to socially irresponsible behavior, which could result in less severe financial impacts.

Accordingly, we expect that the negative reaction of stakeholders to controversies becomes even more negative in firms from developed countries compared to firms that come from emerging countries (Garcia et al., 2017; Ting et al., 2019). Thus, we propose the following hypothesis:

> *H2*. The impact of CSI on CFP is more negative for firms headquartered in developed than for firms headquartered in emerging countries.

RESEARCH DESIGN

Sample and Data

Our sample consists of 1,971 listed companies from 52 countries and 17 industries included in the Thomson Reuters – Refinitiv Eikon database for a 10-year period, from 2011 to 2020. Therefore, the sample is composed of 17,703 firm–year observations. This database has gained increasing currency among management, finance, CSR, and IB scholars (Brammer et al., 2021; Cuervo-Cazurra et al., 2022; Cheng et al., 2014; Ioannou & Serafeim, 2012; Nirino et al., 2021) because it provides relevant and systematic standardized ESG data for over 6,000 global listed companies included in S&P 500, Russell 1000, MSCI Europe, FTSE 250, MSCI World Index, and 250 emerging-market companies. We then matched Refinitiv Eikon with firm financial information obtained from DataStream through Thomson Reuters. We drop firms with missing observations on ESG controversies, ESG scores, financial information, and control variables. Table 18.1 shows the distribution of our final sample across countries. Firms headquartered in developed countries represent 72.06% of the sample, while firms headquartered in emerging countries account for the remaining 27.94%. Table 18.2 shows the distribution of

Table 18.1. Distribution of Companies Across Home-country Location.

No.	Advanced Countries	Freq.	Percent	No.	Emerging Countries	Freq.	Percent
1	Austria	108	0.85	1	Bermuda	63	1.27
2	Belgium	162	1.27	2	Brazil	468	9.46
3	Canada	1,433	11.23	3	Cayman Islands	18	0.36
4	Czech Republic	18	0.14	4	Chile	153	3.09
5	Denmark	171	1.34	5	China	450	9.10
6	Finland	198	1.55	6	Colombia	45	0.91
7	France	673	5.28	7	Egypt	45	0.91
8	Germany	530	4.15	8	India	474	3.64
9	Greece	90	0.71	9	Indonesia	180	3.64
10	Hungary	27	0.21	10	Israel	90	1.82
11	Republic of Ireland	180	1.41	11	Republic of S. Korea	665	13.45
12	Italy	180	1.41	12	Kuwait	9	0.18
13	Luxembourg	36	0.28	13	Malaysia	279	5.64
14	Netherlands	225	1.76	14	Mexico	162	3.28
15	Norway	126	0.99	15	Morocco	9	0.18
16	Poland	90	0.71	16	Panama	9	0.18
17	Portugal	63	0.49	17	Peru	9	0.18
18	Spain	261	2.05	18	Philippines	108	2.18
19	Sweden	342	2.68	19	Qatar	9	0.18
20	Switzerland	441	3.46	20	Russia	243	4.91
21	United Kingdom	1,786	14.00	21	Saudi Arabia	45	0.91
22	United States of America	5,617	44.03	22	Singapore	9	0.18
				23	South Africa	360	7.28
				24	Sri Lanka	9	0.18
				25	Taiwan	774	15.65
				26	Thailand	117	2.37
				27	Turkey	117	2.37
				28	Ukraine	9	0.18
				29	United Arab Emirates	18	0.36
	Total	12,757	100.00		Total	4,946	100.00

Source: Authors' original work.

our sample across 17 industries, mostly from manufacturing (39.89%), followed by mining, oil, and gas (10.08%), and information technology (8.70%).

Dependent Variable

The dependent variable of this study is *Firm CFP*, measured using Tobin's *Q* (Tobin, 1969), which is a market-based measure widely applied in management and finance studies when assessing the impact of sustainable practices within businesses (Kim et al., 2018). Tobin's *Q* is computed as the sum of market capitalization and total liabilities, divided by total assets (Cuervo-Cazurra et al., 2022). Firms whose Tobin's *Q* is higher than 1 attract investors' attention since the market value of their investment is greater than its assets; Tobin's *Q* below 1 discourages investors from investing in a firm.

Table 18.2. Distribution of Companies Across Sectors.

No.	Sectors	Freq.	Percent	Cum.
1	Accommodation and food services	306	1.73	1.73
2	Administrative and support and waste management and remediation services	315	1.78	3.51
3	Agriculture, forestry, fishing, and hunting	54	0.31	3.81
4	Arts, entertainment, and recreation	71	0.40	4.21
5	Construction	756	4.27	8.48
6	Finance and insurance	198	1.12	9.60
7	Health care and social assistance	162	0.92	10.52
8	Information	1,541	8.70	19.22
9	Manufacturing	7,061	39.89	59.11
10	Mining, quarrying, and oil and gas extraction	1,784	10.08	69.19
11	Other services (except public administration)	54	0.31	69.49
12	Professional, scientific, and technical services	783	4.42	73.91
13	Real estate and rental and leasing	1,051	5.94	79.85
14	Retail trade	1,188	6.71	86.56
15	Transportation and warehousing	855	4.83	91.39
16	Utilities	1,057	5.97	97.36
17	Wholesale trade	467	2.64	100.00
	Total	17,703	100.00	

Source: Authors' original work.

Independent Variables

The first explicative variable is *CSR*, which is measured through the ESG score variable provided by Refinitiv Eikon, and which is a composite score of the ESG performance pillars. This score is calculated using a weighted average of 34%, 42%, and 24% for ESG performance respectively, based on the latest ESG score calculation methodology (Refinitiv, 2022). The environmental score is composed of 61 items across three areas: resource use, harmful gas emissions, and innovation in controlling environmental degradation. The social score includes 61 items across four areas: workforce quality, emphasis on human rights, organizational community involvement, and product responsibility. Finally, the governance score includes 50 items across three areas: management quality (compensation, independence, and diversity), shareholders' rights, and board composition and strategy. The score has a minimum value of 0 and a maximum of 100.

The second explicative variable is *CSI*, which is measured through the ESG controversy score, a variable provided by Refinitiv Eikon that indicates whether the company experienced any controversies or scandals linked to ESG activities or operations, in a given year (Refinitiv, 2022). The ESG controversies scores are calculated based on 23 different ESG scandals or issue topics (number of controversies published in the media linked to business ethics in general, biodiversity, spills and pollution, environmental consequences of products, child labor issues, human rights issues, working conditions, health and safety issues, etc.). The score has a minimum value of 0 and a maximum of 100.

Moderating Variables

To investigate the moderation effects of the country's level of development on the relationship between financial performance and both CSR and CSI, we use the variable *Advanced countries*, which is a dummy assuming value 1 for firms headquartered in advanced countries, and 0 for firms headquartered in emerging countries. This country classification refers to the year 2020 and it is based on the World Economic Situation and Prospects report (United Nations, 2020), which is developed by UNCTAD. As shown in Table 18.1, the number of firms headquartered in advanced countries is equal to 12,757, that is, 72.06% of the sample, while the number of firms in emerging countries is equal to 4,946, that is, the remaining 27.94% of the sample.

In order to check the robustness of our result we also employ, as an alternative variable, the *sustainable development index (SDI)*, which provides a more direct proxy for the country's sensitivity to sustainable development issues (Zhang et al., 2019). The SDI includes five indicators: education, life expectancy, income, CO_2 emissions, and material footprint. Specifically, the SDI is calculated as the quotient of the two following indexes: (1) a "Development Index" calculated as the geometric mean of the education index, the life expectancy index, and the modified income index; and (2) the "Ecological Impact Index" calculated as the average overshoot of CO_2 emissions and material footprint vis-á-vis their per-capita planetary boundaries, indexed on a natural exponential scale (Hickel, 2020).

Control Variables

Based on previous studies (e.g., Li et al., 2019; Nirino et al., 2021; Wang & Sarkis, 2017), we include a few control variables. First, *Firm size* is considered because larger firms could employ more CSR practices and have higher financial performance. The natural logarithm of total assets (instead of the absolute value) is used as the measure for firm size to decrease the standard variation that the largest companies could generate in the data sample (Wang & Sarkis, 2017). Second, we control for the *Firm liquidity* which can influence strategic decisions on sustainable practices (Li et al., 2019). The level of liquidity is based on the current ratio which is determined by current assets and current liabilities. Finally, we control for *Years, Industries,* and *Geo-Area*. Year fixed-effects are used to consider temporal variations, industries, and geographic area fixed-effects are used to consider industry structure and geographic region since they may result in differences in profitability (Cuervo-Cazurra et al., 2022). Table 18.3 shows the measurement items and the description of the variables.

Methodology

In studying the impact of CSR and CSI controversies on CFP, we employed a panel design with generalized least squares estimation. This approach allows correcting for the presence of autocorrelation and heteroscedasticity in longitudinal data (Wooldridge, 2000). Further, including a 1-year time lag for all explanatory and control variables mitigates the likelihood of bias that could arise from reverse

Table 18.3. Measurement Items and Description of Variables.

Dependent Variable	*Description*
Firm financial performance – Tobin's *Q*	Market cap + total liabilities, divided by total assets (Surroca et al., 2010)
Independent Variables	*Description*
CSR – ESG score	ESG score measure a company's relative ESG performance, commitment, and effectiveness across 10 main themes (emissions, environmental product innovation, human rights, shareholders, etc.) based on publicly reported data. The score has a minimum value of 0 and a maximum of 100
CSI – ESG controversies	ESG controversy score is calculated based on 23 ESG controversy topics and measures a company's exposure to ESG controversies and negative events reflected in global media. The score has a minimum value of 0 and a maximum of 100
Advanced countries (Firm home-country location)	A dummy variable assuming value 1 for Advanced countries, and 0 for Emerging countries. This country classification is taken for year 2020 by the World Economic Situation and Prospects report (United Nations, 2020) which is produced by UNCTAD
SDI	SDI includes five indicators: education, life expectancy, income, CO_2 emissions, and material footprint. The SDI is calculated as the quotient of the two figures: (1) a "Development Index" calculated as the geometric mean of the education index, the life expectancy index, and the modified income index; and (2) "Ecological Impact Index" calculated as the average overshoot of CO_2 emissions and material footprint vis-á-vis their per-capita planetary boundaries, indexed on a natural exponential scale (Hickel, 2020)
Control Variables	*Description*
Firm size	Natural logarithm of total assets
Firm liquidity	Natural logarithm – current assets divided current liabilities
Year fixed effect	
Industry fixed effect	
Geographic area fixed effect	

Source: Authors' original work.

causality and endogeneity (Tamayo-Torres et al., 2019), and gives time for their effect to be reflected in the CFP.

RESEARCH FINDINGS

Descriptive Statistics

Before presenting the results of the regression analysis implemented to test the hypotheses developed in the first part of the chapter, we present the descriptive statistics and the correlation between the variables adopted (Table 18.4). On a scale of 0 to 100, the average *CSR* and *CSI* scores are 51.00 and 88.03, respectively.

Table 18.4. Descriptive Statistics and Pairwise Correlations.

Variables	(1)	(2)	(3)	(4)	(5)	(6)	(7)
(1) *Firm financial performance*	1.000						
(2) *CSR*	0.004	1.000					
(3) *CSI*	0.025	−0.226	1.000				
(4) *Advanced countries*	0.037	0.110	−0.074	1.000			
(5) *SDI*	−0.006	−0.052	−0.020	0.769	1.000		
(6) *Firm size*	−0.198	0.446	−0.341	0.031	−0.009	1.000	
(7) *Firm liquidity*	0.035	−0.191	0.106	−0.052	−0.011	−0.262	1.000
Observations	17,703	17,703	17,703	17,703	15,815	17,703	17,703
Mean	1.877	51.000	88.039	0.765	26.074	22.660	0.594
Std. Dev.	1.925	20.892	25.487	0.423	8.708	1.417	0.463
Min.	0.106	0.550	0.540	0.000	3.740	14.318	−2.137
Max.	79.036	94.681	100.000	1.000	77.415	27.496	5.853

Source: Authors' original work.

As shown in Table 18.4, the variables do not show high correlation, suggesting that there are no evident multicollinearity problems (Allen, 1997). Also, variables such as *Advanced countries* and *SDI* are highly correlated (0.76), which means that those two variables could be used interchangeably. *Financial performance* is also positively correlated with the *Firm liquidity* but negatively correlated with *Firm size*. Furthermore, to rule out any multicollinearity issues, we calculated the variance inflation factors. The results range from 1.30 (model 1) to 1.43 (model 5), and they are therefore well below the threshold accepted by the previous literature (Hair et al., 1995).

Econometric Findings

Table 18.5 reports the results of our econometric analyses. Model 1 presents the impact of *CSR* on *CFP*, while model 2 shows the impact of *CSI* on *CFP*, with *control variables* included. Thus models 1 and 2 are two restricted versions of model 3 that integrates the effects of *CSR, CSI,* and *Advanced countries*, on *CFP* in one model. Finally, in addition to these direct effects, models 4 and 5 include the interaction term of *CSR* and *CSI* with *Advanced countries. H1* is tested with model 4, which presents the moderating effect of the firms' country of origin (advanced vs emerging) on the relationship between *CSR* and *CFP*. The coefficient of the interaction term is positive and significant (β=0.011; p<0.05); this indicates that *CSR* has a more positive effect on financial performance for companies headquartered in developed countries than emerging countries. *H2* is tested in model 5, which shows the moderating effect of *Advanced countries* on the relationship between *CSI* and *CFP*. As the coefficient of the interaction term is negative and significant (β=−0.002; p<0.05), this indicates that CSI has more negative effects on financial performance for companies headquartered in developed countries compared to companies that are headquartered in emerging countries.

Table 18.5. Main Regression Analysis.

	(1)	(2)	(3)	(4)	(5)
VARIABLES	Tobin's *Q*	Tobin's *Q*	Tobin's *Q*	Tobin's *Q*	Tobin's *Q*
CSR (t−1)	0.011***		0.011***	0.009***	0.011***
	(0.001)		(0.001)	(0.001)	(0.001)
CSI (t−1)		−0.002***	−0.001**	−0.001**	0.002
		(0.001)	(0.001)	(0.001)	(0.001)
Advanced countries (t−1)			0.059	−0.066	0.438
			(0.676)	(0.679)	(0.693)
*CSR * Advanced countries (t−1)*				0.003**	
				(0.002)	
*CSI * Advanced countries (t−1)*					−0.004**
					(0.002)
Firm size (t−1)	−0.405***	−0.335***	−0.413***	−0.416***–	−0.414***
	(0.012)	(0.012)	(0.013)	(0.013)	(0.013)
Firm liquidity (t−1)	0.090**	0.096***	0.096***	0.095***	0.096***
	(0.037)	(0.037)	(0.037)	(0.037)	(0.037)
Year fixed effect	Yes	Yes	Yes	Yes	Yes
Industry fixed effect	Yes	Yes	Yes	Yes	Yes
Geographic area fixed effect	Yes	Yes	Yes	Yes	Yes
Constant	10.154***	9.292***	10.389***	10.523***	10.088***
	(0.331)	(0.347)	(0.724)	(0.727)	(0.734)
Observations	17,703	17,703	17,703	17,703	17,703
Number of ID	1,971	1,971	1,971	1,971	1,971
Chi square statistics	2112.05***	1999.69***	2119.42***	2127.44***	2129.71***

Standard errors in parentheses.
*** $p<0.01$, ** $p<0.05$, and * $p<0.1$.

To further explore and gain more insights on the positive/negative signs of the moderation effects, we plotted the results of the interaction terms. Fig. 18.1 clearly shows that the relationship between *CSR* and *CFP* is positive for firms' home-country location in both advanced and emerging countries, but higher for firms headquartered in advanced countries. This confirms that the impact of CSR on financial performance is more positive for companies headquartered in developed than in emerging countries, thus fully confirming our *H1*. Conversely, Fig. 18.2 shows that the relationship between *CSI* and *Firm financial performance* is more negative for firms headquartered in advanced countries, while for firms headquartered in emerging countries, it is not only less negative, but it also tends to have a positive trend. Thus, our *H2* is fully supported.

Robustness Checks

We performed additional tests to check the robustness of our results and provide additional evidence in Table 18.6. More specifically, we employed *SDI* as a proxy for the overall sustainability development of the countries. As regards the relationship between CSR and financial performance, the coefficient of the

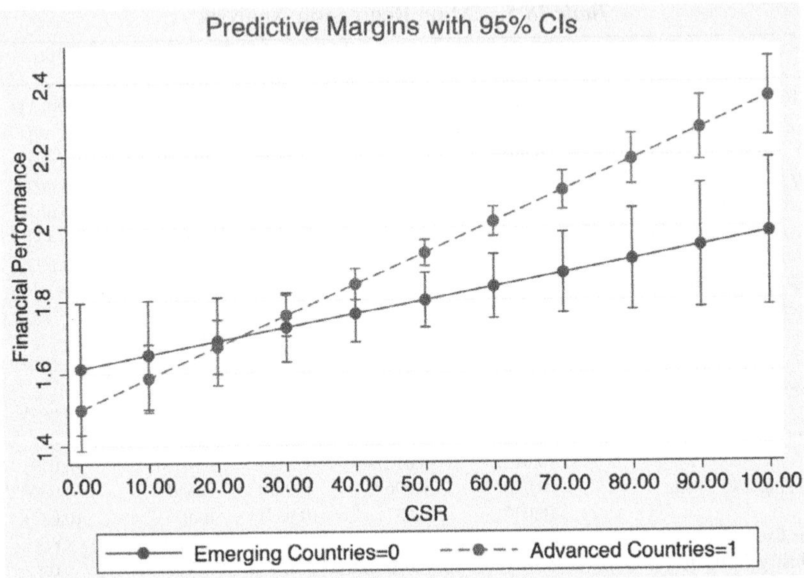

Fig. 18.1. Plot of the Interaction Between CSR and Advanced Countries Performed in Model 4 of Table 18.5. *Source*: Authors' original work.

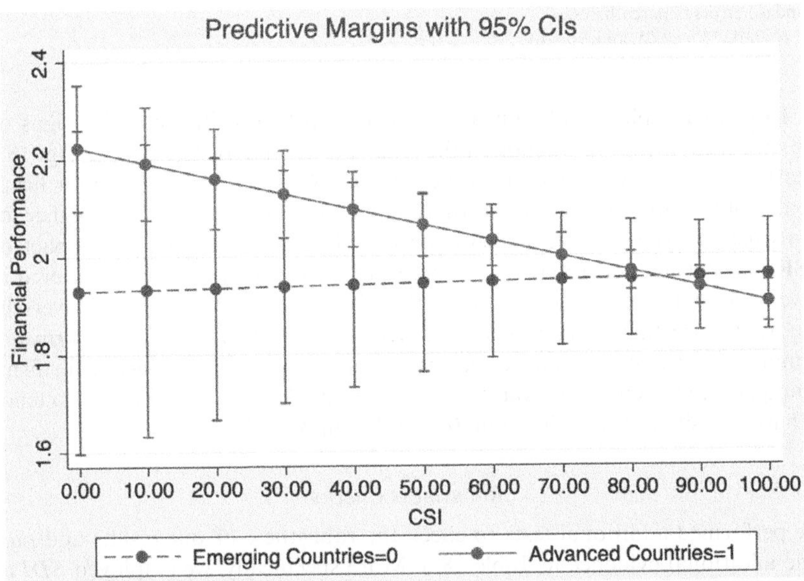

Fig. 18.2. Plot of the Interaction Between CSI and Advanced Countries Performed in Model 5 of Table 18.5. *Source*: Authors' original work.

Table 18.6. Robustness Check.

	(1)	(2)	(3)	(4)	(5)
VARIABLES	Tobin's *Q*	Tobin's *Q*	Tobin's *Q*	Tobin's *Q*	Tobin's *Q*
CSR (t−1)	0.011***		0.013***	0.008***	0.012***
	(0.001)		(0.001)	(0.003)	(0.001)
CSI (t−1)		−0.002***	−0.002**	−0.002**	0.005***
		(0.001)	(0.001)	(0.001)	(0.002)
SDI (t−1)			−0.075*	−0.080**	−0.052
			(0.040)	(0.040)	(0.040)
CSR SDI (t−1)*				0.001*	
				(0.000)	
CSI SDI (t−1)*					−0.002***
					(0.000)
Firms size (t−1)	−0.405***	−0.335***	−0.434***	−0.435***	−0.435***
	(0.012)	(0.012)	(0.014)	(0.014)	(0.014)
Firm liquidity (t−1)	0.090**	0.096***	0.110***	0.110***	0.112***
	(0.037)	(0.037)	(0.039)	(0.039)	(0.039)
Year fixed effect	Yes	Yes	Yes	Yes	Yes
Industry fixed effect	Yes	Yes	Yes	Yes	Yes
Geographic area fixed effect	Yes	Yes	Yes	Yes	Yes
Constant	10.154***	9.292***	13.162***	13.276***	12.548***
	(0.331)	(0.347)	(1.318)	(1.320)	(1.329)
Observations	17,703	17,703	15,815	15,815	15,815
Number of ID	1,971	1,971	1,971	1,971	1,971
Chi square statistics	1999.69***	2112.05***	1815.45***	1823.86***	1831.16***

Standard errors in parentheses.
*** $p<0.01$, ** $p<0.05$, and * $p<0.1$.

interaction term is positive and significant ($\beta=0.001$; $p<0.1$), thus confirming that CSR has a more positive effect on financial performance for companies headquartered in countries with higher SDI. As regards the relationship between CSI and firm financial performance, the interaction term is negative and significant ($\beta=-0.002$; $p<0.01$), thus suggesting that CSI has more negative effects on financial performance for companies headquartered in countries with higher sustainability index (i.e., developed countries' firms) than in emerging countries' firms.

DISCUSSION AND CONCLUSION

The literature examining the relationship between firm's financial performance and CSR and CSI has progressed through multiple perspectives, beginning with the shareholder view and then shifting to the stakeholder and legitimacy views. We build on the stakeholder and legitimacy perspectives and claim that the CSR/CSI–financial performance relationship differs for firms headquartered within developed versus emerging countries. We argue that the main relationship depends on different contextual factors, and in particular on the level of economic development of the country, which affects the sensitivity of stakeholders toward sustainability (Goerzen et al., 2023). More specifically, our argument is that, compared to advanced countries, emerging economies suffer from lower levels of environmental and social awareness and weaker regulations and regulatory enforcement (Brammer et al., 2021). Thus, we assume that emerging country contexts might place fewer demands on companies to adhere to superior CSR practices and less importance on the negative consequences arising from CSI. Accordingly, our findings confirm a more positive relationship between CSR and CFP and a more negative relationship between CSI and CFP for firms headquartered in developed than in emerging markets.

Our findings contribute in advancing the stakeholder and legitimacy perspectives within the realm of the relationship between CSR/CSI and CFP. On the one hand, using the economic development of the companies' home-country as a factor reflecting the sensitivity of stakeholders toward sustainability (Aouadi & Marsat, 2016; Cuervo-Cazurra et al., 2022; Cuervo-Cazurra et al., 2021; Naeem et al., 2022; Nirino et al., 2021), we enrich the stakeholder perspective by providing a more comprehensive understanding of the country-level contingencies affecting the relationship between CSR/CSI and financial outcomes (Tamayo-Torres et al., 2019; Wang & Sarkis, 2013). On the other hand, we enhance the legitimacy perspective by showing the importance of aligning corporate behavior with societal expectations and regulatory frameworks to maintain legitimacy and achieve positive financial outcomes while avoiding negative ones.

We also see some practical implications of our findings. Even if the effect of CSR is stronger in firms headquartered within advanced countries, we find a positive effect also in firms headquartered in emerging countries. Hence, also the practitioners of emerging countries should nurture and foster a CSR culture in their focal firms, by giving the utmost care to issues raised by external stakeholders and being proactive in devising engagement mechanisms with social groups to enhance their reputation and, in turn, firms' financial performance. At the same time, practitioners in developed countries should pay extra attention to CSI, by adopting strategies to prevent and minimize the probability that these accidents occur in order to avoid the negative effect on performance.

The results of the study should be seen in the light of the following limitations, which however open future research opportunities. First, the study focuses on large, listed companies. Therefore, our results cannot be generalized to all types of companies. Although we control for geographical area fixed effects, industry fixed effects, and country fixed effects, future studies could disentangle contingent effects related to firm size and/or industry-specific characteristics. Second, CSR practices – like any other form of strategy – could differ across companies. At the same time, CSI controversies can be of different types. So, it is important that future studies account for the different forms of CSR and CSI and analyze their effect on the financial performance of the companies. Third, it would be interesting to investigate the effect of specific categories of the CSR indicators on the financial performance.

REFERENCES

Aguilera, R. V., Rupp, D. E., Williams, C. A., & Ganapathi, J. (2007). Putting the S back in corporate social responsibility: A multilevel theory of social change in organizations. *Academy of Management Review*, *32*(3), 836–863. https://doi.org/10.5465/amr.2007.25275678

Allen, M. (1997). The problem of multicollinearity. In *Understanding regression analysis*. Springer. https://doi.org/10.1007/978-0-585-25657-3_37

Alshehhi, A., Nobanee, H., & Khare, N. (2018). The impact of sustainability practices on corporate financial performance: Literature trends and future research potential. *Sustainability*, *10*(2), 494. https://doi.org/10.3390/su10020494

Aouadi, A., & Marsat, S. (2016). Do ESG controversies matter for firm value? Evidence from international data. *Journal of Business Ethics*, *151*(4), 1027–1047. https://doi.org/10.1007/s10551-016-3213-8

Arun, T., Girardone, C., & Piserà, S. (2022). ESG issues in emerging markets and the role of banks. In Nguyen, D. C. (Eds.), *Handbook of banking and finance in emerging markets* (pp. 321–344). Edward Elgar Publishing. https://doi.org/10.4337/9781788972708.00023

Atan, R., Alam, M. M., Said, J., & Zamri, M. (2018). The impacts of environmental, social, and governance factors on firm performance: Panel study of Malaysian companies. *Management of Environmental Quality*, *29*(2), 182–194. https://doi.org/10.1108/MEQ-03-2017-0033

Awaysheh, A., Heron, R. A., Perry, T., & Wilson, J. I. (2020). On the relation between corporate social responsibility and financial performance. *Strategic Management Journal*, *41*(6), 965–987. https://doi.org/10.1002/smj.3096

Bahadori, N., Kaymak, T., & Seraj, M. (2021). Environmental, social, and governance factors in emerging markets: The impact on firm performance. *Business Strategy and Development*, *4*(4), 411–422. https://doi.org/10.1002/bsd2.198

Bhaskaran, R. K., Ting, I. W. K., Sukumaran, S. K., & Sumod, S. D. (2020). Environmental, social and governance initiatives and wealth creation for firms: An empirical examination. *Managerial and Decision Economics*, *41*(5), 710–729. https://doi.org/10.1002/mde.3134

Brammer, S., & Pavelin, S. (2006). Corporate reputation and social performance: The importance of fit. *Journal of Management Studies*, *43*(3), 435–455. https://doi.org/10.1111/j.1467-6486.2006.00597.x

Brammer, S., Nardella, G., & Surdu, I. (2021). Defining and deterring corporate social irresponsibility: Embracing the institutional complexity of international business. *Multinational Business Review*, *29*(3), 301–320. https://doi.org/10.1108/MBR-02-2021-0011

Campbell, L. (2007). Why would corporations behave in socially responsible ways? An institutional theory of corporate social responsibility. *The Academy of Management Review*, *32*(3), 946–967. https://doi.org/10.5465/amr.2007.25275684

Carroll, A. B., & Shabana, K. M. (2010). The business case for corporate social responsibility: A review of concepts, research and practice. *International Journal of Management Reviews*, *12*(1), 85–105. https://doi.org/10.1111/j.1468-2370.2009.00275.x

Cheng, B., Ioannou, I., & Serafeim, G. (2014). Corporate social responsibility and access to finance. *Strategic Management Journal*, *35*(1), 1–23. https://doi.org/10.1002/smj.2131

Christensen, J. H., Kanikicharla, K. K., Aldrian, E., An, S. I., Albuquerque Cavalcanti, I. F., de Castro, M., Dong, W., Goswami, P., Hall, A., Kanyanga, J. K., Kitoh, A., Kossin, J., Lau, N. C., Renwick, J., Stephenson, D. B., Xie, S. P., Zhou, T., Abraham, L., Ambrizzi, T., … Zou, L. (2013). *Climate phenomena and their relevance for future regional climate change*. In Climate change 2013 the physical science basis: Working Group I contribution to the fifth assessment report of the Intergovernmental Panel on Climate Change (Vol. 9781107057999, pp. 1217–1308). Cambridge University Press. https://doi.org/10.1017/CBO9781107415324.028

Chung, G. (2021). *ESG has so far been a rich world phenomenon. But emerging markets are catching up*. https://www.institutionalinvestor.com/article/b1s9yy7gmmp1pj/ESG-Has-So-Far-Been-a-Rich-World-Phenomenon-But-Emerging-Markets-Are-Catching-Up

Cuervo-Cazurra, A., Dieleman, M., Hirsch, P., Rodrigues, S. B., & Zyglidopoulos, S. (2021). Multinationals' misbehavior. *Journal of World Business*, *56*(5), 101244. https://doi.org/10.1016/j.jwb.2021.101244

Cuervo-Cazurra, A., Purkayastha, S., & Ramaswamy, K. (2022). Variations in the corporate social responsibility–performance relationship in emerging market firms. *Organization Science*, *34*(4), 1626–1650. https://doi.org/10.1287/orsc.2022.1639

De Lucia, C., Pazienza, P., & Bartlett, M. (2020). Does good ESG lead to better financial performances by firms? Machine learning and logistic regression models of public enterprises in Europe. *Sustainability*, *12*(13), 5317. https://doi.org/10.3390/su12135317

Delmas, M. (2001). Stakeholders and competitive advantage: The case of ISO 14001. *Production and Operations Management*, *10*(3), 343–358. https://doi.org/10.1111/j.1937-5956.2001.tb00379.x

Delmas, M., & Toffel, M. W. (2004). Stakeholders and environmental management practices: An institutional framework. *Business Strategy and the Environment*, *13*(4), 209–222. https://doi.org/10.1002/bse.409

Duque-Grisales, E., & Aguilera-Caracuel, J. (2021). Environmental, social and governance (ESG) scores and financial performance of multilatinas: Moderating effects of geographic international diversification and financial slack. *Journal of Business Ethics*, *168*, 315–334. https://doi.org/10.1007/s10551-019-04177-w

Earnhart, D., & Lizal, L. (2010). Effect of corporate economic performance on firm-level environmental performance in a transition economy. *Environmental and Resource Economics*, *46*(3), 303–329. https://doi.org/10.1007/s10640-009-9342-0

Endrikat, J., Guenther, E., & Hoppe, H. (2014). Making sense of conflicting empirical findings: A meta-analytic review of the relationship between corporate environmental and financial performance. *European Management Journal*, *32*(5), 735–751. https://doi.org/10.1016/j.emj.2014.03.001

Epstein, M. J., & Buhovac, A. R. (2014). *Making sustainability work: Best practices in managing and measuring corporate social, environmental, and economic impacts*. Berrett-Koehler Publishers.

Esfahbodi, A., Zhang, Y., & Watson, G. (2016). Sustainable supply chain management in emerging economies: Trade-offs between environmental and cost performance. *International Journal of Production Economics*, *181*, 350–366. https://doi.org/10.1016/j.ijpe.2016.07.012

Flammer, C., Hong, B., & Minor, D. (2018). *Corporate Governance and the rise of integrating corporate social responsibility criteria in executive compensation: Effectiveness and implications for firm outcomes*. SSRN. http://dx.doi.org/10.2139/ssrn.2831694

Freeman, E. R. (1984). *Strategic management: A stakeholder approach* (Pitman Series in Business and Public Policy, 1st ed.). Harpercollins College Div.

Friede, G., Busch, T., & Bassen, A. (2015). ESG and financial performance: Aggregated evidence from more than 2000 empirical studies. *Journal of Sustainable Finance and Investment*, *5*(4), 210–233. https://doi.org/10.1080/20430795.2015.1118917

Frooman, J. (1999). Stakeholder influence strategies. *Academy of Management Review*, *24*(2), 191–205. https://doi.org/10.5465/amr.1999.1893928

Gao, Y., Zhang, D., & Huo, Y. (2018). Corporate social responsibility and work engagement: Testing a moderated mediation model. *Journal of Business Psychology*, *33*(5), 661–673. https://doi.org/10.1007/s10869-017-9517-6

Garcia, A. S., Mendes-Da-Silva, W., & Orsato, R. J. (2017). Sensitive industries produce better ESG performance: Evidence from emerging markets. *Journal of Cleaner Production*, *150*, 135–147. https://doi.org/10.1016/j.jclepro.2017.02.180

Gartenberg, C. M., Prat, A., & Serafeim, G. (2018). Corporate purpose and financial performance. *Organization Science*, *30*(1), 1–18. http://dx.doi.org/10.2139/ssrn.2840005

Goerzen, A., Sartor, M., Brandl, K., Steger, U., & Terjesen, S. (2023). Widening the lens: Multilevel drivers of firm corporate social performance. *Journal of International Business Studies*, *54*(1), 42–60. https://doi.org/10.1057/s41267-022-00575-x

Gras, D., & Krause, R. (2020). When does it pay to stand out as stand-up? Competitive contingencies in the corporate social performance–corporate financial performance relationship. *Strategic Organization*, *18*(3), 448–471. https://doi.org/10.1177/1476127018805252

Grewatsch, S., & Kleindienst, I. (2017). When does it pay to be good? Moderators and mediators in the corporate sustainability–corporate financial performance relationship: A critical review. *Journal of Business Ethics*, *145*, 383–416. https://doi.org/10.1007/s10551-015-2852-5

Hair, J. F., Anderson, R. E., Tatham, R. L., & Black, W. C. (1995). *Multivariate data analyses with readings*. Prentice Hall.

Harjoto, M. A., Hoepner, A. G. F., & Li, Q. (2021). Corporate social irresponsibility and portfolio performance: A cross-national study. *Journal of International Financial Markets, Institutions and Money*, *70*, 101274. https://doi.org/10.2139/ssrn.3683170

Hickel, J. (2020). The sustainable development index: Measuring the ecological efficiency of human development in the anthropocene. *Ecological Economics*, *167*, 106452. https://doi.org/10.1016/j.ecolecon.2019.106452

Hoang, T. (2018). The role of integrated reporting in raising awareness of environmental, social and corporate governance (ESG) performance. In S. Seifi & D. Crowther (Eds.), *Stakeholders, governance and responsibility* (Developments in Corporate Governance and Responsibility, Vol. 14, pp. 47–69). Emerald Publishing Limited. https://doi.org/10.1108/S2043-052320180000014003

Hull, C. E., & Rothenberg, S. (2008). Firm performance: The interactions of corporate social performance with innovation and industry differentiation. *Strategic Management Journal*, *29*(7), 781–789. https://doi.org/10.1002/smj.683

Ioannou, I., & Serafeim, G. (2012). What drives corporate social performance? The role of nation-level institutions. *Journal of International Business Studies*, *43*(9), 834–864. https://doi.org/10.1057/jibs.2012.29

Javed, M., Rashid, M. A., & Hussain, G. (2016). When does it pay to be good – A contingency perspective on corporate social and financial performance: Would it work? *Journal of Cleaner Production*, *133*, 1062–1073. https://doi.org/10.1016/j.jclepro.2016.06.092

Khanna, T., & Palepu, K. G. (2010). *Winning in emerging markets: A road map for strategy and execution*. Harvard Business Press. https://doi.org/10.1017/CBO9781107415324.004

Kim, K. H., Kim, M., & Qian, C. (2018). Effects of corporate social responsibility on corporate financial performance: A competitive-action perspective. *Journal of Management*, *44*(3), 1097–1118. https://doi.org/10.1177/0149206316634058

Klassen, R. D., & McLaughlin, C. P. (1996). The impact of environmental management on firm performance. *Management Science*, *42*(8), 1199–1214. https://doi.org/10.1287/mnsc.42.8.1199

Kolk, A., & van Tulder, R. J. M. (2010). International business, corporate social responsibility and sustainable development. *International Business Review*, *19*(2), 119–125. https://doi.org/10.1016/j.ibusrev.2009.10.007

Landi, G., & Sciarelli, M. (2019). Towards a more ethical market: The impact of ESG rating on corporate financial performance. *Social Responsibility Journal*, *15*(1), 11–27. https://doi.org/10.1108/SRJ-11-2017-0254

Li, J., Haider, Z. A., Jin, X., & Yuan, W. (2019). Corporate controversy, social responsibility and market performance: International evidence. *Journal of International Financial Markets, Institutions and Money*, *60*, 1–18. https://doi.org/10.1016/j.intfin.2018.11.013

Linnenluecke, M. K. (2022). Environmental, social and governance (ESG) performance in the context of multinational business research. *Multinational Business Review*, *30*(1), 1–16. https://doi.org/10.1108/MBR-11-2021-0148

Lopez-de-Silanes, F., McCahery, J. A., & Pudschedl, P. C. (2020). ESG performance and disclosure: A cross-country analysis. *Singapore Journal of Legal Studies*, 217–241. https://doi.org/10.1017/S0218538520000105

Lu, W., Chau, K. W., Wang, H., & Pan, W. (2014). A decade's debate on the nexus between corporate social and corporate financial performance: A critical review of empirical studies 2002–2011. *Journal of Cleaner Production*, *79*, 195–206.

Manrique, S., & Martí-Ballester, C.-P. (2017, October). Analyzing the effect of corporate environmental performance on corporate financial performance in developed and developing countries. *Sustainability, MDPI*, *9*(11), 1–30. https://doi.org/10.3390/su9112045

Margolis, J. D., & Walsh, J. P. (2003). Misery loves companies: Rethinking social initiatives by business. *Administrative Science Quarterly*, *48*(2), 268–305. https://doi.org/10.2307/3556659

Marsat, S., & Williams, B. (2011, May). *CSR and market valuation: International evidence* [Conference presentation]. In International conference of the French Finance Association (AFFI). Accessed on January 16, 2023. Available at SSRN: https://ssrn.com/abstract=1833581 or http://dx.doi.org/10.2139/ssrn.1833581.

Melinda, A., & Wardhani, R. (2020). The effect of environmental, social, governance, and controversies on firms' value: Evidence from Asia. In W. A. Barnett & B. S. Sergi (Eds.), *Advanced issues in the economics of emerging markets* (International Symposia in Economic Theory and Econometrics, Vol. 27, pp. 147–173). Emerald Publishing Limited. https://doi.org/10.1108/S1571-038620200000027011

Mobius, M., & Ali, U. (2021). ESG in emerging markets: The value of fundamental research and constructive engagement in looking beyond ESG ratings. *Journal of Applied Corporate Finance*, *33*(2), 112–120. https://doi.org/10.1111/jacf.12466

Naeem, N., Cankaya, S., & Bildik, R. (2022). Does ESG performance affect the financial performance of environmentally sensitive industries? A comparison between emerging and developed markets. *Borsa Istanbul Review*, *22*(2), S128–S140. https://doi.org/10.1016/j.bir.2022.01.004

Napier, E., Knight, G., Luo, Y., Chen, J., & Zhang, S. (2023). Corporate social performance in international business. *Journal of International Business Studies*, *54*, 61–77. https://doi.org/10.1057/s41267-022-00584-w

Nardella, G., Surdu, I., & Brammer, S. (2023). What happens abroad, stays abroad? Exploring how corporate social irresponsibility in domestic and international markets influences corporate reputation. *Journal of World Business*, *58*(4), 101420. https://doi.org/10.1016/j.jwb.2022.101420

Nirino, N., Santoro, G., Miglietta, N., & Quaglia, R. (2021). Corporate controversies and company's financial performance: Exploring the moderating role of ESG practices. *Technological Forecasting and Social Change*, *162*, 120341. https://doi.org/10.1016/j.techfore.2020.120341

Orlitzky, M., Schmidt, F. L., & Rynes, S. L. (2003). Corporate social and financial performance: A meta-analysis. *Organization Studies*, *24*(3), 403–441. https://doi.org/10.1177/0170840603024003910

Orsato, R. J., Garcia, A., Mendes-Da-Silva, W., Simonetti, R., & Monzoni, M. (2015). Sustainability indexes: Why join in? A study of the corporate sustainability index (ISE) in Brazil. *Journal of Cleaner Production*, *96*, 161–170. https://doi.org/10.1016/j.jclepro.2014.10.071

Parmigiani, A., Klassen, R. D., & Russo, M. V. (2011). Efficiency meets accountability: Performance implications of supply chain configuration, control, and capabilities. *Journal of Operations Management*, *29*(3), 212–223. https://doi.org/10.1016/j.jom.2011.01.001

Rajesh, R., & Rajendran, C. (2020). Relating environmental, social, and governance scores and sustainability performances of firms: An empirical analysis. *Business Strategy and the Environment*, *29*(3), 1247–1267. https://doi.org/10.1002/bse.2447

Refinitiv. (2022). *Refinitiv (Thomson Reuteres) ESG company scores*. https://www.refinitiv.com/content/dam/marketing/en_us/documents/methodology/refinitiv-esg-scores-methodology.pdf

Reverte, C., Gómez-Melero, E., & Cegarra-Navarro, J. G. (2016). The influence of corporate social responsibility practices on organizational performance: Evidence from eco-responsible Spanish firms. *Journal of Cleaner Production*, *112*, 2870–2884. https://doi.org/10.1016/j.jclepro.2015.08.045

Russo, M. V., & Fouts, P. A. (1997). A resource-based perspective on corporate environmental performance and profitability. *Academy of Management Journal*, *40*, 534–559. https://doi.org/10.2307/257052

Sachin, N., & Rajesh, R. (2022). An empirical study of supply chain sustainability with financial performances of Indian firms. *Environment, Development and Sustainability*, *24*(5), 6577–6601. https://doi.org/10.1007/s10668-021-01717-1

Shi, W., & Veenstra, K. (2021). The moderating effect of cultural values on the relationship between corporate social performance and firm performance. *Journal of Business Ethics*, *174*(1), 89–107. https://doi.org/10.1007/s10551-019-04297-2

Suchman, M. C. (1995). Managing legitimacy: Strategic and institutional approaches. *The Academy of Management Review*, *20*(3), 571–610. https://doi.org/10.2307/258788

Surroca, J., Tribo, J. A., & Waddock, S. (2010). Corporate Responsibility and Financial Performance: The role of Intangible Resources. *Strategic Management Journal*, *31*(5), 463–490. http://www.jstor.org/stable/40587594

Tamayo-Torres, I., Gutierrez-Gutierrez, L., & Ruiz-Moreno, A. (2019). Boosting sustainability and financial performance: The role of supply chain controversies. *International Journal of Production Research*, *57*(19), 6190–6205. https://doi.org/10.1080/00207543.2018.1562248

Ting, I. W. K., Azizan, N. A., Bhaskaran, R. K., & Sukumaran, S. K. (2019). Corporate social performance and firm performance: Comparative study among developed and emerging market firms. *Sustainability*, *12*(1), 26. https://doi.org/10.3390/su12010026

Tobin, J. (1969). A general equilibrium approach to monetary theory. *Journal of Money, Credit and Banking*, *1*(1), 15–29. https://doi.org/10.2307/1991374

United Nations. (2020). *World economic situation and prospects*. https://doi.org/10.18356/ee1a3197-en

Wang, Z., & Sarkis, J. (2013). Investigating the relationship of sustainable supply chain management with corporate financial performance. *International Journal of Productivity and Performance Management*, *62*(8), 871–888. https://doi.org/10.1108/ijppm-03-2013-0033

Wang, Z., & Sarkis, J. (2017). Corporate social responsibility governance, outcomes, and financial performance. *Journal of Cleaner Production*, *162*, 1607–1616. https://doi.org/10.1016/j.jclepro.2017.06.142

Wei, Z., Shen, H., Zhou, K. Z., & Li, J. J. (2017). How does environmental corporate social responsibility matter in a dysfunctional institutional environment? Evidence from China. *Journal of Business Ethics*, *140*, 209–223. https://doi.org/10.1007/s10551-015-2693-2

Wooldridge, J. M. (2000). *Econometric analysis of cross section and panel data*. MIT Press. https://doi.org/10.1017/CBO9780511815478

Zhang, D., Morse, S., & Ma, Q. (2019). Corporate social responsibility and sustainable development in China: Current status and future perspectives. *Sustainability*, *11*(16), 4392. https://doi.org/10.3390/su11164392

Zhao, X., & Murrell, A. J. (2016). Revisiting the corporate social performance–financial performance link: A replication of Waddock and Graves. *Strategic Management Journal*, *37*(11), 2378–2388. https://doi.org/10.1002/smj.2458

CHAPTER 19

THE ROLE OF NGOs IN THE TRANSITION TOWARDS MORE SUSTAINABLE AND INNOVATIVE AGRI-FOOD GVCs

Cyntia Vilasboas Calixto Casnici[a], Germano Glufke Reis[b], David Schulzmann[a,c], Marina Papanastassiou[a,d] and Jeremy Clegg[a]

[a]*Department of International Business, Leeds University Business School, UK*
[b]*Federal University of Parana, Management School, Brazil*
[c]*Aalborg University Business School, Aalborg University, Denmark*
[d]*Carnegie Mellon University in Qatar, Qatar*

ABSTRACT

Multinational enterprises (MNEs) in the agri-food sector are continuously transforming their global value chains (GVCs) to address sustainable development challenges of food security (SDG2) and climate change (SDG13). However, the central role of non-governmental organisations (NGOs) in (re) creating GVCs across multilevel stakeholders through innovative approaches to solve sustainability challenges remains under investigated. This explorative study investigates how international NGOs influence the transformation of large-scale industrial animal agriculture to a more sustainable (cell-based) agri-food GVC. The authors conducted a case study on the Good Food Institute (GFI), an international NGO, that has been an active player in the transition

Walking the Talk? MNEs Transitioning Towards a Sustainable World
Progress in International Business Research, Volume 18, 373–392

ISSN: 1745-8862/doi:10.1108/S1745-886220240000018023

to alternative sources of protein to solve animal-based agriculture sustainability issues. The results show that an international NGO can contribute to the transition to a more sustainable GVC and can enhance the GVC's innovation capabilities.

Keywords: Cultivated meat; MNEs; innovation; international NGOs; sustainable GVCs; SDGs

1. INTRODUCTION

Studies on the influence of NGOs have focussed mostly on NGOs' activism and pressure strategies, the development of standards and guidelines (e.g. animal welfare and sustainability standards), their interfaces with private and public sectors, or see NGOs as one stakeholder among many others that influence organisations (Kourula & Laasonen, 2010). However, their role in reshaping MNE-led GVCs in order to solve complex sustainability challenges – such as in conventional animal-based meat chain – has been overlooked. International NGOs have proven to have outgrown their historically constrained role and now increasingly address issues of global concern that can be more effective in advocating for sustainability improvements in global scope, and mediate multinationals' market and non-market activity (Fee & Gray, 2013; Lambell et al., 2008; Teegen et al., 2004).

In addition, the GVC literature has mostly addressed the governance exerted by major MNEs in leading and shaping GVCs. However, Ponte (2014) calls for research that expands the view that GVC governance relies only on interfirm coordination led by MNEs. This chapter addresses the idea that NGOs may have an important role in this regard, especially in nascent high-tech industries such as in the case of the alt-protein landscape. Broadly, alt-proteins (e.g. plant-based and cell-based proteins) reflect the role of innovation in substantially transforming industrial animal-based agriculture for the better, reducing its environmental footprints and animal suffering (Gerber et al., 2013; Nobre, 2022; Stephens et al., 2018). To this end, the adoption of alt-proteins (in substitution to animal-based meat) can support the progress of the United Nations' Sustainable Development Goals (UN's SDGs) 2 and 13 of food security and climate change, respectively (Nobre, 2022). Thus, in the chapter, we aim to understand how international NGOs can have a leading role in reshaping the agri-food GVCs through innovation, by focussing in the case of the meat GVC, towards sustainability. We thus, explore the following research question: *How do international NGOs influence innovation and impact the transition of large-scale industrial animal agriculture to a more sustainable agri-food GVC?*

We answer this question through an extensive case study of an international NGO, the GFI, which has been an active player towards the transition to alternative sources of protein to solve animal-based agriculture sustainability issues. In-depth interviews were conducted with representatives of GFI's subsidiaries from all over the world. Our results add to the international business (IB) literature that explores the NGOs and MNEs' partnerships in sustainably addressing environmental and social problems (Ghauri et al., 2021; Liu et al., 2020) by showing the new roles and scopes of action that international NGOs can take in the transition to more sustainable GVCs by participating actively in the innovation process.

2. LITERATURE REVIEW

2.1. Global Food Systems, GVCs, and Innovation

Food systems have garnered increasing concerns in relation to their impact on food security and environmental sustainability (Hoek et al., 2021; Mergos & Papanastassiou, 2017). According to a recent study by Crippa et al. (2021), food systems were estimated in 2015 to be responsible for 34% of total greenhouse gas (GHG) emissions, being meat production, especially cattle, by far the most important cause of these emissions. OECD/FAO projected that meat consumption will increase by 29% by 2029 (compared to the base period of 2017–2019) with Brazil, China, the European Union, and the United States to account for 60% of meat production by 2029 (OECD/FAO, 2020). According to IPCC (2020), dietary changes could reduce global CO_2 emissions by up to eight billion tonnes per year. Moreover, meat production is also related to other important issues such as deforestation, zoonoses spread and amplification, antibiotic resistance, and animal suffering (Kelly et al., 2021). These problems are likely to have a greater negative impact on the planet in the coming years, as the demand for meat advances. Therefore, it is evident that there is a need to decarbonise this industry and mitigate its effect on climate change in coming years.

At the same time, food systems are contained in complicated GVCs that are mostly led by MNEs, creating a close link between trade and foreign direct investment. The global agri-food industry is becoming 'increasingly dominated by value-chain relationships in which leading firms exercise vertical coordination' (Humphrey & Memedovic, 2006, p. 7) with the top 100 agri-food and beverage MNEs accounting for one-third of global food production and more than one-half of the technological activities. The impact of MNEs in shaping agri-food GVCs and in particular the industrial part of the GVC, that is, the food and drink sectors are then strongly confirmed (Narula & Wahed, 2017; Scoppola, 2021).

Furthermore, there is extant literature on the mapping of agri-food GVCs where the emphasis is on upgrading as it reflects how MNEs' innovative activities are organised among global and local stakeholders (Lee et al., 2012; Lee & Gereffi, 2021). Recent work by Ambos et al. (2021) further analyses how innovation is organised in MNE-led GVCs and raises attention towards an interactive and open model of innovation where connectivity between MNEs and local suppliers, in an interdependent mode, generates knowledge and innovation in GVCs (Strange & Humphrey, 2019). However, in recent years it has been acknowledged that innovation in GVCs extends beyond binary relations between MNEs and suppliers and more complex ecosystems are developed where different stakeholders are involved (Papanastassiou et al., 2020). The richness of the GVC innovation ecosystem is, thus, becoming evident where issues of sustainability are raised and need to be addressed (Golgeci et al., 2021).

2.2. GVCs and the Role of International NGOs

Agri-food MNEs, in particular, have been pressed by international and local regulatory bodies, NGOs, and civil society, to adopt more sustainable approaches to their production (Tran et al., 2013). Moreover, there is an increasing consumer demand for new dietary practices (vegans, vegetarians, flexitarians, and climatarians) reflecting a transition to more sustainable consuming behaviour (Taghikhah et al., 2021). In response to these developments, we are currently witnessing the growth of the so-called 'meat substitutes', that are alternative sources of protein, aiming at producing animal meat or meat-like products, without the need of raising and slaughtering animals. This is a compelling case of innovation that may dramatically change the conventional meat production value chain. These products may be plant-based or obtained through the in vitro cultivation of animal cells (GFI, 2021; Godfray et al., 2019). Although plant-based meat products have already gained track in the global food marketplace with an increase in retail sales by 27% in 2020 in the United States (Plant Based Foods Association, 2021), it should be noted that they are not actually animal meat. Plant-based alternatives are produced from vegetable proteins and basically are aimed at 'mimicking' meat appearance, taste, and texture – they even look like fresh raw meat – using plants as building blocks (e.g. peas). Cultivated meat, in turn, is a highly promising product since it is truly animal-based meat; it involves animal cell cultivation in a bioreactor (from cattle, chicken, fish, etc.) and is identical to conventional meat.

Gereffi and Fernandez-Stark (2016) recognise the importance of chain governance in shaping GVCs' structure and dynamics. Gereffi et al. (2001, p. 4) define governance as 'non-market coordination of economic activity'. As in most agri-food GVCs, the meat chain has a high degree of concentration with a set of major downstream global producers and supermarket chains leading and coordinating upstream players (Gereffi & Fernandez-Stark, 2016). Its governance is highly hierarchical, that is, buyer-driven, with meat MNEs and major retailers influencing upstream actors in several aspects, from efficiency management to codes of conduct adoption. In this context, the role of NGOs in animal-based GVCs is mostly in line with the compliance-based model (Lund-Thomsen & Lindgreen, 2014) which assumes that NGOs exert pressure on firms to improve issues related to sustainability, human rights, and working conditions, among others; as a result, it is expected that MNEs develop codes of conduct and guidelines for more socially and environmentally responsible operations. Firms–NGO partnerships may arise in this context as well, as a means for solving key sustainability issues (Bitzer & Glasbergen, 2015; Mousavi & Bossink, 2020). It also involves NGOs setting their own sustainability guidelines and standards that may eventually be disseminated throughout the chain (Kourula & Laasonen, 2010). Reis and Molento (2020), for instance, show that emerging country meat-processing multinationals incorporate several animal welfare standards of international NGOs while they internationalise to more developed countries. They may also engage in advocacy and operational activity (filling local voids) along the chain (Teegen et al., 2004) and have active and political initiatives in NGO–business–government relationships aiming for sustainability development. It should be noted that the agri-food GVC mostly involves well-established commodities trade and global relationships between players (and regions) that perform distinct value-adding activities and have unequal value capture capacities; most importantly, it involves power asymmetries between actors of the chain (Dallas et al., 2019).

NGOs provide several guidelines and standards addressing sustainability and animal welfare, such as the Free Farming and the Sustainable Agriculture Network Sustainable Agriculture Standard (Hajjar et al., 2019; Ransom, 2007). In sum, NGOs frequently engage in normative, advocacy, pressure, political, and advisory activity in the case of animal-based agriculture. However, their role in generating new technology and embracing innovation is still underexplored, as this is the domain where usually MNEs play a central role.

3. THE CONTEXT

Cultivated meat, also known as cell-based meat and cultured meat (Reis, Heidemann, Borini, et al., 2020), is an alternative meat production process that does not require the raising and slaughtering of animals. Animal cells are instead cultivated and multiplied in a bioreactor to produce edible meat (Heidemann et al., 2020). As the technology rapidly improves (Bonny et al., 2015) more cultivated meat MNEs – meat cultivation start-ups and culturing system suppliers – have flourished worldwide (Froggatt & Wellesley, 2019). In addition, research on consumer attitudes has shown that cultivated meat is likely to achieve great acceptance. In Brazil, 63.6% of the participants of a survey stated they would eat cell-based meat (Valente et al., 2019); moreover, 45% of Brazilian consumers are likely to try and 31.5% highly likely to try cultivated products (Bryant & Krelling, 2020). In the United Kingdom, consumers are 40% likely to try cultivated meat (Szejda et al., 2021). These outcomes suggest that this innovation has a high potential in terms of reshaping meat production and consumption systems.

This innovation may dramatically improve the conventional meat production GVC and contribute effectively to climate change mitigation and improve food security concerns. For instance, Tuomisto and Teixeira de Mattos (2011) indicated that 99% less land would be needed to produce cultivated meat while GHG emissions might be 78–96% lower depending on the product (beef, poultry, etc.). Swartz (2021), in turn, identified that cultivated meat could decrease global warming impacts by 17%, 52%, and 85–92% compared to conventional meat production. Furthermore, it also prevents other severe issues related to intensive meat production such as increasing antibiotics resistance, zoonoses spread, and animal suffering. As shown in a systematic literature review conducted by Nobre (2022), alternative sources of protein, like cultivated meat, can also help the advancement of environmental and health-related SDGs (UN, 2015). Table 19.1 shows how cultivated meat can help to improve sustainability in the meat chain and how the transition is likely to support the progress of the UN's SDGs.

Consequently, while sustainability improvements in the meat chain basically involve incremental innovation, for example, reducing water waste and pollution (Gerber et al., 2013) and normative improvement, for example, adopting animal

Table 19.1. How the Transition to Cultivated Meat Innovation Relates to the UN's SDGs.

How cultivated meat can improve sustainability in the meat chain	Environmental UN's SDGs that relate to cultivated meat's sustainability improvements
Enhanced food security and sustainable agriculture	Goal 2. End hunger, achieve food security and improved nutrition, and promote sustainable agriculture
Reduction of GHG emissions	Goal 6. Ensure availability and sustainable management of water and sanitation for all
Less land use	Goal 7. Ensure access to affordable, reliable, sustainable, and modern energy for all
Better water use (less waste and pollution that affect biodiversity and aquatic species)	Goal 12. Ensure sustainable consumption and production patterns
Less energy use	Goal 13. Take urgent action to combat climate change and its impacts
Biodiversity preservation (terrestrial and aquatic)	Goal 14. Conserve and sustainably use the oceans, seas, and marine resources for sustainable development
	Goal 15. Protect, restore, and promote sustainable use of terrestrial ecosystems, sustainably manage forests, combat desertification, and halt and reverse land degradation and halt biodiversity loss
How cultivated meat can improve sustainability in the meat chain	Health UN's SDGs that relate to cultivated meat's sustainability improvements
Control of pathogen contamination	Goal 3. Ensure healthy lives and promote well-being for all at all ages
Reduction of zoonotic diseases	
Risk mitigation of cardiovascular, cancer, and obesity diseases	
Reduction of pandemic social problems in low- and high-income nations	

Sources: Nobre (2022) and UN (2015).

welfare standards, the transition to cultivated meat will lead to radical innovation (Dahlin & Behrens, 2005), since it has the potential to promote a breakthrough change in the meat value chain as a whole. As some aspects of this technology are still under development, and new players (e.g. start-ups, regulatory agencies, meat-processing MNEs, equipment suppliers, and investors) are still coming on board, these developments pave the way for the emergence of novel NGOs' role and activity in promoting innovation.

4. METHODOLOGICAL APPROACH

4.1. Data Collection

In this qualitative single case study research (Eisenhardt & Graebner, 2007), we are focussing on an international NGO, the GFI. GFI was founded in 2016 and it aims to support the development of alternative proteins around the globe, by creating and sharing knowledge about cultivated meat, plant-based meat, and fermentation. Apart from the headquarters in the United States, GFI currently operates in five different regions namely, Brazil, Europe, Israel, India, and

Asia Pacific (GFI, 2022). Furthermore, GFI has initiatives in three key areas: (1) advancing science and technology research in alternative proteins; (2) mediating business relationships, engaging companies, and investors internationally in order to advance alt-protein innovation and scale the value chain; and (3) advocating for policy-making and public research funding.

Therefore, we interviewed key decision makers in GFI headquarters and in all international branches to obtain a global perspective of the activities developed by the NGO. Our data collection covered at least three respondents in each GFI unit, reflecting the structural division and area of expertise of their personnel, that is – Science and Technology, Policy, and Industry (Corporate Engagement). Therefore, we will present a single case study within an embedded unit of analysis suitable for theory building (Yin, 2018).

Interviews were conducted by two of the researchers of this study during May to early June 2022, through MS Teams. They were our main source of data. We decided to have semi-structure interviews (Wengraf, 2001), since there was a need to guide the conversation towards innovation and sustainability, but open-ended questions to allow them to express their opinions and insights. Some executives were identified and approached through LinkedIn. For others, we followed the snowball sampling strategy by GFI members in order to facilitate access to respondents within the same location and across expertise (Emmel, 2013).

The 19 interviews lasted 8 hours and 54 minutes and we were able to generate 137 pages of transcript. Two interviews were composed of more than one member since they shared responsibilities. We used the Trint software to support the transcription of the MS Teams recordings (consented by all interviewees) of the 21 respondents and for the language translation from two respondents who were more comfortable speaking in their non-English native language, that is, in the case of Brazil. Since the respondents were working closely together within their region and within their roles across the regions, they were knowledgeable of each other's circumstances. This cross-functional and cross-regional duplicated knowledge allowed us to determine to have reached data saturation (Marshall et al., 2013). Table 19.2 shows the interview summary table split by region, number of interviews, and the period in which the interviews were conducted.

Secondary data were collected from February to June 2022. It included annual reports, specific studies and publications shared by GFI members, access to their databases, GFI's newsletter, McKinsey's reports, main business newspapers, such as *Forbes* and the *Economist*. It created a database of 41 documents. The insights from the secondary data were important to create a general view of the organisation, especially because most part of the interviewees were hired in the last 24 months due to the expansion of GFI worldwide.

Since the notions of validity and reliability are applied in quantitative research and are not suitable for qualitative research, we applied a variety of alternative techniques in order to increase trustworthiness through credibility, transferability, dependability, and confirmability (Hajro et al., 2017; Lincoln & Guba, 1985; Shah & Corley, 2006).

We carried out an extensive in-depth analysis based on regional industry reports, slides, and press releases to enable corroboration and triangulation of the evidence (Yin, 2018). The archival material allowed us to have a broader range

Table 19.2. Interviews Summary.

Region	Number of Interviewees	Period
United States (HQ)	4 (2 interviewees from Corporate Engagement) US-POL = Policy Coordinator US-SCI = Lead Scientist for Cultivated Meat US-CE-1 = University Innovation Manager US-CE-2 = Senior Investment Engagement Specialist	May 2022
Brazil	4 (2 members from the Policy team) BR-POL-1= Vice-President Public Policy BR-POL-2= Policy Analyst BR-SCI= Alternative Proteins Science and Technology Vice-President BR-CE= Corporate Engagement Director	May and June 2022
Singapore	4 (2 members from the Corporate Engagement) SG-POL= Policy Specialist SG-SCI= Science and Technology Specialist SG-CE-1= Corporate Engagement Specialist 1, for start-ups and universities SG-CE-2= Corporate Engagement Specialist 2, for MNEs	June 2022
India	3 (1 from each pillar) IN-POL= Senior Policy Specialist IN-SCI= Science and Technology Specialist IN-CE= Senior Innovation Specialist	May 2022
Europe	3 (1 from each pillar) EUR-POL= Policy Manager EUR-SCI= Science Technology Manager EUR-CE= Corporate Engagement Manager	May 2022
Israel	3 (1 from each pillar) IS-POL= Director of Strategic Alliances IS-SCI= Senior Scientist IS-CE= Business Analyst	May 2022

Source: Authors.

of GFI's activities over the last five years, complement the other data, and reduce the potential bias from the interviews (Eisenhardt & Graebner, 2007).

In a nutshell, we developed a case study protocol and interview guide to structure the data collection and analysis. Finally, to establish an appropriate understanding of our key constructs, we used multiple sources of evidence, mainly internal documents published by GFI, but also newspaper articles and industry reports (Zhao et al., 2020).

4.2. Data Analysis

The data analysis followed a data-driven inductive approach best suited to obtain rich and detailed descriptions that enable linking causal explanations and adding context to theory generation (Birkinshaw et al., 2011). Two researchers jointly analysed and proof-read the individual transcripts and began the data analysis by marking texts and paragraphs line-by-line to generate initial first-order codes guided by our research questions (Miles et al., 2014). We analysed GFI's stakeholders, individual goals, achievements, past challenges that were overcome, current challenges that are still being faced, and future agenda in addressing the current challenges in driving forward a sustainable value chain with cultivated meat. Most of the first-order codes were in vivo codes that allowed to convey the original words and expressions used by the respondents (Corbin & Strauss, 2015). The first-order coding process was extended to six secondary data reports to complement and confirm the initial first-order codes and therefore strengthen the data quality. Each interview or document was broken down in order to deliberate the significance to support the themes resulting in 24 first-order codes (Gioia et al., 2013).

In the next step, we identified relationships across the primary and secondary first-order codes and consolidated them into second-order themes related to GFI's activities in influencing and forming the cultivated meat value chain. This process resulted in six different second-order codes. After we have formed the aggregated second-order codes, we have further condensed them into four different overarching theoretical themes. The coding process required the co-authors multiple times to iterate between the collected data and previous studies, leading us to search for additional literature and reshaping our research question (Miles et al., 2014).

5. RESULTS

Our results demonstrate how GFI has had a global influence in advancing innovation and orchestrating the transition towards a more sustainable agri-food GVC. We have identified themes based on their research activities in the science and technology field, advocacy for new governmental regulations, and their engagement with new smaller and established larger industrial players in the meat industry. The three themes relate to the development of the cultivated meat industry, identifying and articulating regulatory issues, and liabilities in the nascent value chain. Table 19.3 depicts the results that have emerged during this inductive and iterative data analysis process.

Table 19.3. Data Presentation.

First Order Codes	Second Order Themes	Constructs
Capacity building Talent development of relevant technical staff Disseminate funding opportunities to relevant academics Created academic alternative protein course Form student groups on campuses to create talent pipeline	Education and research infrastructure	
Build a regional alternative protein community Connect different stakeholders Counselling some firms that are now operating in the market Collaboration with big companies and set up partnerships and formalised collaborations Facilitate relationships across the value chain and foster connections	NGO's engagement	Developing the industry

(*Continued*)

Table 19.3. (*Continued*)

First Order Codes	Second Order Themes	Constructs
Increase awareness about alternative proteins	Raising awareness	
Acceptance of cultivated meat and terminology		Articulating market and regulatory issues
Support fair, efficient regulation, and safety of alternative proteins	Contribute institutional development	
Provide government and industry advisory		
Achieving a common voice in the industry		
Support firms in talking to the public sector		
Alleviating the bottlenecks by means of internal analyses of the industry, filling research gaps, and proposing solutions and fund research	Chain development on a global basis	
Proof of concept showing the feasibility of this product by launching manufacturing units		Liabilities in the value chain
Labelling	Reduce bottlenecks	
Managing to create infrastructure for scale up		
Bringing costs down (scale) and making the products better (taste)		
Strengthening food security		
Funding – lack of public investment		

Source: Authors.

5.1. Developing the Industry's GVC

GFI has been leading a proactive role in the development of the cultivated meat industry by providing resources and influencing educational institutions to establish new alternative protein programmes and promote research funding. Moreover, GFI has been active in building up a cultivated meat network through stakeholder engagement across the value chain and uniting them in an alternative protein community.

5.1.1. Education and Research Infrastructure

Cultivated meat is a nascent industry that has yet to develop the capacity in the workforce and research infrastructure to keep up with current demands and facilitate future growth. GFI recognises the need for capacity building as highlighted by the Vice President for Science and Technology:

> [...] we don't have ... people trained or ... experts right in the field [across industry and academia]. So, it's super important to develop capacities ... (and influence) universities to developing courses and lectures ... and other events.' (BR-SCI)

GFI has acted upon the need and has started internally developing educational materials that have been used within educational institutions and with other stakeholders in order to educate them about alternative proteins.

> I provide the courses on alternative proteins in a several Israeli universities (and outside of Israel) and I ... also provide the lectures, develop some resources in the field. (IS-SCI)

The format for disseminating cultivated meat knowledge can be through in-person lectures at educational institutions or through online webinars that are accessible to different interested stakeholders to get involved.

The collaboration with universities is not limited to lectures but aims to develop the educational field on alternative proteins and specifically on cultivated meat. This includes building up educational programmes on alternative protein that is contributing to satisfy the increasing demand for a larger talent pool. The University Innovation Manager explains:

> GFI is activating a new generation of scientists and students to see themselves working in the alt protein field, collaborating as a networked team, and hastening a global protein transition in our lifetimes … we are focused on really building scientific and engineering education and training pathways and advancing scientific research. (US-CE)

One way of establishing collaboration with universities has been successful through a global Alt-Protein Project that currently includes 16 universities and will have grown to 32 universities across the United States, Europe, Asia, and Africa. The Alt-Protein Project has been a means to develop student groups on campus and link them to an academic ecosystem.

Moreover, to promote cultivated meat topics in educational institutes, GFI is working closely with international funding bodies to advocate for funding opportunities in the area of cultivated meat and disseminate them to relevant groups. The funding opportunities are necessary to further develop this nascent area and strengthen commercial feasibility. Besides education, GFI has developed a robust influence on several stakeholders. The NGO plays a role in the industry as a catalyser for new opportunities for MNEs and start-ups as well as an integrator for different actors. As an independent body, its collaborations throughout the value chain legitimise the engagement towards a more sustainable industry.

5.1.2. NGO's Engagement

GFI has a strong influence in the current commercial developments of cultivated meat together with small firms and MNEs, globally. In the Brazilian context, GFI has a strong relationship with large MNEs that are now increasing their portfolio with alternative proteins as well as supporting the creation of new start-ups as indicated by the Corporate Engagement Director:

> We are involved in the JBS and BRF cultured meat projects and several other projects. We helped to create the alternative protein sector here in Brazil. We arrived here in 2017 and in 2019 we launched here Fazenda Futuro …. So, I provide the sector with information so that it can make the decision in the most assertive manner. I'm never going to tell a MNE to go on that route [acquisition of start-ups] or go this way [internal business unit] …. I show all the options for him to build a strategic plan … if he wants to know how to build a business unit, for example, I give him the inputs to do it. (BR-CE)

The current landscape of commercial players globally is small, specialised, and decentralised. To promote more innovation within cultivated meat, GFI has actively supported start-up growth in the industry. One example of facilitating early commercial development and financially supporting entrepreneurs is explained by the Senior Innovation Manager in India:

> My role is focused on building out the early-stage entrepreneurial ecosystem for plant-based, cultivated and fermentation derived meat, eggs and dairy … from ideation up until the time that start-ups launch in the market …. The one (start-up) in cultivated beef (is) MyoWorks and

we've just supported them on sort of understanding space opportunities as well as supporting them on the grant. We are on (board) of a (grant funding programme) committee to evaluate progress. (IN-CE)

The engagement is not limited to support individual start-ups and entrepreneurs with specific advisory and funding support, but also to facilitate interconnectivity between the different companies. The development of an ecosystem helps them to utilise each other's expertise within and across different regions.

As engagement requires an interface with different actors, the need to master regulatory issues was pointed out. Investors will feel comfortable to provide additional funding when they can perceive that the products will be available to consumers in the near future. Moreover, consumers would help advocate for a more sustainable source of protein once they understand the benefits and it is safe to buy.

5.2. Articulating Market and Regulatory Issues

5.2.1. Raising Awareness

One of GFI's challenges was raising awareness about alternative proteins. In the beginning, it was difficult for several employees in the organisation to explain to regulators, potential investors, and consumers what kind of technology was involved in the process of producing cultivated meat as illustrated by the Corporate Engagement Manager in Europe:

Understand organisations in Europe that focus on alternative proteins. Bringing the companies together and facilitate partnerships. Guiding companies to focus on top priorities and raising awareness, educating, engaging. (EUR-CE)

It was necessary to educate stakeholders and show the benefits of mitigating climate change. As the Policy Specialist in Singapore mentioned, his work was connected to include alternative proteins in discussions about sustainability at several events in Asia:

[...] parts of my work that I do is to raise awareness, definitely to try to put all the proteins on the global agenda because everybody keeps talking about global emissions. (SG-POL)

One of the reasons that increasing awareness on alternative proteins was difficult worldwide is related to the fact that there was not a clear definition of it. Especially for cultivated meat, a common terminology would be helpful for its acknowledgement and acceptance. Terms like cultured, clean, and lab-grown are widely employed in newspaper articles (GFI, 2021). Encouraging a universal nomenclature can be important for the industry and research development, as well as funding from regulators as mentioned by the Science and Technology Manager in Europe:

Another issue is around the definition of alternative proteins. So, especially in Europe, the definition of alternative proteins is generally clumped based on fermentation, cultivated, and plant-based but also insects, and it also generally refers to alternative proteins for animal feed; this is something that the EU funding generally includes in scope. (EUR-SCI)

In order to make sure that regulators will be creating a positive regulatory framework for alternative proteins, GFI works closely with several health agencies around the world and state-level institutions when required. It provides resources and counselling to them and supports a deeper understanding on this innovative industry.

5.2.2. Contribute to Institutional Development

Policy is one of the main pillars that sustains GFI's activities. Supporting the establishment of a fair and efficient regulation is a gigantic task that specialists in every unit are contributing to increasing the understanding of alternative proteins and building a global regulatory framework. As indicated by the Policy Director in the United States, it is a defensive work:

> [...] to make sure that these products are regulated fairly and efficiently and safely, sort of on an even playing field with their conventional counterparts. (US-POL)

GFI wants to ensure that alternative proteins are considered in the policy discussion worldwide regarding climate change mitigation and global health (GFI, 2020). In order to actively influence different stakeholders to consider different angles of this new industry, GFI provides government and industry advice by commenting on current studies, developing internal resources and even industry reports. We could see the broad scope of GFI's advisory coverage by the involvement of different members in India, Brazil, and Israel:

> [...] then it comes to government advice, for example, we've provided comments on multiple calls for comment, for public comment in committee as well as in the policy space. (IN-POL)

In order to boost the association of different stakeholders, GFI recognises that it is important to achieve a common voice in the industry when talking to the public sector.

5.3. Liabilities in the GVC

5.3.1. Chain Development on a Global Basis

The distinguished group of scientists at GFI tries to figure out the most complex bottlenecks for the development of alternative proteins. After analysing the industry in depth, they support filling research gaps by proposing solutions and funding additional investigations. The lead scientist in the United States focussed on identifying those bottlenecks and how to alleviate them:

> My role is mostly focused on identifying and articulating the technical and economic bottlenecks that are facing the cultivated meat industry. Coming up with potential solutions in terms of how we go about alleviating those bottlenecks. Research or training programs, that's the bottleneck that we really need to solve here. (US-SCI)

Furthermore, the industry now needs to prove that it is feasible to produce and cultivate meat in large scale in proper manufacturing units and not in university laboratories. After the approval for selling cultivated meat in Singapore, it is expected to see meat companies (traditional MNEs in the meat industry) investing in the construction of pilot facilities in the area (GFI, 2021). One of the largest multinational companies in the world in the agri-food industry signalled that it would go to market by 2024, which accelerated the confidence of supply chain in reducing the costs and reaching consumers soon as declared by one of Customer Engagement Analysts in Singapore:

> The biggest one is kind of proof of concept. And I think that was that good. The good meat facility just launched in Singapore a few days ago and that's really trying to show kind of conviction that the cost can come down because the biggest criticism of cultivated meat is that will never be kind of reached price parity. (SG-CE-2)

5.3.2. Reduce Bottlenecks

During the interviews, the current challenges that the industry was facing was clear. GFI members had suggested in the future agenda some relevant barriers need to be surpassed in order to move ahead in scaling up the production of cultivated meat. One of the next steps in the forefront of the industry is labelling. There is a concern about the impact of damaging labelling restrictions on cultivated meat as the plant-based meat is already facing in relevant markets such as the European Union and the United States. Therefore, they are also investing in advocacy strategies:

> [...] one person is focusing on labelling. So, ensuring that there is no sort of like harmful labelling restrictions. (EUR-POL)

The key challenge in this value chain is creating infrastructure for scaling up. It requires not only a large amount of investment, but also government support and incentives. In order to reduce the costs, it is necessary to scale up infrastructure and technology in competitive regions.

> [...] if we don't start building factories now to produce cultivated meat, then we can all work on developing the best formulation, bringing down media prices. But we're not going to be able to actually produce anything at scale. And then we're also kind of not addressing the challenge. (EUR-CE)

In addition, the products should taste better and be affordable to consumers. They need to be convenient for most people to buy them and to be successful. One of the ways to do that is by working with the existing supply chain for meat related products. The SciTech Specialist from Europe emphasised the priorities required to develop economic viable production:

> [...] bringing costs down and making the products better. Those are the two core priorities for us, since basically that's all we're trying to do on our science and technology work is help the industry scale and help them improve actual biomimetic kind of mimicry of these products, so they genuinely taste the same meat and cost the same or less. (EUR-SCI)

A secure global food system is a recognised interest of national governments. Once commercialisation of cultivated meat is viable, it will be proved that advances in food technology by producing meat from animal cells can reduce climate impact (GFI, 2022). Finally, it will only be possible to propel this new industry forward with additional research funding. Public investment in this area is limited in comparison to electric vehicles or renewable energy.

6. DISCUSSION

It is evident that the nascent cultivated meat technology represents an opportunity to examine how innovation can lead to the development of a more sustainable meat chain. Aligned with the interactive and open model of innovation advocated by Ambos et al. (2021), the cultivated meat case, sheds light on how an international NGO can also have a lead role in the development and transition to a new and more sustainable GVCs through engagement in innovation.

As shown in our results, this role may have a broad scope, involving the simultaneous coordination of a diversity of initiatives, such as investing in technology development, lobbying, facilitating firms' engagement with the new chain, showing active presence in regulatory initiatives, education, and knowledge sharing, among others. Moreover, these activities involve the orchestration of several stakeholders, like start-ups, MNEs, other NGOs, universities, research institutions, government bodies, and venture capital firms as mentioned by interviewees. In addition, it was also observed that with the support of their overseas subsidiaries, the studied NGO has also an extensive geographic scope of influence addressing both, local issues (regulatory and local developments) and the articulation of the chain on a global basis, for example, facilitating MNEs and start-ups interaction. These efforts aim to develop technological capabilities at not only a national level but also ultimately attempt to leverage the development of the cultivated meat industry at a global level.

In particular, three main aggregate themes were identified: developing the industry, articulating market and regulatory issues, and addressing liabilities in the value chain. The first theme involves the development of technological capabilities that are required for the transition; thus, there are efforts to advance cultivated meat products and process upgrading (Lee & Gereffi, 2015; Morrison et al., 2008). As shown in the literature, upgrading in GVCs requires engaging firms, investing in infrastructure, and developing human capital (Gereffi & Fernandez-Stark, 2016). It also requires access to education and training, funding and finance, and the participation of research institutions as catalysts. In this case, we could note that GFI initiatives focus both at the firm level (supporting single start-up development and collaborating with MNE) and industry-level (connecting different stakeholders and stimulating sectorial development, e.g., bioreactor suppliers' development) compared to MNEs which focus only on the firm level for upgrading. For instance, in some regions, the NGO's subsidiaries may focus on developing courses or garnering initial government funding while, in others, initiatives may focus on advanced technology development efforts (e.g. developing bioreactors with greater capacities).

As for the second theme, articulating market and regulatory issues, since regulation is a major bottleneck for the cultivated meat to advance (Stephens et al., 2018), several NGO initiatives are focussed on securing the approval of cultivated meat products for production and sales in other countries than Singapore, especially in larger markets such as Europe and the United States. Successful approvals are likely to affect the cultivated meat GVC shape reflecting in which markets downstream firms, especially meat MNEs that have invested in the technology (e.g. BRF, JBS, and Tyson), are going to operate.

Finally, the third theme, addressing liabilities in the value chain, shows how NGOs can have a governance role in GVC, that is, assuring resource and activity flows, as well as interfirm relationships, through formal and informal coordination mechanisms (Gereffi & Lee, 2012). The NGO may adopt a GVC perspective to identify technology and capabilities development bottlenecks on a global basis (e.g. creating conditions for scaling up production, reducing costs, and improving product quality worldwide). However, different from the governance exerted by MNEs in conventional agri-food systems, which are largely dependent on buyer–suppliers power asymmetries, we contend that NGO governance is relational (Gereffi & Fernandez-Stark, 2016). Thus, the governance benefits actors' interactions and it depends on complex information sharing and collaborative knowledge development; such relationships tend to rely more intensively on ingredients such as reputation, access to knowledge, and trust. Therefore, we can infer that NGO governance may address GVC's market specificities as conventional MNEs, however, it will also target the expansion of industry needs. In order to figure out and clear up bottlenecks, GFI not only shares information among stakeholders, but also creates the bridge for their interaction to further industry advancement.

7. CONCLUSION

Our content analysis provided fine-grained details on how an NGO can influence innovation in GVCs and contribute to the transition to a more sustainable (cell-based) agri-food chain in addressing SDG2 and SDG13. This role is commonly led by MNEs, known for their innovative nature and large resources to affect value chains, whereas institutions and non-firm players act as partners in co-shaping it to support social and environmental evolution (De Marchi et al., 2020). However, as postulated by Ambos et al. (2021), an interactive and open innovation model should be the basis for understanding innovation in GVCs. Therefore, our research further extends Ambos et al.'s (2021) work, as it shows that international NGOs, like GFI, may have an extended protagonist role in facilitating and creating innovation in GVCs such as in the context of alternative proteins.

MNEs are counting on GFI's expertise to evaluate their activities in this field, when developing an internal business unit or acquiring a promising start-up. Education and raising awareness about cultivated meat to academic, business, and governmental stakeholders were frequently mentioned during the interviews as key activities for GFI. Furthermore, regulators and investors seek counselling based on GFI's independence. The transparency and open access for published research also provide reliability to its developed resources and databases. Therefore, our findings suggested a significant change in the cultivated meat GVC's governance, which is orchestrated by an NGO. GFI coordinates the transformation to a more sustainable agri-food sector (Gereffi, 2001; Ponte, 2014) by facilitating relationships across the value chain and fostering connections to enhance innovation. Different from MNE's governance model, which is focussed on firm level upgrading, the NGO's ultimate aim is to develop a strong alternative protein community.

This study is not without limitations. The challenges and future agenda for the development of this innovative and sustainable industry described in this chapter are based on qualitative data with method-related limitations to generalisation.

Even though we covered all the regions and at least three GFI respondents per unit supplemented by secondary data, our findings focus exclusively on GFI's perspective on its role and activities. Further studies must also incorporate different stakeholders along the alternative protein value chain such as MNEs, startups, regulators, and investors to provide contrasting perspectives.

REFERENCES

Ambos, B., Brandl, K., Perri, A., Scalera, V. G., & Van Assche, A. (2021). The nature of innovation in global value chains. *Journal of World Business*, *56*(4), 101221.

Birkinshaw, J., Brannen, M. Y., & Tung, R. L. (2011). From a distance and generalizable to up close and grounded: Reclaiming a place for qualitative methods in international business research. *Journal of International Business Studies*, *42*(5), 573–581.

Bitzer, V., & Glasbergen, P. (2015). Business–NGO partnerships in global value chains: Part of the solution or part of the problem of sustainable change? *Current Opinion in Environmental Sustainability*, *12*, 35–40.

Bonny, S. P., Gardner, G. E., Pethick, D. W., & Hocquette, J. F. (2015). What is artificial meat and what does it mean for the future of the meat industry? *Journal of Integrative Agriculture*, *14*(2), 255–263.

Bryant, C., & Krelling, F. (2020). *Alternative proteins in Brazil: Nomenclature for plant based & cultured meat*. OSF Preprints.

Corbin, J., & Strauss, A. (2015). *Basics of qualitative research: Techniques and procedures for developing grounded theory* (J. Corbin & A. Strauss, Eds., 4th ed.). SAGE Publications.

Crippa, M., Solazzo, E., Guizzardi, D., Monforti-Ferrario, F., Tubiello, F. N., & Leip, A. (2021). Food systems are responsible for a third of global anthropogenic GHG emissions. *Nature Food*, *2*(3), 198–209.

Dahlin, K. B., & Behrens, D. M. (2005). When is an invention really radical?: Defining and measuring technological radicalness. *Research Policy*, *34*(5), 717–737.

Dallas, M. P., Ponte, S., & Sturgeon, T. J. (2019). Power in global value chains. *Review of International Political Economy*, *26*(4), 666–694.

De Marchi, V., Di Maria, E., Golini, R., & Perri, A. (2020). Nurturing international business research through global value chains literature: A review and discussion of future research opportunities. *International Business Review*, *29*(5), 101708.

Eisenhardt, K. M., & Graebner, M. E. (2007). Theory building from cases: Opportunities and challenges. *Academy of Management Journal*, *50*(1), 25–32.

Emmel, N. (2013). *Sampling and choosing cases in qualitative research: A realist approach*. SAGE Publications.

Fee, A., & Gray, S. J. (2013). Transformational learning experiences of international development volunteers in the Asia-Pacific: The case of a multinational NGO. *Journal of World Business*, *48*(2), 196–208.

Froggatt, A., & Wellesley, L. (2019). *Meat analogues: Considerations for the EU* [Research Paper]. Energy, Environment and Resources Department, The Royal Institute of International Affairs Chatham House. Retrieved March 11, 2019, from https://www.chathamhouse.org/2019/02/meat-analogues

Gerber, P. J., Steinfeld, H., Henderson, B., Mottet, A., Opio, C., Dijkman, J., Falcucci, A., & Tempio, G. (2013). *Tackling climate change through livestock – A global assessment of emissions and mitigation opportunities*. Food and Agriculture Organization of the United Nations (FAO). http://www.fao.org/3/i3437e/i3437e.pdf

Gereffi, G., Humphrey, J., Kaplinsky, R., & Sturgeon, T. J. (2001). Introduction: Globalisation, value chains and development.

Gereffi, G., & Lee, J. (2012). Why the world suddenly cares about global supply chains. *Journal of Supply Chain Management*, *48*(3), 24–32.

Gereffi, G., & Fernandez-Stark, K. (2016). *Global value chain analysis: A primer*. The Duke Center on Globalization, Governance & Competitiveness. http://dukespace.lib.duke.edu/dspace/bitstream/handle/10161/12488/2016-07-28_GVC% 20Primer%202016_2nd%20edition.pdf

Gioia, D. A., Corley, K. G., & Hamilton, A. L. (2013). Seeking qualitative rigor in inductive research: Notes on the Gioia methodology. *Organizational Research Methods*, *16*(1), 15–31.

GFI. (2020). *2020 state of industry report cultivated-meat*. https://gfi.org/resource/cultivated-meat-eggs-and-dairy-state-of-the-industry-report

GFI. (2021). *Seeding radical change*. https://gfi.org/wp-content/uploads/2021/07/2021-GFI-Impact-Report-GFI-supporters.pdf?fbclid=IwAR1HozEzRA1ScsspZgTwJoSog_jq5G9AID9WX-7HywMYl_tyqSpMR4H3VEIg

GFI. (2022). *2021 state of industry report cultivated-meat and seafood*. https://gfieurope.org/wp-content/uploads/2022/04/2021-Cultivated-Meat-State-of-the-Industry-Report.pdf

Ghauri, P., Strange, R., & Cooke, F. L. (2021). Research on international business: The new realities. *International Business Review*, *30*(2), 101794. https://doi.org/10.1016/j.ibusrev.2021.101794

Godfray, H. C. J., Springmann, M., Sexton, A., Lynch, J., Hepburn, C., & Jebb, S. (2019). *Meat: The future series alternative proteins*. World Economic Forum. Retrieved April 5, 2019, from http://www3.weforum.org/docs/WEF_White_Paper_Alternative_Proteins.pdf

Golgeci, I., Makhmadshoev, D., & Demirbag, M. (2021). Global value chains and the environmental sustainability of emerging market firms: A systematic review of literature and research agenda. *International Business Review*, *30*(5), 101857.

Hajjar, R., Newton, P., Adshead, D., Bogaerts, M., Maguire-Rajpaul, V. A., Pinto, L. F., McDermott, C. L., Milder, J. C., Wollenberg, E., & Agrawal, A. & Agrawal, A. (2019). Scaling up sustainability in commodity agriculture: Transferability of governance mechanisms across the coffee and cattle sectors in Brazil. *Journal of Cleaner Production*, *206*, 124–132.

Hajro, A., Gibson, C. B., & Pudelko, M. (2017). Knowledge exchange processes in multicultural teams: Linking organizational diversity climates to teams' effectiveness. *Academy of Management Journal*, *60*(1), 345–372.

Heidemann, M. S., Molento, C. F. M., Reis, G. G., & Phillips, C. J. C. (2020). Uncoupling meat from animal slaughter and its impacts on human–animal relationships. *Frontiers in Psychology*, *11*, 1824. https://doi.org/10.3389/fpsyg.2020.01824

Hoek, A. C., Malekpour, S., Raven, R., Court, E., & Byrne, E. (2021). Towards environmentally sustainable food systems: Decision-makingfactors in sustainable food production and consumption. *Sustainable Production and Consumption*, *26*, 610–626.

Humphrey, J., & Memedovic, O. (2006). *Global value chains in the agri-food sector*. https://agris.fao.org/agris-search/search.do?https://news.un.org/en/story/2021/03/1086822

IPCC. (2020). *Climate change and land: An IPCC special report on climate change, desertification, land degradation, sustainable land management, food security, and greenhouse gas fluxes in terrestrial ecosystems*. https://www.ipcc.ch/site/assets/uploads/sites/4/2020/02/SPM_Updated-Jan20.pdf

Kelly, R. F., Jennings, A., Hunt, J., Hamman, S. M., Mazeri, S., Nkongho, E. F., Ngwa, V. N., Tanya, V., Sander, M., Ndip, L., & Bessell, P. R. (2021). The epidemiology of bacterial zoonoses in pastoral and dairy cattle in Cameroon, Central Africa. *Zoonoses and Public Health*, *68*(7), 781–793.

Kourula, A., & Laasonen, S. (2010). Nongovernmental organizations in business and society, management, and international business research: Review and implications from 1998 to 2007. *Business & Society*, *49*(1), 35–67.

Lambell, R., Ramia, G., Nyland, C., & Michelotti, M. (2008). NGOs and international business research: Progress, prospects and problems. *International Journal of Management Reviews*, *10*(1), 75–92.

Lee, J., & Gereffi, G. (2015). Global value chains, rising power firms and economic and social upgrading. *Critical Perspectives on International Business*, *11*(3–4), 319–339.

Lee, J., & Gereffi, G. (2021). Innovation, upgrading, and governance in cross-sectoral global value chains: The case of smartphones. *Industrial and Corporate Change*, *30*(1), 215–231.

Lee, J., Gereffi, G., & Beauvais, J. (2012). Global value chains and agrifood standards: Challenges and possibilities for smallholders in developing countries. *Proceedings of the National Academy of Sciences*, *109*(31), 12326–12331.

Lincoln, Y. S., & Guba, E. G. (1985). *Naturalistic inquiry*. SAGE Publications.

Liu, S. Y. H., Napier, E., Runfola, A., & Cavusgil, S. T. (2020). MNE–NGO partnerships for sustainability and social responsibility in the global fast-fashion industry: A loose-coupling perspective. *International Business Review*, *29*(5), 101736. https://doi.org/10.1016/j.ibusrev.2020.101736

Lund-Thomsen, P., & Lindgreen, A. (2014). Corporate social responsibility in global value chains: Where are we now and where are we going?. *Journal of Business Ethics*, *123*(1), 11–22.

Marshall, B., Cardon, P., Poddar, A., & Fontenot, R. (2013). Does sample size matter in qualitative research?: A review of qualitative interviews in is research. *Journal of Computer Information Systems*, *54*(1), 11–22.

Mergos, G., & Papanastassiou, M. (2017). *Food security and sustainability*. Palgrave Macmillan.

Miles, M. B., Huberman, A. M., & Saldaña, J. (2014). *Qualitative data analysis: A methods sourcebook* (3rd ed.). SAGE Publications.

Morrison, A., Pietrobelli, C., & Rabellotti, R. (2008). Global value chains and technological capabilities: A framework to study learning and innovation in developing countries. *Oxford Development Studies*, *36*(1), 39–58.

Mousavi, S., & Bossink, B. (2020). Corporate–NGO partnership for environmentally sustainable innovation: Lessons from a cross-sector collaboration in aviation biofuels. *Environmental Innovation and Societal Transitions*, *34*, 80–95.

Narula, R., & Wahed, M. S. (2017). The dominant presence of MNES in agro-food GVCs: Implications for the developing countries. In G. Mergos & M. Papanastassiou (Eds.), *Food security and sustainability* (71–88). Palgrave Macmillan.

Nobre, F. S. (2022). Cultured meat and the sustainable development goals. *Trends in Food Science & Technology*, *24*, 140–153.

OECD FAO. (2020). *Agricultural outlook 2020–2029*. https://doi.org/10.1787/1112c23b-en

Papanastassiou, M., Pearce, R., & Zanfei, A. (2020). Changing perspectives on the internationalization of R&D and innovation by multinational enterprises: A review of the literature. *Journal of International Business Studies*, *51*(4), 623–664.

Plant Based Foods Association. (2021). *Retail sales data*. Retrieved September 09, 2021, from https://plantbasedfoods.org/marketplace/retail-sales-data/

Ponte, S. (2014). The evolutionary dynamics of biofuel value chains: From unipolar and government-driven to multipolar governance. *Environment and Planning A*, *46*(2), 353–372.

Ransom, E. (2007). The rise of agricultural animal welfare standards as understood through a neo-institutional lens. *International Journal of Sociology of Agriculture and Food*, *15*(3), 26.

Reis, G. G., Heidemann, M. S., Borini, F. M., & Molento, C. F. M. (2020). Livestock value chain in transition: Cultivated (cell-based) meat and the need for breakthrough capabilities. *Technology in Society*, *62*, 101286.

Reis, G. G., & Molento, C. F. M. (2020). Emerging market multinationals and international corporate social responsibility standards: bringing animals to the fore. *Journal of Business Ethics*, *166*(2), 351–368.

Scoppola, M. (2021). Globalisation in agriculture and food: The role of multinational enterprises. *European Review of Agricultural Economics*, *48*(4), 741–784.

Shah, S. K., & Corley, K. G. (2006). Building better theory by bridging the quantitative–qualitative divide. *Journal of Management Studies*, *43*(8), 1821–1835.

Stephens, N., Di Silvio, L., Dunsford, I., Ellis, M., Glencross, A., & Sexton, A. (2018). Bringing cultured meat to market: Technical, socio-political, and regulatory challenges in cellular agriculture. *Trends in Food Science & Technology*, *78*, 155–166.

Strange, R., & Humphrey, J. (2019). What lies between market and hierarchy? Insights from internalization theory and global value chain theory. *Journal of International Business Studies*, *50*(8), 1401–1413.

Swartz, E. (2021). *Anticipatory life cycle assessment and techno-economic assessment of commercial cultivated meat production*. Good Food Institute. https://gfi.org/wp-content/uploads/2021/03/cultured-meat-LCA-TEA-policy.pdf

Szejda, K., Bryant, C. J., & Urbanovich, T. (2021). US and UK consumer adoption of cultivated meat: A segmentation study. *Foods*, *10*(5), 1050.

Taghikhah, F., Voinov, A., Shukla, N., Filatova, T., & Anufriev, M. (2021). Integrated modeling of extended agro-food supply chains: A systems approach. *European Journal of Operational Research*, *288*(3), 852–868.

Teegen, H., Doh, J. P., & Vachani, S. (2004). The importance of nongovernmental organizations (NGOs) in global governance and value creation: An international business research agenda. *Journal of International Business Studies*, *35*(6), 463–483.

Tran, N., Bailey, C., Wilson, N., & Phillips, M. (2013). Governance of global value chains in response to food safety and certification standards: The case of shrimp from Vietnam. *World Development*, *45*, 325–336.

Tuomisto, H. L., & Teixeira de Mattos, M. J. (2011). Environmental impacts of cultured meat production. *Environmental Science &Technology*, *45*(14), 6117–6123.

UN. (2015). *Transforming our world: The 2030 agenda for sustainable development (A/RES/70/1)*. United Nations. Retrieved July 20, 2022, from https://sustainabledevelopment.un.org/post2015/transformingourworld/publication

Valente, J. D. P. S., Fiedler, R. A., Sucha Heidemann, M., & Molento, C. F. M. (2019). First glimpse on attitudes of highly educated consumers towards cell-based meat and related issues in Brazil. *PloS One*, *14*(8), e0221129.

Wengraf, T. (2001). *Qualitative research interviewing*. SAGE Publications. https://dx.doi.org/10.4135/9781849209717

Yin, R. K. (2018) *Case study research and applications: Design and methods* (6th ed.). SAGE Publications.

Zhao, S., Tan, H., Papanastassiou, M., & Harzing, A. W. (2020). The internationalization of innovation towards the South: A historical case study of a global pharmaceutical corporation in China (1993–2017). *Asia Pacific Journal of Management*, *37*, 553–585.

INDEX

Printed and bound by CPI Group (UK) Ltd, Croydon, CR0 4YY

17/09/2026

14964138-0002